1st International Conference on Animal Production for Food Sustainability (ICAPFS 2018)

The Future, Challenges, and Strategy for Animal Production

IOP Conference Series: Earth and Environmental Science
Volume 287

Padang, Indonesia
10 - 11 October 2018

ISBN: 978-1-5108-9329-0
ISSN: 1755-1307

Printed from e-media with permission by:

Curran Associates, Inc.
57 Morehouse Lane
Red Hook, NY 12571

Some format issues inherent in the e-media version may also appear in this print version.

Institute of Physics
Dirac House, Temple Back
Bristol BS1 6BE UK

Phone: 44 1 17 929 7481
Fax: 44 1 17 920 0979

techtracking@iop.org

Additional copies of this publication are available from:

Curran Associates, Inc.
57 Morehouse Lane
Red Hook, NY 12571 USA
Phone: 845-758-0400
Fax: 845-758-2633
Email: curran@proceedings.com
Web: www.proceedings.com

Table of contents

Volume 287

International Conference on Animal Production for Food Sustainability
10–11 October 2018, Padang, Indonesia

Preface

Papers

Animal Production and Technology

Feed Nutrition

Social Economic

IOP Conf. Series: Earth and Environmental Science **287** (2019) 011001 doi:10.1088/1755-1315/287/1/011001

Preface

Assalamualaikum wr.wb.
Dear all of the scientists, delegates, participants, ladies, and gentlemen,

Alhamdulillah...as the host of The 1st International Conference on Animal Production for Food Sustainability (ICAPFS), we do impress, thankful, and present a high appreciation for your participation in joining the ICAPFS 2018 in Padang, West Sumatra-Indonesia. As an international meeting for researchers, educators, professionals, and students, the meeting will attract more than 100 participants with common but diverse interests in the future of animal production.

The theme of 1st ICAPFS 2018 is "The Future, Challenges, and Strategy for Animal Production". The livestock sector globally is highly dynamic. In developing countries, it is evolving in response to rapidly increasing demand for livestock products, largely driven by human population growth, income growth, and urbanization. In the future, production will increasingly be affected by competition for natural resources, particularly land and water, competition between food, feed, fuel and fiber by the need to operate in a carbon-constrained economy. Developments in breeding, nutrition and animal health will continue to contribute to increasing potential production and further efficiency and genetic gains. Demand for livestock products in the future could be heavily moderated by socio-economic factors such as human health concerns and changing socio-cultural values. There is considerable uncertainty as to how these factors will play out in different regions of the world in the coming decades.

On behalf of 1st ICAPFS 2018 committee and all associates, we wish all of the participants having a great achievement of success and fulfill the expectation as well as enjoying the interaction with all scientists participating in Conference.

Thank you, invited speakers and delegates, thank you for choosing to come to this conference and to "Minang" Padang-West Sumatra, Indonesia. We hope the arrangements we have put in place meet with your requirements. We wish you fruitful deliberations and intellectually and socially rewarding stay in Padang, West Sumatra.

We are looking forward to meeting you all in the future Congress to continue.

Terimakasih, Thank you

Rusmana W.S. Ningrat
Chairman of 1st ICAPFS 2018

IOP Conf. Series: Earth and Environmental Science **287** (2019) 011002 doi:10.1088/1755-1315/287/1/011002

Peer review statement

All papers published in this volume of *IOP Conference Series: Earth and Environmental Science* have been peer reviewed through processes administered by the proceedings Editors. Reviews were conducted by expert referees to the professional and scientific standards expected of a proceedings journal published by IOP Publishing.

Genetic polymorphisms of the OLR1 and DGAT1 genes associated with milk components in Holstein Friesian dairy cattle under an intensive management in Central Java

Anneke Anggraeni

Indonesian Research Institute for Animal Production, PO Box 221, Ciawi, Bogor, Indonesia.
*Corresponding author e-mail: ria.anneke@yahoo.co.id.

Abstract. Oxidized Low Density Lipoprotein Receptor 1 (OLR1) gene serves to encode the binding vascular surface receptors and to degrade the oxidized low-density lipoprotein. In dairy cattle a nucleotide mutation at 3'UTR location of the OLR1 gene results in C allele of which related to higher milk fat content than A allele. Diacylglycerol O-acyltransferase 1 (DGAT1) gene has a K232A mutation of which K allele also associated with higher milk fat content against A allele. This study was aimed to study the association of variant genotypes of the OLR1 and DGAT1 genes on milk fat and other milk components in Holstein Friesian (HF) dairy cattle. The target of this research was to obtain genotypes of the OLR1 and DGAT1 gene affecting highly to milk fat content and other milk components in domestic HF cattle. Animals observed were Holstein Friesian (HF) lactating cows for 53 heads kept intensively at one government dairy station at BBPTU Baturraden, Central Java. Base mutations at 3'UTR location of the OLR1 gene and at K232A mutation of the DGAT1 gene were identified by PCR-RFLP (polymerase chain reaction - restriction fragment length polymorphism) techniques. Allele frequency of the base mutation of each gene was analyzed by PopGen 32 program. Milk components of fat, protein, solid non-fat (SNF) and lactose were generated from daily milk test for morning, afternoon and average of both. Study of the association of variant genotypes of the OLR1 and DGAT1 genes on individual milk component was analyzed by GLM considering the fixed effects of lactation period (1-3, 4-6), calving season (1-4 mo., and 5-8 mo.) and calving year (2011, 2012). Duncan Multiple Range Test was used to test significant differences of the averages among subclasses. The OLR1 gene resulted in A allele and C allele of whose the frequencies were respectively 0.457 and 0.543, while the DGAT1 gene resulted in A allele and K allele with the frequencies respectively 0.426 and 0.575. Variant genotypes of the OLR1 gene gave a significant different on milk fat production ($P<0.05$). Genetic polymorphism of the OLR1 gene had a significant effect on milk fat content in the noon, the highest was for CC genotype, followed by AC, and the lowest was AA, with the fat contents of CC and AC genotypes higher to AA one by 4.87% and 1.38% respectively. However variant genotypes of the OLR1 gene had no significant effect on other milk components ($P>0.05$). Similarly, variant genotypes of the DGAT1 gene consisting of genotype AK and genotype KK did not significantly affect on all of milk components. It was concluded that there was a fairly good control of the OLR1 gene to fat content in HF cows. The base mutation at the 3'UTR location of the OLR1 gene can be considered as an initial information in developing a genetic assisted selection (GAS) technique for milk fat content in domestic HF cows.

Keywords – genetic polymorphisms, OLR1, DGAT1, Holstein Friesian, milk components.

1. Introduction

Production of milk and other milk components are economically valuable traits for dairy cattle industry. To be able to improve the quality of milk, it is necessary to increase milk fat content and other milk components such as protein, dried matter, and lactose. High milk contents can increase the selling price of fresh milk from dairy cows. Based on the provisions of the Indonesian National Standard for fresh milk from dairy cattle it is required, among others, to contain a minimum level of milk components for fat by 3%, protein by 2.8%, and non-fat dried-matter by 7.8% [1]. The quality of milk produced by dairy cows therefore needs to pay attention. Milk production and the quality of milk in dairy cattle are a quantitative trait that is controlled by a large number of quantitative trait loci (QTL) and influenced by the environment [2] [3] [4]. The biosynthesis of milk fat in a cow's body is a complex series of processes, passing various pathways controlled by major genes with high genetic variation [5].

Genetic improvement of milk quality is usually done through quantitative selection, but in accelerating the genetic response obtained, can be supported by molecular-based selection, namely through exploration of SNP variants from major genes or genome exploration. Identification of genetic polymorphisms in major genes that have a large influence on the milk components of dairy cattle can be used to accelerate the results of conventional selection programs. The Candidate gene strategy has been proposed by direct search for quantitative trait loci (QTL). Genetic variation in a gene, affecting the physiological pathways and phenotypes, further the proportion of genetic and phenotypic variation would likely be to affect the breeding strategy for improvement of important traits [4] [6]. Milk fat contents are controlled by a number of major genes, including SCD1 gene (stearoyl-CoA desaturase 1), ACACA gene (Acetyl-CoA Carboxylase Alpha), OLR1 gene (oxidized low density lipoprotein receptor 1), and DGAT1 gene [2] [3] [4] [5] [7].

Oxidized low density lipoprotein receptor 1 (OLR1) gene functions to encode oxidized low density lipoprotein receptors such as lectins, which are proteins that bind and degrade oxidized low density lipoproteins [8]. This gene in dairy cattle is located on chromosome 5 (BTA5) at intervals between 106 - 108 cM consisting of 5 exons with the function of encoding 279 amino acids of protein. It is known that there is a significant influence of the marka rs109019599 or g.8232CNA or C223A at the non-translational region or 3'UTR of this gene on milk fat content and milk fat production [2] [3] [9]. CC and AC genotypes consistently produce higher levels of milk fat against the AA cows [2] [10] [11].

Diacylglycerol acyltransferase 1 (DGAT1) gene is one of candidate genes that has a very high potency to become a gene marker in increasing milk fat content in dairy cattle. The DGAT1 gene encodes a microsomal enzyme by using diacylglycerol and fatty acyl coA as the terminal of substrates catalyzes and committed in triacylglycerol biosynthesis. The DGAT1 gene is also involved in triacylglycerol metabolism such as intestinal fat absorption, lipoprotein assembly, adipose tissue formation and lactation [4]. This gene is located on bovine chromosome 14 with a total length of 14,117 bp consisting of two flanking regions, 17 exons and 16 introns. This gene encodes Acyl-CoA: diacylglycerol acyl transferase 1 as essential enzyme to role in the synthesis of triglycerides with the catalyst of the reaction of diacylglycerol and fatty acids [12]. A non-conservative substitution of lysine with alanine (K232A) at positions 10,433 and 10,434 in exon 8 showed an association to milk fat content and milk components.

This study was aimed to study the association of variant genotypes of the OLR1 and DGAT1 genes on milk fat and other milk components in domestic Holstein Friesian (HF) dairy cattle. The target of this research is to obtain genotypes of the OLR1 and DGAT1 gene affecting highly to milk fat content and other milk components in domestic HF cattle.

2. Materials and methods

The research materials were Holstein Friesian (HF) Lactating cows kept at BBPTU Baturraden, Purwokerto, Subdistrict. Banyumas District, Central Java. A total number of HF cows for 53 heads were observed at the satus of lactation months of 1-8 months, lactation periods of 1-6, and calving years of 2011-2012. Data of milk production were collected from daily milk testing generated at morning, afternoon, and average of both of milking times per lactating cow.

All of those HF cows as samples were collected their fresh blood. Blood samples were taken for DNA analysis collected from jugular vein using a 21 G X 1½ sized vacutainer or 10 ml syringe containing heparin anticoagulant substance. A total of 1 to 2 ml of blood samples was stored in a 10 ml tube and added 8 ml of ethanol (EtOH) absolute containing 1 mM EDTA.

Milk component was obtained using a formol titration technique where p is the amount of NaOH for titration on milk samples and q is the amount of NaOH for blank titration. Components of milk (%) analyzed include fat content, protein, lactose, and non-fat dried-matter.

2.1. DNA extraction

DNA extraction using 0.2% NaCl, ethylene diamide tetra acetate (EDTA), 10% sodium diodecyl sulfate (SDS), sodium tris EDTA (STE 1.0x), proteinase k 5.0 mg / ml, phenol, CIAA (24 chloroform: 1 isoamyl alcohol), absolute ethanol, 70% ethanol and 80% EDTA (TE).

2.2. Genotyping OLR1 gene

Genotyping the OLR1 gene included briefly DNA extraction activities, amplification by PCR (polymerase chain reaction) technique from base fragments in untranslational region or 3'UTR of the OLR1 gene. The RFLP (restriction polymerase chain reaction) method was applied to cut DNA amplicons as PCR products. The primary pairs namely forward: 5'-TCCCTAACTTGTTCCAAGTCCT-3' and reverse: 5'-GCTTTCTTCTAG GCATTGTGAG-3' [9]. The PCR product (143 bp) was cut with the restriction enzyme PstI which recognized the C * TGCAG as a restriction base.

2.3. Genotyping DGAT1 gene

PCR analysis using DNA samples and premix PCR. Premix PCR consisted of 9.35 µl sterile distilled water, 0.3 µl triposfat deoxy nucleotide (dNTP), 3.0 µl buffer (5.0x), 1.0 µl MgCl2, 0.05 µl DNA polymerase taq enzyme and 0.3 µl primer. The primary used namely forward 5's-GCACCATCCTCTTCCTCAAG-3' and reverse 5'-GGAAGCGCTTTCGGATG-3' [12]. Analysis of RFLP using PCR and premix RFLP products. The premix consisted of EaeI restriction enzymes [13], sterile distilled water and buffer.

2.4. Data analysis

The frequency of a certain genotype was obtained by comparing the total number of that genotype cows to all genotypes obtained, while the Hardy-Weinberg equilibrium (HWE) was tested by Chi square analysis. To analyze gene frequency and H-W equilibrium were conducted by Gen Pop 3.2. Program.

To study of the association of variant genotypes of individual of the OLR1 and DGAT1 genes on each milk component observed was analyzed by the General Linear (GLM) Model by considering the fixed effects of lactation period (1-3, 4-6), calving season (1-4 mo., and 5-8 mo.) and calving year (2011, 2012). Duncan The Multiple Range Test is the difference between the averages among subclasses. Analysis was conducted by using SAS Program Package ver. 9.1 [14].

3. Results and discussion

3.1. Genotyping genes

Genetic polymorphisms of the OLR1 and DGAT1 gene can become potential molecular markers for selection on milk fat content and other milk components as important economic traits in dairy cattle [2] [10] [12] [13] [15] [18]. DNA amplification of both gene OLR1 and DGAT1 gene using PCR techniques were successfully performed on all of blood samples of the observed HF cows for the total 53 heads.

3.1.1. OLR1

Gene Genotyping of the 3 'non-coding area of the OLR1 gene using PCR-RFLP with PstI restriction enzymes recognized the changes on base A to C at the 6-base cutting site from C *TGCAG. The mutation is thought to be a silent mutation. If there no cutting PCR product (146 pb) was declared as AA genotype, whilst three base fragments (146 bp, 120 bp, and 26 bp) found was identified as AC genotype, and if two fragments (120 bp and 26 bp) found was identified as CC genotypes [9]. The genotyping results in 53 HF cattle were found to result in three genotypes namely AA, AC and CC genotypes by the respective frequencies of 0.2075, 0.5283, and 0.2642; while the frequencies of A and C alleles were respectively 0.4574 and 0.5426. Chi-square test showed that the A and C alleles of the OLR1 gene had $\chi2$ calculation = 0.1740 < $\chi2$ P0.05 = 0.6765. This showed that the OLR1 locus was in the Hardy Weinberg (H-W) equilibrium. Similar results were obtained in HF cattle in Iran [11].

Variant genotypes of the OLR1 gene at the same base mutation of HF cattle in Iran that found the genotype frequencies were almost similar for AA, AC and CC genotypes, namely 0.22, 0.50, and 0.28 respectively; whilst the frequencies of A and C alleles were 0.47 and 0.53 respectively [11]. Almost similar frequencies for A and C alleles were also reported in HF cattle in USA, namely 0.46 and 0.54 [2], and HF cattle in Poland, namely 0.43 and 0.57 [9]. Another investigation at the same base markers g.8232CNA in the untranslation region of 3' UTR region of the OLR1 gene in Red-and-White strain Polish obtained genotype frequencies from the most consecutively for CC cattle (0.53), AC cattle (0.34), and AA cattle (0.13) [16]; so that the dominant for C allele (0.7) against A allele (0.3). CC genotype had the highest fat percentage but AA genotype for the lowest, while AC genotype for intermediate (P<0.05) [17]. The CC cows produced more milk fat yield against the AC genotypes AC (P<0.1) and AA ones (P<0.01). CC and AC cows had a higher milk protein content than AA cows P<0.01). These associations revealed, the SNP has the potential to be considered as a marker in marker-assisted selection.

3.1.2. DGAT1 gene

Amplification product obtained on the DGAT1 gene had a base fragment length of 411 pb. Genotyping DGAT 1 gene by the EaeI restriction enzyme successfully cut the mutation site of the DGAT1 gene. The position of the base mutation occurred on the 8th exon, at the position of the 10th base of 433 which caused the change of adenine to guanine (A → G) or the transition mutation at the 10th base of 434. This caused the changes in adenine base to cytosine one (A → C) or in the form of transversion mutation at GC*GGCC bases (Kong et al., 2007). The mutation resulted in changes in amino acids from lysine (K = AAG) to alanine (A = GCG) for having a significant effect on milk fat content and other milk components of dairy cows [12].

Genotyping results of the DGAT1 in the HF cows observed resulted in A and and K alleles. K allele was identified when the base fragment was not cut off by the EaeI restriction enzym, resulting in a base fragment of 411 bp. Conversely, the A allele was obtained when the base fragment was cut by the enzyme which resulted in the GC*GGCC mutation site, so that two base fragments were obtained by the fragment lengths of 203 bp and 208 bp. Genotyping DGAT1 gene in the observed HF lactating cows resulted in only two types of genotypes, namely KK genotype (411 bp), and KA genotype (203 bp, 208 bp and 411 bp), without AA genotype (203 bp and 208 bp). Genotype frequency of AK cow (0.8679) was dominant to that of KK genotype (0.1321), while frequency of K allele (0.5745) was higher than that of A allele (0.4255). The Chi Square test showed that for A and K alleles had $\chi2$ calculation = 25.0734 > $\chi2$ P0.05 = 0.00001. This shows that the DGAT1 gene at the observed locus was in H-W equilibrium. If Hardy-Weinberg equilibrium is constant from by generations indicates the accurances of selection, mutation, migration, and genetic drift [6].

Based on the results of studies on dairy cows of the Slovak Spotted breed and its crossbreds [18] frequency of genotype homozygotes for AA genotype (frequency 0.790), whilst only 2 were homozygotes for KK genotype in DGAT1 K232A locus (frequency 0.035). The rest of cows were AK heterozygotes (frequency 0.175). Further it was described allele frequenciest for A allele and K allele were succesively 0.877 and 0.123. Hardy-Weinberg equilibrium was found that proved no significant

IOP Conf. Series: Earth and Environmental Science **287** (2019) 012001 doi:10.1088/1755-1315/287/1/012001

deviation of equilibrium in DGAT1 at K232A locus. It was assumed that the frequenciest of A allele and AA (AK) genotypes were most probably caused by artificial selection of dairy products, which negatively influenced the occurrence and persistence of K allele.

3.2. Non genetic effects on milk components

Averages of milk fat content and other milk components as generated from milk daily testing in the morning, afternoon and both of the obserbed HF cows by the total number 53 heads were presented in Table 1. Examination of the influence of non-genetic factors included lactation period (lactation 1, 2), calving season (1-4 months, 5-8 months) and calving yeas (2011, 2012) on individual milk component showed that the morning milk fat content was significantly affected (P <0.05) by lactation period, calving season and calving year. So these three factors also had significant effect on the averages of individual milk component. Fat contentes in the morning and average tests were higher for the 2nd lactation period to the 1st one by 5.67% and 3.64%; 2nd season (months 5-8) to the 1st one (January-April) by 4.15% and 3.60%; and calving year of 2012 to the 2011 one by 3.25% and 0.77%. Instead milk protein in the afternoon was significantly affected by the calving season and calving year. For HF cows in the 2nd calving season to that 1st one produced a higher protein content by 0.47%; while the calving year 2012 to that of 2011 produced a higher protein by 0.76%.

Table 1. Average of milk fat content and other milk components based on non-genetic factor and milking time in Holstein Friesian cattle.

Milk component/ Factor	Variable	Cow (hd.)	Morning	Noon	Both
Fat					
Lact. Period	1	39	4.321±0.380*	4.420±0.255	4.372±0.223*
	2	14	4.566±0.248	4.491±0.250	4.531±0.185
Calv. season	1-4	16	4.262±0.393*	4.346±0.212	4.306±0.188*
	5-8	37	4.439±0.343	4.479±0.261	4.461±0.223
Calv. year	2011	45	4.364±0.374*	4.450±0.264	4.409±0.236*
	2012	8	4.506±0.296	4.374±0.180	4.443±0.135
Protein					
Lact. Period	1	39	3.067±0.188	3.019±0.116	3.045±0.128
	2	14	3.015±0.100	3.006±0.121	3.011±0.056
Calv. season	1-4	16	3.086±0.211	3.006±0.110*	3.047±0.128
	5-8	37	3.039±0.149	3.020±0.121	3.032±0.108
Calv. year	2011	45	3.034±0.171	3.012±0.120*	3.026±0.116
	2012	8	3.160±0.124	3.035±0.102	3.099±0.077
Lactose					
Lact. Period	1	39	4.514±0.271	4.463±0.176	4.490±0.188
	2	14	4.444±0.160	4.440±0.193	4.444±0.089
Calv. season	1-4	16	4.512±0.258	4.430±0.167	4.478±0.163
	5-8	37	4.488±0.245	4.465±0.180	4.478±0.172
Calv. year	2011	45	4.461±0.238*	4.452±0.184*	4.456±0.169*
	2012	8	4.689±0.215	4.405±0.157	4.589±0.120
Non-fat dried-matter					
Lact. Period	1	39	8.322±0.458	8.225±0.308	8.275±0.321
	2	14	8.201±0.275	8.184±0.330	8.190±0.152
Calv. season	1-4	16	8.338±0.449	8.183±0.298	8.263±0.291
	5-8	37	8.270±0.410	8.228±0.320	8.251±0.290
Calv. year	2011	45	8.235±0.410	8.206±0.319	8.222±0.290*
	2012	8	8.603±0.338	8.261±0.279	8.434±0.205
Average		53	8.290±0.419	8.215±0.311	8.254±0.287

Note: * at the same column showed significantly different (P <0.05).

However all of lactose and non-fat dried-matter were not affected by non-genetic factors, the exception was for morning lactose content that was significantly affected by the calving year. For lactose content in the morning test day was higher in 2012 (4.47%) against the previous year.

3.3. Effect of variant genotypes on milk fat content and milk components

3.3.1. Effect of OLR1 gene

The results of the GLM analysis in Table 2 showed that the genetic polymorphism of the OLR1 gene had a significant effect on the level of milk fat content of HF cows milking in the afternoon, from the highest were for CC genotype cows, followed by AC cows, and the lowest AA cows. The fat contents of CC and AC cows were higher to the AA ones by 4.87% and 1.38% respectively. However, the fat contents in the morning and both (morning and noon) were not significantly (P> 0.05) influenced by the genotypes of the OLR1 gene. Other milk components, including fat, lactose, and non-fat dried-matter were also not significantly affected by the OLR1 gene.

Table 2. Average of milk fat content and other milk components based on variant genotype of the OLD1 and DGAT1 gene and milking time in HF cattle.

Milk component/ Factor	Geno-type	Cow (hd.)	Morning	Noon	Both
			Fat (%)		
OLRI-genotipe	AA	11	4.301±0.235	4.351±0.190	4.326±0.175
	AC	28	4.430±0.342	4.411±0.225*	4.423±0.219
	CC	14	4.363±0.484	4.563±0.313*	4.466±0.259
DGAT-genotipe	AK	46	4.401±0.368	4.448±0.256	4.427±0.218
	KK	7	4.281±0.343	4.377±0.242	4.331±0.260
			Protein (%)		
OLRI-genotipe	AA	11	3.052±0.100	3.044±0.119*	3.050±0.073
	AC	28	3.063±0.160	3.001±0.115	3.034±0.125
	CC	14	3.033±0.231	3.024±0.121	3.031±0.121
DGAT-genotipe	AK	46	3.063±0.172	3.013±0.118	3.040±0.117
	KK	7	2.987±0.149	3.037±0.115	3.014±0.092
			Lactose (%)		
OLRI-genotipe	AA	11	4.511±0.166	4.498±0.185	4.508±0.116
	AC	28	4.530±0.265	4.437±0.174	4.484±0.198
	CC	14	4.414±0.257	4.465±0.192	4.441±0.137
DGAT-genotipe	AK	46	4.508±0.248	4.454±0.180	4.482±0.172
	KK	7	4.413±0.239	4.480±0.183	4.449±0.147
			Non-fat dried-matter (%)		
OLRI-genotipe	AA	11	8.313±0.279	8.285±0.325	8.300±0.195
	AC	28	8.336±0.438	8.179±0.303	8.250±0.334
	CC	14	8.181±0.476	8.231±0.328	8.209±0.265
DGAT-genotipe	AK	46	8.314±0.420	8.207±0.313	8.263±0.294
	KK	7	8.131±0.406	8.261±0.320	8.199±0.249
Average		53	8.290±0.419	8.215±0.311	8.254±0.287

The CC genotype cows produced a higher fat level in the noon compared to the AA ones were reported in some previous studies [2] [9] [10]. The highest milk fat content (3.31%) was produced by CC cows, the moderate one by the AC cows (3.17%), and the lowest by the AA cows (2.99%) [10]. AC and CC genotypes produced more fat than AA genotype [11]. In Holstein-Friesian Polish cattle (Red-and-White strains), where both AC and CC cows produced higher milk fat yield per lactation than AA cows, namely 280.19 ± 73.20 kg and 270.81 ± 59.43 kg vs. 249.32 ± 40.22 kg [16].

Variant genotype of the OLR1 gene also had a significant effect (P <0.05) on milk protein content in the afternoon of which AA genotype produced higher protein against AC and CC genotypes. Significant association between C allele and milk protein content from the 3'UTR region of the OLR1 gene seemed that this mutation base could be a potential gene assisted selection (GAS) for molecular selection for the increasing milk fat content program in domestic HF dairy cattle.

3.3.2. Effect of DGAT1 gene

Variant genotypes of the DGAT1 gene in related to the occasion of a base mutation at the base site of GC*GGCC base sequences the HF cows observed cows showed that both of AK and KK genotypes had no significant effect on all milk components. In the other side no AA genotype was found, so the effect of variant genotypes could not be compared among the three genotype. A non-conservative substitution of lysine by alanine (K232A) at the positions 10,433 and 10,434 in exon 8 in the DGAT1 gene can be considered for molecular selection for future breeding programs [19]. KK genotype had the greatest effect on all milk contents, while AA genotype had the greatest effect on milk yield.

4. Conclusion

Genotyping of the 3'UTR region of the OLR1 gene by the restriction enzyme Pst1 resulted two alleles (A and C), three genotypes (AA, AC, and CC), and in H-W equilibrium. While genotyping non conservative substitution of lysine with alanine (K232A) of the DGAT1 gene produced two genotypes (AK and KK). The CC and AC genotypes of the OLR1 gene resulted higher milk fat content to the AA one. This SNP of the OLR1 gene can be initially considered as a candidate gene assisted selection (GAS) for increasing milk fat content in domestic HF cattle.

5. References

[1] BSN (Badan Standarisasi Nasional) untuk Susu Segar - Bagian 1: Sapi. SNI 4141.1. 2011. Jakarta.

[2] Khatib H., S.D. Leonard, V. Schutzkus, W. Luo, and Y.M. Chang. 2006. Association of the OLR1 gene with milk composition in Holstein dairy cattle. J. Dairy Sci. 89: 1753-1760.

[3] Wang, X., Li, T., Zhao, H.B., Khatib, H., 2013. Short communication: amutation in the 3′ untranslated region diminishesmicroRNA binding and alters expression of the OLR1 gene. J. Dairy Sci. 96: 6525–6528.

[4] Mohammed, S.A., A. Siham, Rahamtalla, S. Salah, A. Ahmed, Elhafiz, M. Bakhiet, Dousa, and M. Khalid. 2015. DGAT1 Gene in Dairy Cattle. Review Article. Global Journal of Animal Scientific Research. 3(1):191-198.

[5] Bouwman, A.C., H. Bovenhuis, M.H.P.W. Visker, and J.A.M. van Arendonk. 2011. Genomewide association of milk fatty acids in Dutch dairy cattle. BMC Genetics 12(43): 1-12. https://doi.org/10.1186/1471-2156-12-43.

[6] Anggraeni A, C. Talib, S.A. Asmarasari, T. Herawati, and E. Andreas. 2017. Genetic polymorphisms of IGF1, GH, and OPN genes in crosses Peranakan Ongole cattle based on birth type in Central Java. JITV 22(4): 165-172. DOI: http://dx.doi.org/10.14334/jitv.v22i4.1625.

[7] Kęsek, M., T. Szulc, and A.Z. Steciwko. 2014. Genetic, physiological and nutritive factors affecting the fatty acid profile in cows' milk – a review. Animal Science Papers and Reports 32 (2): 95-105.

[8] Kataoka, H., Kume, N., Miyamoto, S., Minami, M., Murase, T., Sawamura, T., Masaki, T., Hashimoto, N., and Kita, T. 2000. Biosynthesis and post- translational processing of lectin-like oxidized low density lipoprotein receptor-1 (LOX-1). N -linked glycosylation affects cell-surface expres- sion and ligand binding. J Biol Chem 275, 6573–6579.

[9] Komisarek, J. and Z. Dorynek. 2009. Effect of ABCG2, PPARGC1A, OLR1 and SCD1 gene polymorphism on estimated breeding values for functional and production traits in Polish Holstein-Friesian bulls. J. Appl. Genet. 50: 125-132.

[10] Soltani-Ghombavan M., S. Ansari-Mahyari, G.R. Ghorbani, and M.A. Edriss. 2013. Association of a polymorphism in 3' un-translated region of OLR1 gene with milk fat and protein in dairy cows. Arch. Tierz. 56, 32-40.
[11] Mashhadi, M.H. 2017. A research on association between SCD1 and OLR1 genes and milk production traits in Iranian Holstein dairy cattle. Iranian J. of Applied Animal Sci. 7(2): 243-248.
[12] Winter A., Krämer W., Werner F.A.O., Kollers S., Kata S., Durstewitz G., Buitkamp J., Womack J.E., Thaller G., Fries R. 2002. Association of a lysine-232/alanine polymorphism in a bovine gene encoding acyl-CoA: diacylglycerol acyltransferase (DGAT1) with variation at a quantitative trait locus for milk fat content. Proceedings of the National Academy of Sciences of the United States of America, 99, 9300–9305.
[13] Kong HS, Oh JD, Lee JD, Yoon DH, Choi YH, Cho BW, Lee HK, Jeon GJ. 2007. Association of sequence variations in DGAT1 gene with economic traits in Hanwoo (Korea cattle). Asian-Aust. J. Anim. Sci. 20 (6): 817-820.
[14] SAS®/STAT Software, Release 9.1. SAS Institute, Inc., Cary, NC. USA.
[15] Molee, A, N. Duanghaklang, and P. Na-Lampang. 2011. Effects of Acyl-CoA:diacylglycerol acyl transferase (DGAT1) gene on milk production traits in crossbred Holstein dairy cattle. Trop Anim Health Prod DOI 10.1007/s11250-011-9959-1.
[16] Luczak, I. K and E.C. Piątkowska. 2018. Polymorphism in the OLR1 gene and functional traits of dairy cattle. Veterinarski Arhiv 88 (2): 171-177.
[17] Ghombavani, M.S., S.A. Mahyar, and M.A. Edriss. 2013. Association of a polymorphism in the 3' untranslated region of the OLR1 gene with milk fat and protein in dairy cows. Archiv Tierzucht 56 (32): 328-334.
[18] Lešková, L., M. Bauer, P. Chrenek, Z. Lacková, J. Soročinová, V. Petrovič, and G. Kováč. 2013. Detection of DGAT1 gene polymorphism and its effect on selected biochemical indicators in dairy cows after calving. Acta Vet. Brno 13(82): 265–269. doi:10.2754/avb201382030269.
[19] Patel, R.K., J.B. Chauhan, K.J. Soni, and K.M. Singh. 2009. Genotype and allele frequencies of DGAT1 gene in Indian Holstein bulls. Curr Trends Biotechnol Pharm. 3: 386-389.

Production performance of laying Japanese quail that given citric acid as acidifier

Astriana Napirah*, Hamdan Has, Amrullah Pagala, and La Ode Nafiu

Faculty of Animal Science, Universitas Halu Oleo, Kendari, Southeast Sulawesi, Indonesia.
*Corresponding author e-mail: astriana_napirah@yahoo.com.

Abstract. This research aimed to study the effect of citric acid on production performance of laying Japanese quail. One hundred of 12 weeks laying Japanese quails were divided into 4 treatments and 5 replication based on completely randomized design. The treatments were citric acid addition level, they were 0, 0,3, 0,6, and 0,9% of citric acid. This research used a self-mixing feed that contained 2800 kcal/kg metabolizable energy and 17,5% crude protein. The parameters measured were feed consumption, egg weight, egg production, and feed conversion ratio. The obtained data during 4 weeks measurement were analyzed using variance analysis and continued with Duncan's Multiple Range Test. The result showed that addition of citric acid on Japanese quail feed did not give any significant effect (P>0,05) on feed consumption and egg weight, but the addition of citric acid gave a significant effect (P<0,05) on egg production and feed conversion ratio. During 4 weeks, addition 0,6% citrit acid showed egg production above 75%. It can be concluded that addition of 0,6% citric acid showed a better egg production and feed conversion ratio performance.
Keywords – laying Japanese quail, citric acid, acidifier.

1. Introduction

The Japanese quail is one of poultry which have been farmed in Indonesia. Many Japanese quail are farmed because it has a high productivity and does not require extensive land for maintenance. In addition, quail meat and eggs also contain good nutrition [1] [2].

In an effort to optimize the productivity of poultry, in recent years many experiments has carried out. The addition of the feed additive have been much done. One of the ingredients of the feed additive that has the potential to promote poultry productivity are organic acids.

Organic acids, are known as acidifier, has a growth enhancing effect [3]. In poultry, organic acids could improved body weight gain, feed convertion ratio, and also increasing the villus height of small intestine [4]. Application the organic acids can also boost poultry immunity [5] and gut health [6]. In addition, giving of organic acids could increase production [7] and quality of eggs in laying hens [8].

One of the organic acids that can be used in the poultry feed is citric acid. Citric acid can improve growth, feed conversion [9] and health of poultry [10]. Citric acid also could increase phosphorus utilization in poultry diet [11]. Based on the description before, then this research was conducted to study the effect of citric acid on the production performance of laying Japanese quails.

2. Methods

2.1. Materials

One hundred of 12 weeks laying Japanese quails were housed and handling during 5 weeks of experiment. A self-mixing feed that contained 2800 kcal/kg metabolizable energy and 17,5% crude protein was used. The feed was composed of corn, rice bran and commercial RK24 concentrates. During experiment, feed and water were given ad libitum.

2.2. Experimental design

Laying Japanese quails were divided into 4 groups of treatment and each treatment was replicated 5 times. The treatments used in this experiment were 4 addition level of citric acid:

P0 = 0% citric acid/kg feed
P1 = 0,3% citric acid/kg feed
P2 = 0,6% citric acid/kg feed
P3 = 0,9% citric acid/kg feed

Before the data retrieval was performed, the feed adaptation was carried out for one week. After that, the data retrieval than performed during 4 weeks.

2.3. Measured parameters

The measured parameter in this experiment were feed consumption, egg weight, egg production, and feed conversion ratio. Feed consumption was measured every day by calculating the difference between given and leftover feed. Quail eggs were collected on the morning and afternoon. Every afternoon, eggs weight were weighed using analytical scale. Egg production was expressed as a percentage (%) and calculated by comparing the number of eggs produced by the number of quails reared [12]. Meanwhile the feed conversion ratio was calculated by comparing the amount of consumed feed with weights of eggs produced in a certain time.

2.4. Data analysis

The data obtained were analyzed using analysis of variance based on completely randomized design. If a significant effect found, the analysis would be continued using Duncan's Multiple Range Test.

3. Results and discussion

Feed consumption, egg weight, egg production and feed conversion ratio of laying Japanese quail given citric acid were shown in Table 1.

Table 1. Feed consumption, egg weight, egg production and feed conversion ratio of laying Japanese quail given citric acid.

Parameters	Treatments			
	P0	P1	P2	P3
Feed consumption (g/quail/day)	21,87	21,90	21,95	21,89
Egg weight (g)	10,06	10,09	10,06	10,05
Egg production (%)	68,71^{a}	70,00^{a}	77,00^{b}	71,43^{a}
Feed conversion rasio	3,16^{a}	3,10^{a}	2,83^{b}	3,03^{a}

Means within a line with no common superscript differ significantly (P<0,05).

Feed consumption and egg weight were not affected (P>0,05) by citric acid addition on feed. Similar things have also been reported. Giving of organic acids, butyric, fumarate, and lactate acid, did not affect feed consumption of broiler [4]. Addition of organic acids mixture (formic acid and salt of butyric, propionic and lactic acids) had no effect on feed intake and egg weight of laying hens [7].

Quail feed consumption ranged from 21,87 to 21.95 g/quail/day. Feed consumption obtained in this study are in accordance with [13] explanation. According to [13], laying quail consumption during production ranged from 19,8 to 22 g/quail/day.

Quail egg weight obtained in this study ranged from 10,05 to 10,09 g. These values did not differ greatly with some previous researches. Quail egg weights ranged from 10,73 to 10,92 g with an average 10,86 g [14], Japanese quail egg weight is 10,34 g [15]. This result showed that addition citric acid up to 0,9% in feed doesn't give any negative effect on quail eggs weight. It means that citric acid could maintain the process of egg formation lasts well. Citric acid did not contain harmful compounds that can reduce the weight of quail eggs.

In the other hand, giving citric acid in quail feed showed a significant effect (P<0,05) on egg production and feed conversion ratio. Feed conversion ratio was found lower in addition 0,6% citric acid (P2). Similar thing also found in egg production. It became higher in addition 0,6% citric acid. That addition 0,5% citric acid could improve broiler feed conversion ratio [9]. That organic acid supplementation could improve the laying hens production [7].

Egg production and feed conversion ratio obtained in this study were in accordance with [13] explanation. Laying Japanese quail given a diet that contain 18% crude protein would have egg production about 71,9% and feed conversion ratio about 3,3 [13]. In this study, egg production and feed conversion of laying Japanese quail given 0,6% citric acid were 77,00% and 2,83 respectively. These values are more optimal.

Citric acid as an organic acid have a capability of maintaining the balance of microorganism in feed and digestive tract. Dietary organic acid could inhibit the microorganism growth in feed and consequently could maintain the balance of digestive tract microorganisms [8]. Further, it also could modify the pH of intestines.

Citric acid's ability in maintaining the pH balance and microorganisms in digestive tract might be one of the causes of the feed digestion process become more optimal. Feed conversion could be improved. Further, egg production was also becoming more optimal.

4. Conclusion

Giving citric acid up to 0,9% in quail feed did not affect the feed consumption and egg weight, but in the other hand, giving 0,6% citric acid could improve production of laying quail and its feed conversion ratio. It can be concluded that addition of 0,6% citric acid showed a better egg production and feed conversion ratio performance.

5. References

[1] Genchev, A., G. Mihaylova, S. Ribarski, A. Pavlov, and M. Kabakchiev. 2008. Meat quality and composition in Japanese quails. Trakia Journal of Science. 6(4): 72 – 82.

[2] Genchev, A. 2012. Quality and composition of Japanese quail eggs (*Coturnix japonica*). Trakia Journal of Science. 10(2): 91 – 101.

[3] Luckstadt, C., N. Senkoylu, H. Akyurek, and A. Agma. 2004. Acidifier – A modern alternative for anti-biotic free feeding in livestock production, with special focus on broiler production. Veterinarija ir zootechnika. 27(94): 91 – 93.

[4] Adil, S., T. Banday, G. A. Bhat, M. S. Mir., and M. Rehman. 2010. Effect of dietary supplementation of organic acids on performance, intestinal histomorphology, and serum biochemistry of broiler chicken. Veterinary Medicine International. Volume 2010: 1 – 7.

[5] Abdel-Fattah, S. A., M. H. El-Sanhoury, N. M. El-mednay, and F. Abdel-Azeem. 2008. Thyroid activity, some blood constituents, organ morphology and performance of broiler chicks fed supplemental organic acids. International Journal of Poultry Science. 7(3): 215 – 222.

[6] Paul, S. K., G. Halder, M. K. Mondal, and G. Samanta. 2007. Effect of organic acid salt on the performance and gut health of broiler chicken. The Journal of Poultry Science. 44: 389 – 395.

[7] Soltan, M. 2008. Effect of dietary organic acid supplementation on egg production, egg quality and some blood serum parameters in laying hens. International Hournal of Poultry Sciences. 7(6): 613 – 621.

[8] Youssef, A. W., H. M. A. Hassan, H. M. Ali, and M. A. Mohamed. 2013. Effect of probiotics, prebiotics and organic acids on layer performance and egg quality. Asian Journal of Poultry Science. 2013: 1 – 10.

[9] Islam, M. Z., Z. H. Khandaker, S. D. Chowdhury, and K. M. S. Islam. 2008. Effect of citric acid on the performance of broilers. J. Bangladesh Agril. Univ. 6(2): 315 – 320.

[10] Chowdhury, R., K. M. S. Islam, M. J. Khan, M. R. Karim, M. N. Haque, M. Khatun, and G. M. Pestis. 2009. Effect of citric acid, avilamycin, and their combination on the performance, tibia ash, and immune status of broilers. Poultry Science. 88: 1616 – 1622.

[11] Boling-Frankenbach, S. D., J. L. Snow, C. M. Parsons, and D. H. Baker. 2001. The effect of citric acid on the calcium and phosporus requirements of chicks fed corn-soybean meal diets. Poultry Science. 80: 783 – 788.

[12] North, M. O. 1984. Breeder Management in Commercial Chicken Production Manual. The Avi Publishing Company. Inc. Westport, Connecticut. pp. 298 – 321.

[13] Shanaway, M. M. 1994. Quail Production System. A Review. Food and Agriculture Organization of The United Nations. Rome.

[14] Soliman, F. N. K., R. E. Rizk, and J. Brake. 1994. Relationship between shell porosity, shell thickness, egg weight loss, and embryonic development in Japanese quail eggs. Poultry Science. 73: 1607 – 1611.

[15] Song, K. T., S. H. Choi, and H. R. Oh. 2000. A comparison of egg quality of pheasant, chukar, quail and guinea fowl. Asian Australasian Journal of Animal Sciences. 13(7): 986 – 990.

Effect of FSH dosage on the number and quality of Pesisir cattle embryos

F L Syaiful, T Afriani dan E Purwati
Faculty of Animal Science, Andalas University, Padang, West Sumatera, Indonesia.
Email: ferrylismanto5@gmail.com.

Abstract. This study investigates the effect of FSH supplementation on the resulting number and quality of super-ovulated Pesisir cattle embryos. This study aims investigates the effect of FSH supplementation on the resulting number and quality of super-ovulated Pesisir cattle embryos. After progesterone therapy fifteen Pesisir cattle were each given five injections of FSH followed by AI with Simmental semen. Embryos were flushed out on the sixth to eighth day after AI and counted and graded under a microscope. Cows given 16 ml and 12 ml of FSH produced the highest percentage of A quality embryos (22% and 20% respectively) while no high-quality embryos were collected from those given 14 ml FSH. 12 ml FSH treatment resulted in the highest number of B quality embryos (20%) while a14 ml dose produced less than 7%. One-third of the embryos resulting from 14 ml FSH were C grade but 12 and 16 ml dosages resulted in no embryos of this grade. The highest numbers of degraded embryos were found from cows given 12 ml (60%), while the lowest from cows given 14 ml (33.33%). The percentage of the recovered ovum that remained unfertilized from cows treated with 16 ml, 14 ml, and 12 ml were 33.3%, 66.7%, and 0% respectively. While 14ml dosages of FSH resulted in the highest number of embryos, the higher embryo from a 16 ml FSH suggests that this dosage will produce superior results.
Keywords – FSH, superovulation, embryos, Pesisir Cattle.

1. Introduction

Pesisir cattle are a local Indonesian cattle breed raised in West Sumatera, especially in South Pesisir district, in coastal areas including Padang Pariaman regency, South Pesisir and Agam regencies [4].

Pesisir cattle are raised for beef [16]. These cattle have a high ability in converting low-quality feed into meat [7] and are able to adapt well to traditional low intensity raising methods and are resistant to diseases and parasites [1].

However, Pesisir cattle numbers have recently been declining sharply and livestock productivity has decreased due to limited natural resources. The grassed areas in areas that support Pesisir cattle have declined to reduce the amount of feed available and farmers have been switching to imported cattle instead [7]. However, this local breed still has great potential to produce hybrids with Simmental cattle to increase the supply of meat for community nutrition needs and provide income to farmers in South Pesisir [3].

Reproductive biotechnology, which is advancing rapidly. such as Artificial Insemination (AI) and Embryo Transfer (ET) is one way to rapidly increase the population of these cattle and increase livestock productivity. AI has already been used extensively in Pesisir cattle but ET use is still limited. ET, however, may be more effective to increase the livestock population and also provides opportunities for embryo manipulation.

Reproductive biotechnology provides a way to overcome the challenges of meeting productivity gain targets without destroying local biological resources by overcoming the constraints of small-scale production and the low productivity of local native livestock. High-quality embryos have the potential to improve livestock population numbers and improve overall genetic quality so increasing production and profitability. This study investigates the feasibility of using super-ovulation with FSH hormone to increase corpus luteum numbers and quality of embryos in Pesisir cows to pass on the genetic advantages of superior quality cows to more hybrid offspring so increasing the productivity of the population.

2. Research methods

2.1. Research materials

Research materials used were progesterone, CIDR, superovulation *Follicle Stimulating Hormone* (FSH), lubricating gel, povidone-iodine, alcohol, cotton, tissue, lactate ringer medium, physiological NaCl, gentamicin, calf serum, lidocaine, 15 coastal cattle tail, and frozen Simmental semen.

Research equipment used were Petri dishes, micropipettes, Pasteur pipettes, pipettes, a Bunsen burner, Eppendorf microtubes size (1.5 ml; 0.5 ml and 0.2 ml), vortex, micro centrifuge, PCR machine, camera to record electrophoresis gel results, , tissue, cotton, aluminum foil, stereo microscope, Foley catheter, silicone tubing and syringe, embryo filter, stirring rod, IA gun and CIDR applicator.

2.2. Super-ovulation

Fifteen Pesisir cows divided into three groups of five cows were implanted with CIDR (containing progesterone hormone) in front of the cervix for 11 days. From day 10 the nine that had produced Corpus luteum were injected with FSH in the morning and afternoon for 3 days. Each group received a different dosage of hormone: 12 ml FSH + 200 mg GnRH, 14 ml FSH + 200 mg GnRH and 16 ml FSH + 200 mg GnRH. On day 3 FSH injections were accompanied by an injection of Capriglandin for estrus detection in the morning and afternoon. Artificial Insemination (AI) was performed after visible signs of estrus. Embryo collections were performed on days 6 to 8 after AI.

2.3. Embryo flushing

On days 6-8 after AI embryos were harvested/flushed as follows; donor cattle were placed in a special cage with a 10-20 cm elevated front. Then a 2% lidocaine chloride epidural anesthesia was administered between the last sacrum and the first coccygeal bone. A cervical dilator was used to ensure clear flow of fluid from the uterus.

Next, a Foley catheter attached to an AI gun was used to flush out each cornu in turn. After the catheter was placed in the uterus the balloon was inflated with a 20ml syringe to seal off the inside of the cornua to prevent the escape of fluid or embryo into the abdominal cavity when flushing. The AI gun was removed and a Y shaped hose connected to the Foley catheter, one tube connected to the lactate ringer/antibiotic flushing media and the other into a container equipped with an embryo filter to receive the fluid flushed from the uterus.

The collection bottle was then removed along with the upper part of the filtered vial containing any filtered embryos collected, then the number and quality status of the embryo were evaluated.

2.4. Evaluation of fresh embryo

Evaluation of the embryos was performed under a stereomicroscope at 40x magnification. The embryos were counted and graded into 5 groups according to quality (grade), namely: A, B, C, Dg (*degenerated*), Uf (*unfertilized*).

3. Results and discussions

3.1. Number of embryos

The number of embryos collected at various doses of FSH and GnRH is shown in Table 1.

Table 1. FSH Dose and number of embryos in Pesisir Cows 7 days after AI.

Treatment	Number of Embryos and Ovum per Cows
12 ml FSH	1.67±0.58
14 ml FSH	5.00±3.00
16 ml FSH	3.00±1.00
Total	3.22±2.17

There was a statistically significant difference between the number of an embryo for each dose level of FSH ($P < 0.05$) with a dose of 14ml FSH producing the largest number of embryos. This large number of embryo is because supplementation of FSH and GnRH can stimulate the ovaries to grow and mature the follicles. [5] explains that the function of FSH is to stimulate follicular growth and mature de Graff follicles and together with LH stimulates estrogen release. [14] affirm that the number of embryos recovered is highly determined by the success of the super-ovulation process triggered by the FSH hormone.

Application of gonadotropin hormone at the time of follicular wave increases the super-ovulation response. This response will be higher if the hormone is administered on the day before or on the day of follicular waves than one or two days afterward.

It this study, the number of embryos was smaller than the number of corpus lutetium observed [15]. This is because not all the embryos were recovered during flushing or were not visible under the microscope. Embryos may be destroyed when the blastomeres separate from the pellucid zone. Possibly, some embryos were lost into the abdominal cavity if too much flushing fluid was inserted. [14] suggest that the low acquisition of embryos is largely determined by the success of the superovulation process by the FSH hormone. A large and hanging uterus can prevent complete closure of the catheter balloon so that the flushing fluid can seep to the other side. When too many oocytes are produced the fimbriae may fail to capture them all.

3.2. Embryo quality

Embryos were graded based on the morphology using the scale developed [14]. The results are displayed in Table 2. Both 16ml and 12 ml dosages of FSH produced similar numbers of grade A embryos but 14 ml dose resulted in none. A similar number of B grade embryos were produced with 12 ml dosage resulting in the highest number. All C grade embryos were from cows given a 14 ml dosage. The largest number of embryos was degraded.

Differences in the quality of embryo at various doses of FSH indicate that in superovulation does not produce all oocytes simultaneously hence fertilization is not simultaneous either and embryos end up at different levels of development. [4] states that variations in the stage of embryonic development within the same cycle can affect the life of the embryo. The optimal time for insemination is important because silent estrus and different cycle lengths in cattle can lead to the less optimal timing of insemination and in lower fertility rates. Unfertilized ovum may occur because the ovum is ovulated too late to be fertilised. A degraded embryo or an unfertilized oocyte may also cause a decrease in the quality of other embryos. The genetic variation of individual cows also affects the quality of the embryo [9] as does environmental influences such as nutrients obtained by livestock also affect the quality of embryos as an adequate intake of nutrients is needed to sustain the life of the embryo optimally.

Table 2. Embryo Quality of Pesisir Cattle after different FSH Dosage.

Hormone	Embryo Quality					Total
	A	B	C	Dg	Uf	
12 ml FSH	1	1	0	0	0	2
+	0	0	0	2	0	2
200 mg GnRH	0	0	0	1	0	1
Average	0.33±0.58	0.33±0.58	0±0	1.00±1.00	0±0	1.67±0.58
14 ml FSH	0	1	3	3	1	8
+	0	0	2	3	0	5
200 mg GnRH	0	0	0	2	0	2
Average	0±0	0.33±0.58	1.67±1.53	2.67±0.58	0.33±0.58	5.00±3.00
16 ml FSH	2	0	0	0	0	2
+	0	0	0	2	1	3
200 mg GnRH	0	1	0	1	2	4
Average	0.67±1.15	0.33±0.58	0±0	1.00±1.00	1.00±1.00	3.00±1.00

From this, it can be seen that the largest number of viable embryos per cow 0.22 was obtained using a 14ml dose of FSH. However, if only high-grade embryos that are likely to result in successful pregnancies after ET are considered a 16 ml dosage produced the best results with 0.33 A and B grade embryo per cow. This number may appear small but if ET is successful then a single cow can be superovulated several times a year resulting in a larger number of high-quality offspring.

4. Conclusion

The highest number of embryos per super-ovulated cow was gained with 14 ml FSH supplementation (0.67) and the lowest number with 12 ml FSH (0.22). However, with 14 ml, no A grade embryos were obtained, and a third were C grade. No clear relationship between FSH dosage and embryo quality was evident. Although a 16 ml FSH dosage resulted in fewer embryo overall (0.33) all of these were A and B grade.

5. References

[1] Adrial. 2010. Potensi sapi Pesisir dan upaya pengembangannya di Sumatera Barat *(Pesisir cattle potential and development efforts in West Sumatra)*, J. Litbang Pertanian, 29(2): 66-72.

[2] Adriani., B. Rosadi., Depison. 2009. Penggunaan follicle stimulating hormone dan pregnant mare serum gonadotrophin untuk superovulasi pada sapi persilangan Brahman. *(Use of follicle stimulating hormone and pregnant mare gonadotrophin serum for superovulation in Brahman cross cows,* Media Peternakan, 32(3): 163-170.

[3] Afriani T., Jaswandi, Defrinaldi, dan Y. E. Satria. 2014. Pengaruh waktu pemberian gonadotropin releasing hormone (GnRH) terhadap jumlah korpus luteum dan kecepatan timbulnya berahi pada sapi Pesisir. (*Effect of time of administration of gonadotropin releasing hormone (GnRH) on the number of corpus luteum and speed of estrus on Pesisir cattle),* J. Peternakan Indonesia. 16(3): 194-197.

[4] Anwar, S. 2004. Kajian keragaman karakter eksternal dan DNA mikrosatelit sapi Pesisir Sumatera Barat. *(Study of external character diversity and microsatellite DNA of West Sumatra Coastal cattle),* Disertasi Pascasarjana IPB, Bogor.

[5] Hafez, B., E.S.E. Hafez. 2000. Reproduction in Farm Animals. 7th Edition. Philadelphia: Lippincott Wiliams and Wilkins.

[6] Hendri., N. Nufus., S. Sulastri. 2014. Jumlah corpus luteum pada ovarium kiri dan kanan sebagai respons superovulasi pada sapi Frieshian Holstein, Limousin dan Simmental. (*The number of corpus luteum in the left and right ovaries in response to superovulation in*

Frieshian, Holstein, Limousin and Simmental cows),. Prosiding Seminar Nasional Ruminansia: 35-37.

[7] Hendri, Y. 2012. Dinamika Perkembangan Sapi Pesisir sebagai Sapi Lokal Sumatera Barat. (*The Dynamics of the Development of Pesisir Cattle as a Local Breed in West Sumatra),* J. Litbang Pertanian. 32(1): 39-45.

[8] Maidaswar. 2007. Efisiensi Superovulasi pada Sapi Melalui Sinkronisasi Gelombang Folikel dan Ovulasi. Thesis. (*Efficiency of superovulation in cows through synchronization of follicular waves and ovulation.*) Bogor: Sekolah Pascasarjana IPB.

[9] Mermillod, P., C. Boccard., C. Wils., A. Massie., F. Dessy.1992. Effect of oviduct-condition medium and of cumulus cells on bovine embryo development *in vitro*, Theriogenology, 37: 256.

[10] Praharani L., E. Triwulaningsih. 2007. Karakteristik bibit kerbau pada agroekosistem dataran tinggi. *(Characteristics of buffalo stock in highland agroecosystems),* Prosiding Seminar dan Lokakarya Nasional Usaha Ternak Kerbau. Pusat Penelitian dan Pengembangan Peternakan, Bogor.

[11] Roheni., Sabran., Hamdan. 2007. Potensi peran dan permasalahan beternak kerbau di Kalimantan Selatan. *(Potential role and challenges of buffalo breeding in South Kalimantan)* Prosiding Seminar dan Lokakarya Nasional Usaha Ternak Kerbau. Pusat Penelitian dan Pengembangan Peternakan, Bogor.

[12] Saili, T., Fatmawati., A.S. Aku. 2014. Gambaran Aktivitas Ovarium Sapi Betina yang di potong pada Rumah Pemotongan Hewan (RPH) Kendari berdasarkan Folikel Dominan dan Corpus Luteum. (*Description of Ovarian Activity in Cows slaughtered at Kendari Slaughterhouse (RPH) based on Dominant Follicles and Corpus Luteum).*Universitas Haluoleo, Kendari.

[13] Saito S. 1997. *Manual on Embryo Transfer of Cattle*. National Livestock Embryo Centre (NLEC) Cipelang and Japan International Cooperation Agency (JICA).

[14] Subchan F.A., R. Handarini., S.W. Siswanti. 2016. Perbedaan waktu penyuntikan hormon FSH terhadap respon superovulasi sapi Angus. (*Influence of the timing of injecting the FSH hormone on the response of Angus cow superovulation)* Jurnal Peternakan Nusantara 2(1): 34-42.

[15] Syaiful.F.L., T. Afriani., E. Purwati. 2017. Respon Berbagai Dosis Hormon FSH dan GNRH Terhadap Jumlah Corpus Luteum dan Embrio Sapi Pesisir. (*Response to Various Dosages of* FSH *Hormones and* GNRH *on the Number of Corpus Luteum and Pesisir Cattle Embryos).* Prosiding Seminar Nasional Pengembangan Peternakan Berkelanjutan Ke-9. Universitas Padjajaran, Bandung. Jatinangor 17 November 2017.

[16] Udin, Z., Hendri., F. Rahim., Jaswandi., Y. Ferina. 2014. Kualitas semen cair sapi Pesisir dalam bahan pengencer yang berbeda. (*The quality of Pesisir cattle semen in different diluents).* Prosiding Seminar Nasional Ruminansia; 30-32.

[17] Victorbuana. 2010. Peluang Usaha Ternak Sapi Potong. (*Beef Cattle Business Opportunities.)* PeluangUsaha.web.id.

[18] Wattimena J. 2011. Pematangan oosit domba secara *in vitro* dalam berbagai jenis serum. (*Maturation of sheep oocytes in vitro in various types of serum.*) J Agrinimal. 1: 22-27.

The relationship between hormone and blood mineral connection with the success of artificial insemination in Bali cow on the low and medium areas Jambi Province

Fachroerrozi Hoesni*, and Firmansyah

Faculty of Animal Science, University of Jambi, Kampus Pinang Masak Jl. Raya Jambi - Muara Bulian KM.15, Mendalo, Jambi, Indonesia.
*Corresponding Author: rozi.hoesni@gmail.com.

Abstract. The purpose of this study was to analyze the relationship between the content of the hormone progesterone and minerals in the blood with the success of artificial insemination in Bali cattle in the medium and lowlands of Jambi Province. The research method used in this study is survey and laboratory. The survey method is used to evaluate the success of AI in the form of service per conception (S/C). The sampling technique used in this research is Stratified Random Sampling which consists of 2 strata, namely: Strata I is bali cattle that are inseminated by inseminators in the medium areas of Jambi Province represented by Tebo District. Stara II is a bali cattle that is inseminated by inseminators in the lowand areas of Jambi Province represented by Tanjung Jabung Barat District. Laboratory research was conducted to measure the blood and mineral blood hormones of Balinese cattle, to measure the content of the hormone progesterone in the blood of Balinese cattle using radioimmunoassay (RIA) technique. The results of the correlation analysis (Pearson Correlation, Kendall's tau_b, and Spearman's rho) obtained a relationship between the content of the hormone progesterone in the blood of cattle with the success of artificial insemination in cattle is very significant. The results of the correlation analysis (Pearson Correlation) obtained from the relationship between blood mineral content in the form of Zn, Se and Mg with the success of IB in cattle in the central cattle area of Jambi Province is very significant.
Keywords – artificial insemination, hormones, minerals, Bali cows.

1. Introduction

The success of IB in cattle is influenced by many factors, including the hormone content and mineral content in the blood of cattle. Progesterone is a hormone that is very important to maintain pregnancy. Progesterone is produced by the corpus luteum. The corpus luteum is formed after the follicle secures the egg, so that in superovulated treated cows it produces more corpus luteum and the progesterone hormone. According to [6] the hormone progesterone is often called pregnancy hormone, because its function is to maintain pregnancy.

In ruminant livestock, according to [17] the adequacy of macro and micro minerals is important to note given that these minerals are absolute and must be in feed so that the health and productivity of livestock are not disturbed. This mineral need increases especially during growth and reproductive development. Mineral elements are very important in the physiological processes of livestock. Elements of macro essential minerals such as Ca, Mg, Na, K, and P are needed to structure the body such as bones and teeth, while micro elements such as Fe, Cu, Zn, Mo, and J function for the activity of enzyme and homonal systems in the body.

This study has the purpose of 1) to analyze the relationship of the content of the hormone progesterone in the blood to the success of artificial insemination in Bali cattle in the medium and lowlands of Jambi Province, and 2) to analyze the relationship of minerals in the blood to the success of artificial insemination in Bali cattle in the region medium and lowlands of Jambi Province.

2. Materials and methods

The research method used in this study is survey and laboratory. The survey method is used to evaluate the success of AI in the form of service per conception (S/C), namely how many times insemination occurs until pregnancy occurs in bali cattle in the medium and lowland areas of Jambi Province.

The sampling technique used in this research is Stratified Random Sampling [7] which consists of 2 (two) strata, namely: Strata I is a Bali cattle breed inseminated by inseminators in the medium land region of Jambi Province represented by Tebo District. Stara II is a bali cattle that is inseminated by inseminators in the lowland areas of Jambi Province represented by Tanjung Jabung Barat District.

To measure the content of the hormone progesterone in the blood of bali cattle using radioimmunoassay (RIA) technique. The RIA technique is one of the most sensitive detection methods, which is based on antigen-antibody interactions. Radioactive labeled antigen (hormone) can be used to detect hormone content in a sample. Furthermore, to analyze the relationship of blood hormone and mineral content with the success of artificial insemination in Bali cattle in the medium and lowland areas of Jambi Province, Pearson Correlation, Kendall's tau_b, and Spearman's rho correlations were used.

3. Results and discussions

3.1. The success rate of Artificial Insemination (AI)

The results of the study in the field obtained data that the success rate of IB in Bali cattle in Jambi Province as reflected by service per conception (S/C) was an average of 1.52 ± 0.32. According to [8], the normal S/C values are in the range 1.6 - 2.1 and according to [4], the ideal S/C values range between 1.6 and 2.0. The lower the S / C value, the more fertile the cow, whereas the high S/C value indicates the low level of fertility of the cow. [3] provides guidance in evaluating the success of the implementation of artificial insemination (AI) by providing a standard value of service per conception (S/C) is 1.6. For each region in Jambi Province, there was a variation in the success rate of AI as presented in the following table.

Table 1. Success rate of AI in cattle in Jambi Province.

No.	Jambi Province Region	Service per Conception (S/C)
1.	Medium Areas (Tebo District)	1,34 ± 0,17^{b}
2.	Low Areas (Tanjung Jabung Barat District)	1,47 ± 0,16^{c}
	Jambi Province	**1,52 ± 0,32**

Description: The mean ± SD in one column followed by different lowercase letters shows the difference.

Then in the analysis of the mean difference test for the success rate of AI in cattle between the medium areas (Tebo District) and the low areas (Tanjung Jabung Barat Distric) there was a significant value = 0.00 <α 0.05, which means there are differences. This value means that the success rate of AI in cattle in the medium areas (Tebo District) is better than the low areas (Tanjung Jabung Barat Distric). [13] conducted research in the highlands and lowlands in West Sumatra with the conclusion that the reproductive efficiency of cattle in IB both service per conception (S/C) and conception rate (RC) in the lowlands is better than S/C and CR in the highlands.

The results of [15] conclude that the results of AI in coastal and mountainous areas are relatively the same, which is 37.5% in coastal areas and 40% in mountainous regions. According to [2] [14], differences in the success of AI in mountainous regions with beaches are thought to be due to differences in altitude which will have a lot of influence on diet, quality of vegetation, wind speed which ultimately stresses the level of livestock which will result in a difference in productivity. The influence of environmental factors on the reproductive appearance of livestock.

Likewise with the results of [12], young cattle that are raised in high topographic areas experience faster puberty compared to young cows that are kept in low topography. The rate of growth of reproductive cells and body cells of cattle for adult sex has the highest value for high topographic regions compared to the medium and low topography.

3.2. Relationship of the progesterone hormone with the success of AI cattle

Progesterone is a hormone that is very important to maintain pregnancy. Progesterone is produced by the corpus luteum. The corpus luteum is formed after the follicle secures the egg, so that in superovulated treated cows it produces more corpus luteum and the progesterone hormone. According to [18], the concentration of the hormone progesterone < 1 nmol/l indicates that there is no reproductive activity in the ovary; between 1-3 nmol/l indicates activity but is doubtful; and concentrations > 3 nmol/l indicate activity in the ovary.

Progesterone is a very important hormone in regulating the normal cycle function of female animal reproduction [5]. In the estrus cycle, namely in the luteal phase, the hormone progesterone inhibits the secretion of the gonadotropin hormone, folicle stimulating hormone (FSH) and luteinizing hormone (LH) in the anterior pituitary gland. These obstacles cause follicles in the ovary to not develop and the hormone estrogen is not produced, so that animals do not show symptoms of estrus [9] [6]. The results of the study found that the progesterone content of cattle in Jambi Province averaged 5.55 ± 1.97 ng/mL, for each region in Jambi Province as presented in the following table.

Table 2. Content of progesterone cattle in Jambi Province.

No.	Jambi Province Region	Cattle Progesterone Hormone Content (ng/mL)
1.	Medium Areas (Tebo District)	5,77 ± 1,86
2.	Low Areas (Tanjung Jabung Barat District)	5,49 ± 2,54
	Jambi Province	**5,55 ± 1,97**

Description: The mean ± SD in one column followed by different lowercase letters shows the difference.

The results of [11] mean that progesterone levels when pregnant Bali cattle or after AI are 15.43 ± 0.50 ng /mL; 17.16 ± 0.34 ng /mL; and 20.78 ± 0.59 ng /mL, for gestational age 30 days, 60 days and 90 days, respectively. Progesterone levels in pregnant Bali cattle increased in line with gestational age, ie from gestational age 30 to 90 days, and statistically it appeared to significantly increase (P <0.05) between gestational age 30th, 60th day and 90th day. According to [6] the hormone progesterone is often referred to as pregnancy hormone, because its function is to maintain pregnancy. In cattle the main source of pregnancy progesterone is the corpus luteum so that the cow including the very dependent corpus luteum or the presence of the corpus luteum as a source of progesterone during pregnancy is very necessary

The results of correlation analysis (Pearson Correlation, Kendall's tau_b, and Spearman's rho) obtained the output that is the relationship between the content of the hormone progesterone in the blood of cattle with the success of artificial insemination in cattle in Jambi Province was very significant at the level of $\alpha = 0.01$. Correlation coefficient (r) between the content of the hormone progesterone in the blood of cattle and the success of artificial insemination in cattle in Jambi Province differed slightly between the three correlation models.

Based on the correlation coefficient, there is a relationship between the content of the hormone progesterone in the blood of cattle and the success of AI in cattle in Jambi Province. These results indicate the higher the content of the hormone progesterone in the blood of cattle, the higher the success of AI in cattle in Jambi Province, or in other words the better the factor of livestock, the higher the success of AI. Physiologically, the hormone progesterone functions to maintain pregnancy in all

IOP Conf. Series: Earth and Environmental Science **287** (2019) 012004 doi:10.1088/1755-1315/287/1/012004

species by stimulating uterine growth and suppressing myometrial contractions [1]. Progesterone is widely used to monitor postpartum ovarian activity, diagnosing early pregnancy in cattle [10].

The results of this study are in line with the research of [16] who found that the percentage of cattle pregnancy is influenced by many things including the estrus quality of AI acceptors which are directly affected by hormonal conditions. There is a positive correlation between the quality of estrus and the percentage of pregnancy, which is the better the estrus state, the greater the percentage of pregnancy. The reproductive process is related to the mechanism of the hormonal system, namely the relationship between hypothalamus-pituitary hormones namely gonadotrophin releasing hormone (GnRH), follicle stimulating hormone (FSH) and luteinizing hormone (LH), ovarian hormones (estrogen and progesterone) and uterine hormones (prostaglandins). If the mechanism of the system runs normally, the percentage of cattle pregnancy will be higher.

3.3. Relationship of blood mineral content to the success of cattle AI

In the study, besides also analyzed the relationship of feed mineral content, also analyzed the relationship of blood mineral content of cattle with the success of IB in cattle in Jambi Province. Research in the field to obtain blood mineral content is shown in the table below.

Table 3. Blood mineral content of cattle in Jambi Province.

No	Blood Minerals	Jumlah
1.	Zn	1,33 ± 0,19
2.	Cu	0,24 ± 0,13
3.	Se	0,61 ± 0,08
4.	Mg	2,97 ± 0,38
5.	Ca	0,01 ± 0,06
6.	P	0,01 ± 0,02

Correlation analysis results (Pearson Correlation) obtained output, namely: The relationship between blood mineral content in the form of Zn, Se and Mg with the success of IB in cattle in Jambi Province was very significant at the level of $\alpha = 0.01$. While the relationship between blood mineral content in the form of Cu, Ca and P with the success of IB in cattle in Jambi Province was not significant at the level of $\alpha = 0.01$ and $\alpha = 0.05$.

4. Conclusion

1. The success rate of AI in cattle in the middle areas is better than the low areas in Jambi Province,
2. There is a relationship between the content of the hormone progesterone in the blood of cattle and the success of AI in cattle in Jambi Province.
3. There is a relationship between blood mineral content in the form of Zn, Se and Mg with the success of AI in cattle in Jambi Province.

5. Reference

[1] Al-Asmakh, M. 2007. Reproductive Functions of Progesterone. Middle East Fertility Society Journal 12 (3): 147-152.

[2] Awemu, E. M., L. N. Nwakolar, and B. Y. Abubakar. .1999. Environmental Influences on Preweaning Mortality and Reproductive Performance of Red Sakoto does. Small Ruminant Research. 34: 161-165.

[3] Direktorat Jenderal Peternakan. 2010. Blue Print Program Swasembada Daging Sapi 2014. Kementrian Pertanian Direktorat Jenderal Peternakan. Jakarta.

[4] Dwiyanto, K. 2012. Optimalisasi Teknologi Inseminasi Buatan Untuk Mendukung Usaha Agribisnis Sapi Perah dan Sapi Potong. Bunga Rampai. Puslitbangnak.

[5] Graham, J. D, L. C. Christine. 1997. Physiological Action of Progesterone in Target Tissues. Endocrine Reviews 18 (4): 502-519.

[6] Hafez, E. S. E. 2000. Reproductuion in Farm Animals. Lea and Febiger, Philadelphia.

[7] Harun Al Rasyid. 1994. Teknik Penarikan Sampel dan Penyusunan Skala. disunting oleh Teguh Krismantoroadji dkk., Program Pascasarjana Universitas Padjadjaran, Bandung.

[8] Jainudeen, M. R. and E. S. E. Hafez. 2008. Cattle and Buffalo dalam Reproduction in Farm Animals. 7th Edition. Edited by Hafez E. S. E. Lippincott Williams & Wilkins. Maryland. USA.

[9] McDonald, L. E. 2000. Veterinary Endocrinology and Reproduction. 3 Ed. London. Bailliere Tindall. Pp 315-367.

[10] Osman, M. M., K. H. M. El Bayomi, A. S. Abdoon, and A. A. S. El Nabtiti. 2012. Using of Progesterone Field Test (on Farm) as a Tool for Early Detection of Pregnancy in Dairy Farms. *SCVMJ* 17(1): 9-18.

[11] Pemayun, T. G. O., I. G. N. B. Trilaksana, dan M. K. Budiasa. 2014. Waktu Inseminasi Buatan Yang Tepat Pada Sapi Bali dan Kadar Progesteron Pada Sapi Bunting. Jurnal Veteriner. Vol. 15 No. 3 September 2014. Hal: 425-430.

[12] Syahwal, S., B. P. Purwanto dan I. G. Permana. 2013. Studi Hubungan Respon Ukuran Tubuh dan Pemberian Pakan Terhadap Pertumbuhan Sapi Pedet dan Dara pada Lokasi yang Berbeda. JITP Vol. 2 No. 3, Januari 2013. Hal: 175-188.

[13] Udin, Z dan. B. P. Zesfin. 2010. Analisis Mineral Utama dan trace Mineral pada Induk Sapi yang di IB di Sumatera Barat. Laporan Penelitian. Universitas Andalas, Padang.

[14] Urdaneta, L. D., G. T. Hernandez, C. M. B. Perez, O. G. Betancourt, F. G. Cossio, M. O. Arce and O. G. Betancourt. 2000. Comparison of Alpine and Nubian Goats for some reproductive traits under dry tropical condition. Small Rum. Res. 36: 91-95.

[15] Utomo, S. 2013. Pengaruh Perbedaan Ketinggian Tempat Terhadap Capaian Hasil Inseminasi Buatan Pada Kambing Peranakan Ettawa. Sains Peternakan Vol. 11 (1), Maret 2013. Hal: 34-42.

[16] Wahyudi, L., Susilawati, T. dan Isnaini, N. 2014. Tampilan Reproduksi Hasil Inseminasi Buatan Menggunakan Semen Beku Hasil Sexing Pada Sapi Persilangan Ongole Di Peternakan Rakyat. Jurnal Ternak Tropika Vol. 15, No.1. Hal: 80-88.

[17] Widhyari, S. D., A. Esfandiari, A. Wijaya, R. Wulansari, S. Widodo, dan L. Maylina. 2015. Tinjauan Penambahan Mineral Zn dalam Pakan Terhadap Kualitas Spermatozoa pada Sapi *Frisian Holstein* Jantan. Jurnal Ilmu Pertanian Indonesia (JIPI), April 2015. Vol. 20 (1): 72 77.

[18] Zdunczyk, S., E. S. Mwaanga, J. M. Tepicht, W. Baranski, and T. Janowski. 2002. Plasma Progesterone Levels and Clinical Findings in Dairy Cows with Post-Partum Anoestrus. Bull. Vet. Inst. Pulawy, 46, pp. 7986.

Differences of antibody titer to avian influenza and hematology profile on local ducks in Central Java

Ismoyowati, D Indrasanti, and I H Sulistyawan

Faculty of Animal Science, Jenderal Soedirman University, Purwokerto, Central Java, Indonesia.
Email: moy.moyowati@gmail.com.

Abstract. Avian Influenza (AI) in poultry farms in Indonesia is still a threat to the sustainability of livestock and farmers. Duck is one of the most heterogeneous natural reservoirs and hosts of AI viruses. This study aims to detect antibody titer AI levels in blood serum and duck eggs as well as to know the different hematology profile of Tegal and Magelang ducks in Central Java. Research method was survey at farmer level. Farmer and duck samples were determined by purposive random sampling of 10 breeders for each region. Selection of respondents and ducks was based on the number of duck owners and duck age. The samples of farmers and ducks taken were determined randomly as many as 10 breeders in each area and from each breeder was selected 6 female ducks for their blood and 10 eggs to be analyzed using antibody titer against AI. The variables measured in this study were antibody titer against AI measured on blood serum and serum egg yolk, with inhibition hemagglutination test and haematological profile. The data obtained were analyzed using T-test. The results showed protective antibodies against AI on blood serum Tegal ducks (seropositive) 58% while Magelang ducks 46%. Titer of serum antibodies of Tegal duck in the range of 22-25 while Magelang ducks 22-26. Titer of egg antibody in Tegal ducks in the range of 20-210 while Magelang ducks 20-27. T-test on AI antibody titer on blood serum and serum of eggs of Tegal duck and Magelang ducks (which have different environment) showed no significant difference (P> 0.05). The result of T-test on leukocyte fraction profile showed that on heterophile, lymphocyte, and H / L both ducks showed a very significant difference (P <0.01). Higher H / L values in Tegal ducks show stress due to heat stress. T-test on red blood cells, white blood cells, hemoglobin, packed cell volume (PCV), total plasma proteins, eosinophils, monocytes, fibrinogen and albumin showed no significant difference (P> 0.05). The study concluded that there was no difference of antibody titer level to AI and hematology profile in Tegal and Magelang ducks despite different maintenance environment. However, Tegal duck was indicative of higher stress with higher H / L values than Magelang duck.
Keywords - Avian Influenza, hematology, Magelang duck, Tegal duck, antibody titer.

1. Introduction

AI epidemic to the poultry declared in January 2004 in Indonesia has significantly decreased each year; 2.751 cases were reported in 2007 and it gradually declined up to 104 by the end of the year [1]. AI breakout ended in Tegal and Magelang in 2016 but vaccination among the breeders was rare. AI infection to ducks in a region may trigger antibody against AI. Ducks have a higher seroprevalence against High Pathogenic Avian Influenza (HPAI) H5N1 than chickens [2] [3]. Duck is one of the most heterogenous natural reservoirs and host for AI with the ability of shedding a mass of virus and mild pathological and clinical symptoms [4] [5]. Accordingly, ducks are concerned to transmit the virus to other ducks or even to other poultry.

In addition to detecting antibody titer to particular disease, examination on blood profile serves as a method to evaluate the physiological condition and health status of the poultry. It may be a good indicator for poultry in different environment and habitat [6]. This research, therefore, used survey method to investigate the in-situ antibody titer and blood profile of Tegal ducks and Magelang ducks from the native habitat in Tegal and Magelang.

1.1. The purpose of the research

The purpose of the research were to detect the level of antibody titer against AI in the ducks' blood serum and eggs, and to observe the different hematological profile between Tegal ducks and Magelang ducks in Central Java farming duck.

2. Materials and methods

2.1. The material of the research

Research materials were two breeds of ducks and the eggs from the farmers. which were collected from duck breeding centers in Central Java that included "Berkah Abadi" farmer breeders group in Pesurungan Lor, Margadana, Tegal; and "Sido Rukun" farmers breeders group in Sawangan, Magelang. Ducks farmers in Tegal as the respondents had a minimum 600 ducks aged 8-15 months. Farmers in Magelang had 150 ducks with the same age range. The selection was based on the preliminary survey that observed a farmer in Tegal generally owned 400 adult female ducks while in Magelang 100 productive ducks. Ten farmers were randomly selected from every region. Blood sample was collected from six hens from each farmer, and 10 eggs were taken from each breeding center to analyze the antibody titer against AI.

2.2. Methods

Survey method was conducted among the farmers. Purposive random sampling was used to select 10 farmers and ducks in each region/respondents and ducks were selected based on the numbers of ducks in the breeders and ducks' age. The measured variables in the study were antibody titer against AI from the blood serum and egg yolk serum, and hematological profile i.e., erythrocyte count, leukocyte count and differential leukocyte, hematocrit value, hemoglobin, total plasma protein, albumin, fibrinogen and the H/L ratio. Serology test was performed to measure antibody titer against AI using hemagglutination inhibition (HI) test according to the procedure of Office International des Epizooties [7] to determine antibody titer against H5N1 virus. The obtained data were subject to T-test.

2.3. Egg yolk serum preparation procedure [8]

The eggs were cracked then the yolks were separated from the white. The egg yolks were paced in a petri dish and the yolk membrane was washed in aquadest. Then, the yolks were stored in a tube to measure the volume, added with Dulbecco solution twice as much as the yolk volume, then homogenized by stirring. As much chloroform as the yolk was added. The solution was stirred to homogenous and formed emulsion, then centrifuged at 1000 x g for 30 minutes at room temperature. Upon centrifugation, the solution was separated into three layers—orange liquid (lecithin) at the bottom, semi-solid yolk emulsion in chloroform in the middle, and the liquid phase of protein serum at the top. The top layer which contained IgY was stored at -20 ° C until analysis was performed.

3. Result and discussion

3.1. Antibody titer against AI in Tegal duck in Tegal region and Magelang Duck in Magelang Region

The propagation pattern of AI virus is generally affected by the season; the prevalence of AI is higher in autumn, winter and rainy seasons. In Indonesia, AI breakout tends to increase during a rainy season because the high humidity is the contributing factor to AI virus breeding and spread in the environment. Wild waterfowl such as Anseriformes (mallard, Muscovy and goose) and Charadriiformes (seagull, oceanic bird and wild bird) are the natural reservoir of type A influenza virus and play important role in the virus ecology and propagation.

Ducks breeding in the breeder centres in Tegal and a fraction of farmers in Magelang used an enclosed ducks house. This model has a lower risk of disease spread than the traditional or nomadic breeding. Nomadic behaviour and a considerably long-distant transporting increase the risk of the ducks as a carrier of AI from one region to another. furthermore, feed poisoning risk during the free-range can be overcome by an enclosed breeding system [9] [10]. Antibody titer in Tegal duck in Tegal regency and Magelang duck in Magelang regency is presented in Table 1.

Table 1. Mean of antibody titer of blood serum and egg serum of Tegal duck in Tegal and Magelang duck in Magelang.

Breeders	Mean of antibody titer			
	Ducks' blood serum (2^n)		Ducks' egg serum (2^n)	
	Tegal	**Magelang**	**Tegal**	**Magelang**
1	4.2	3.5	1.6	2.1
2	3.8	3.9	1.1	1.5
3	3.6	3.4	2.2	2.4
4	3.8	3.7	1	1.7
5	3.9	3.7	1.1	1.5
6	3.7	3.7	2.2	2.4
7	3.5	3.7	1	1.7
8	3.9	3.4	1.1	1.5
9	3.7	3.8	2.2	2.4
10	3.7	3.5	1.1	1.6
Mean	3.80	3.63	1.43	1.86

The antibody is protective against AI when it inhibits the serum which is diluted 1 : 16 (2^4) or log 2^4 using antigen 4 HAU [7]. The antibody protective against AI in the blood serum of Tegal duck and Magelang duck was seropositive; 58% and 46%, respectively. The protective antibody titer in egg serum of Tegal duck and Magelang ducks was seropositive, 11% and 4%, respectively. Antibody titer serum in Tegal duck and Magelang duck was 2^2-2^5 and 2^2-2^6, respectively. Antibody titer egg in Tegal and Magelang duck was 2^0-2^{10} and 2^0-2^7, respectively. The seropositive serum showed immunity in the ducks which may link to the previous exposure to AI in the region. On the other hand, the seropositive egg serum may due to the maternal antibody from the hen. However, the study shows that the majority of the ducks have low antibody titer against AI (Table 1).

Result of HI mean value showed that Tegal duck in Tegal had a slightly higher antibody titer compared to that of Magelang duck in Magelang. It may due to the coastal area of Tegal that allows a more frequent interaction between Tegal duck and the wild waterfowls than that of Magelang duck. Accordingly, several ducks with high immunity had a seropositive antibody titer against AI. However, the result of paired t-test on antibody titer AI on blood serum and egg serum did not show a significant effect. The possible reason for this was the absence of AI vaccine in both area; therefore, the ducks may not survive in the next AI breakout despite the fact that Tegal and Magelang had a high prevalence of AI.

Some contributing factors to the level antibody titer against AI are the history of AI incidence and AI vaccination. AI H5N1 vaccination increased antibody titer in ducks which is transferrable to the eggs produced and ducklings [11]. Infection case in a breeding centre without vaccination follow-up may lead to a poor distribution of antibody titer, i.e. some ducks have a very high antibody titer and the rest have the counterpart level.

A dense population and a free breeding system of waterfowl are the contributing factors to waterfowl as the potential reservoir of H5N1 virus in Indonesia [3]. An enclosed breeding system for Tegal ducks and Magelang ducks would decrease AI incidence compared to the backyard system (free-range around the cage) due to the minimum contact with the ducks or the other poultry [12], particularly for ducks with poor immunity or without AI vaccination. Furthermore, improvement of breeding management is important by grouping the ducks based on age to prevent the contagion [13].

High antibody titer was assumedly due to the body immune system in responding to the continuous, long-term AI exposure from the environment which, by no means, naturally vaccinated the poultry. This process will likely to occur in backyard system due to lack of biosecurity, so AI virus can easily transmit from one poultry to another through direct or indirect contact. Positive HI test showed that in the animal's body exist an antibody as a sign of early AI infection or exposure in the poultry. The gradual exposure may stimulate AI antibody but insufficient to induce the clinical symptoms [14].

3.2. Hematological profile of Tegal duck in Tegal and Magelang duck in Magelang

Hematological profile can be used to identify the clinical health status because blood plays an important role to measure the physiological, pathological and nutritional status of an animal. The change in blood constituent in animal physiological status may due to genetic and non-genetic factors such as age, sex, breeding and management system [15] [16]. The hematological profile in the present study (in situ) included erythrocyte, leukocyte, haemoglobin, packed cell volume (PCV), total protein plasma, heterophile, eosinophil, limphocyte, monocyte, fibrinogen and albumin and H/L ratio. The surveyed hematological profile and the standard hematological profile are presented in Table 2.

Table 2. Mean of hematological profile of Tegal duck and Magelang duck in breeder centres in Tegal and Magelang.

Hematological profile	Mean		Standar (Scalm, 2010)
	Tegal duck	**Magelang duck**	
Erythrocyte (jt/µl)	3.048	3.243	2.01±0.4
Leukocyte (.10^6 / µL)	9.489	9.567	23.4–24.8
Hb (g/dL)	10.893	10.987	11.4±1.6
PCV (%)	33.033	34.967	39±5.1
TPP (g/dl)	3.913^a	3.300^b	-
Heterophile segment (%)	36.83^a	28.23^b	29±1.4
Eosinophil (%)	7.1	7.07	0.2±0.17
Basophil (%)	0	0	2.2±0.3
Limphocyte (%)	50.33^b	59.27^a	66±1.4
Monocyte(%)	6	5.4	2.5±0.3
Fibrinogen (g/dl)	0.460	0.460	-
Albumin (g/dl)	3.597	3.822	-
H/L	0.773^a	0.497^b	-

Note: Values bearing different superscript within column show a highly significant difference (P>0,01) based on t-test

Table 2 shows the different blood parameter of Tegal and Magelang ducks compared to the normal standard [17]. The mean values of erythrocyte, leukocyte, haemoglobin, PCV and albumin in Magelang ducks were higher than those of Tegal ducks. TPP, heterophile, eosinophil, basophil, lymphocyte, monocyte and H/L ratio in Tegal ducks were higher than those of Magelang ducks (Table 2).

The result of t-test of erythrocyte, leukocyte, hemoglobin, packed cell volume (PCV), eosinophil, monocyte, fibrinogen and albumin did not show a significant difference (P>0.05). it indicated that there was no difference in hematological profile in both types of ducks despite the different breeding environment and nutrition in feed. The hematological profile was relatively similar because both ducks are derived from one species Anas plathyrynchos. A significant difference in blood profile is generally attributed to the variation in breed, hormone and nutrition [15]. Although Magelang ducks were given a lower protein feed (17,99%) than that of Tegal ducks (19,25%), the environment of Magelang region has supported the minimum requirement of growth and protein for the ducks. The result of this research was in line with [6] that the contributing factors to hematological value were breed, sex, age and reproductive status, heart rhythm, breeding procedure and nutritional status of the animals.

Table 2 shows that erythrocyte, hemoglobin and PCV of Magelang ducks are biologically higher than those of Tegal ducks. The influential factors may include the different breeding region; Magelang is a highland and Tegal is a coastal area. According to [18], erythrocyte count would increase in a lower temperature and decrease in a higher temperature. In addition, the heat stress in Tegal ducks may affect blood profile, particularly to erythrocyte, hematocrit, hemoglobin and leukocyte [19].

The result of t-test on hematological profile showed that the total protein plasma, heterophile, lymphocyte and H/L ratio of both types of ducks were significantly different ($P<0.01$); the nutrient feed of Tegal ducks was higher than that of Magelang ducks. The different nutrient content profoundly affected the blood profile, particularly the blood protein synthesis [20]. The high protein feed resulted in a higher total protein than that of Magelang duck. Nutrition offered to Tegal ducks was rich in composition and variation, containing 19.25% crude protein and 2914 kcal/g energy, while Magelang ducks consumed 17.99% crude protein and 2801 kcal/g energy. However, the different nutrition did not significantly improve blood profile because of the potential heat stress for Tegal ducks.

A higher H/L ratio in Tegal duck showed a heat stress. The average temperature during the three-month observation (July - September 2017) in Tegal duck breeding centre was 30,53 ± 1,38°C while in Magelang was 28,32 ± 1,00°C. The environment temperature of both regions is above the thermoneutral zone for ducks, i.e., 23-25°C [21]. Heat stress was detected from the decreasing leukocyte count, particularly lymphocyte ($P<0.01$) and the high H/L ratio compared to those of Magelang ducks which were below the Scalm standard [17]. Lymphocyte is the part of leukocyte that plays role in the immunity system. IgG is the main antibody produced by lymphocyte [20] [22]. The result of this research was in line with that of [22] that lymphocyte count decreased during the heat stress, as observed from the increasing H/L ratio. However, the effect of nutrition to Magelang ducks required a further investigation considering the low leukocyte level in the ducks compared to the Scalm standard [17].

4. Conclusions

Tegal ducks reared in Tegal region and Magelang ducks in Magelang region have been exposed to AI. The level of antibody titer to AI in blood serum and egg yolk was relatively similar. Also, Tegal ducks and Magelang ducks showed a relatively similar hematological profile and stress despite being reared in a different environment

5. References

[1] Ministry of Agriculture, Directorate General of Livestock and Animal Welfare. The situation of Avian Influenza (AI) breakout to poultry up to 31 August 2017. 2017. http://ditjennak.pertanian.go.id/artikel.

[2] Henning J, Wibawa H, Morton J, Usman TB, Junaidi A and Meers J. 2010. Scavenging Ducks and Transmission of Highly Pathogenic Avian Influenza, Java, Indonesia. *Emerging Infectious Diseases* **16** (8): 1244-1250.

[3] Hewajuli DA and Dharmayanti NLPI. 2012. The relation between AI and waterfowl to create a genetic diversity and the role of waterfowl as the reservoir in the propagation of AI virus. *Wartazoa* **22** (1): 12-23.

[4] Keawcharoen J, Van Riel D, Van Amerongen G, Bestebroer T, Beyer WE, Lavierent RV, Osterhaus ADME, Fouchier RAM and Kuiken T. 2008. Wild ducks as long-distance vectors of highly pathogenic Avian Influenza virus (H5N1). *Emerg. Infect. Dis.* **14** (4): 600-607.

[5] Ronnmark R and Gisaeus E. 2011. Illustration of the host range of influenza a virus with the natural reservoir of influenza A virus, accidental hosts, and the subtype that have been identified in the different groups http://www.infectionecologyandepidemiology.net/index.php/ Accesed 5 February 2018.

[6] Dalai M, Puspamitra S, Bhattacherjee A, Acharya D, Acharya G and Mohanty PK. 2015. Comparative haematology of Anas platyrhynchos (Anseriformes) and Coturnix coturnix japonica (Galliformes). *J. Entomology and Zoology Studies* **3** (5): 50-53.

[7] OIE. 2000. International Animal Health Code mammals, birds and bees Ninth Edition, Office International Des Epizooties, Paris, France.

[8] Hamal KR, Burgess SC, Pevzner IY and Erf GF. 2006. Maternal Antibodi Transfer from Dams to Their Egg Yolks, Egg Whites, and Chicks in Meat Lines of Chickens. *Poultry Science* **85**: 1364–1372.

[9] Saleh E. 2004. Backyard duck reeding management. Department of Animal Husbandry. Usu Digital Library:1-15.

[10] Indonesian Agency for Agricultural Research and Development. 2012. Beware of ducks in the propagation of avian influenza virus or AI. 19-25 September 2012 Agroinovasi Sinar Tani No.3474 year XLIII.

[11] Ismoyowati, Mufti M, Samsi M and Susanto A. 2013. Maternal Antibody Titer Against Avian Influenza Transferred from Hens to The Eggs and Ducklings. *Animal Production* **15**(3): 189-195.

[12] Hadipour MM, Habibi G and Vosoughi A. 2011. Prevalence of Antibodies to H9N2 Avian Influenza Virus in Backyard Chickens around Maharlou Lake in Iran. *Pakistan Veterinary Journal* **31**: 192-194.

[13] Sarah CH, Manvell RJ, Schulenburg B, Shell W, Wikramaratna PS, Perrins C, Sheldon BC, Brown IA and Pybus OG. 2016. Antibody responses to avian influenza viruses in wild birds braden with age Proceedings of the Royal Society B 283 pp: 1-9.

[14] Elfidasari D, Puspitasari RL and Frisa A. 2014. Detection of antibody due to AI virus subtype H5N1 in domestic waterfowl surrounding the natural conservation in Pulau Dua. *Jurnal Al-Azhar Indonesia Seri Sains and Teknologi* **2** (4): 260-269.

[15] Pandian C, Pandiyan MT, Sundaresan A and Omprakash AV. 2012. Haematological Profile And Erythrocyte Indices In Different Breeds Of Poultry. *International Journal of Livestock Research* **2**(3): 89-92.

[16] Etim NA, Williams ME, Akpabio U and Offiong EEA. 2014. Haematological Parameters and Factors Affecting Their Values. *Agricultural Science* **2** (1): 137-47.

[17] Weiss DJ and Wandrop KJ. 2010. Schalm's Veterinary Hematology. Sixth edition. Wiley-Blackwell. Iowa-USA. Pp.977-986.

[18] Guyton AC and Hall JE. 2008. Lesson book of Medical Physiology. 11th Ed. Irawati et al.. Jakarta (ID): EGC. Translated from Textbook of Medical Physiology.

[19] Harlova HJ, Blaha M, Koubkova J, Draslarova J and Fucikova A. 2002. Influence of heat stress on the metabolic response in broiler chickens. *Scientia Agriculturae Bohemica* **33**: 145–149.

[20] Swenson MJ. 1993. Physiological Properties and Celluler and Chemical Constituent of Blood. In Dukes Physiology of Domestic Animals. 11 Edition. Comstock Publishing Associates a division of Cornell University Press Ithaca and Londion. Pp. 22- 48.

[21] El-Badry ASO, Hassanane MM, Ahmed ES and El-Kholy KH. 2009. Effect of Early-Age Acclimation on Some Physiological, Immunological Responses and Chromosomal Aberrations in Muscovy Ducks During Exposure to Heat Stress. *Global J. Biotechno. & Biochem.* **4** (2): 152-159.

[22] Kusnadi E. 2009. The effect of different stresses on the change of several blood components and biochemicals of poultry. National Seminar of Livestock Technology and Veterinarian 2009. Faculty of Animal Science, Andalas University, Limau Manis Campus Pp: 580-585.

Acknowledgments
The authors express their sincerest gratitude to the Directorate of Research and Community Service, Ministry of Research and Technology and Higher Education of Indonesia for the competitive grant of basic research in universities.

IOP Conf. Series: Earth and Environmental Science **287** (2019) 012006 doi:10.1088/1755-1315/287/1/012006

Prevalence of resistant *Salmonella* spp. isolated from raw meat and liver of cattle in the Wa Municipality of Ghana

R Ekli[1], F Adzitey[1, 2, *], and N Huda[3]

[1] University for Development Studies, Department of Animal Science, Box TL 1882, Tamale, Ghana.
[2] University for Development Studies, Department of Veterinary Science, Box TL 1882, Tamale, Ghana.
[3] Universiti Malaysia Sabah, Faculty of Food Science and Nutrition, Jalan UMS, 88400, Kota Kinabalu, Sabah, Malaysia.
* Corresponding author: Email: adzitey@yahoo.co.uk; Tel: +233249995310.

Abstract. This study was carried out to determine the prevalence of resistant *Salmonella* spp. isolated from meat (beef) and liver of cattle in the Wa, Municipality of Ghana. Isolation of *Salmonella* spp. was done according to the USA-FDA Bacteriological Analytical Manual. Antibiotic susceptibility test was performed using the disc diffusion method and the results interpreted using the CLSI guidelines. Antibiotic residues determination was carried out using Premi®Test Kit. Liver samples 32%(16/50) were more contaminated with *Salmonella* spp. as compared to beef samples 30% (15/50). The *Salmonella* spp. were highly resistant to teicoplanin (96.77%) but susceptible to chloramphenicol (100%), ciprofloxacin (100%), tetracycline (100%), suphamethoxazole/ trimethoprim (100%), amoxycillin/clavulanic acid (93.55%), ceftriaxone (93.55%) and gentamicin (83.87%). The *Salmonella* spp. exhibited 5 antibiotic resistant patterns with the pattern (Tec) being the commonest. Multiple antibiotic index ranged from 0.11 to 0.33. A total of 35 samples (20 beef and 15 liver samples) were examined for antibiotic residues, and 7 (20%) were positive for antibiotic residues; 6 (17%) were found in the liver while 1 (3%) was in the beef. This study revealed that salmonellae are present in beef and liver samples examined and showed high resistance to teicoplanin. Some of the beef and kidney samples harbored antibiotic residues.
Keywords – antibiotics, cattle, drug residues, raw meat, *Salmonella* spp.

1. Introduction

Foodborne disease outbreaks continue to be an important public health problem globally and most of the food safety hazards are caused by foods of animal sources [1]. Bhandare et al. [2] also reported that contaminated raw meat has been identified as one of the main sources of foodborne illnesses. Among the reasons for which meat and meat products are consumed includes their high protein contents, available vitamins, minerals, lipids and savuory sensation. However, due to its high nutrient content which supports the growth of microorganisms, it is classified among the most perishable staple foods [3]. Adeyemo [4] indicated that there are two sources for which microorganisms get into meat that is either from the hide of the animals or from the abattoirs where the animals are slaughtered and processed. Food-borne pathogens such as *Staphylococcus* spp., *Escherichia coli*, *Salmonella* spp., *Shigella* spp., *Bacillus* spp. and *Listeria monocytogenes* have been isolated from meats in Ghana [5,6,7, 8] some of which have been subjected to antimicrobial susceptibility test [9]

According to Fratamico et al. [10] most illnesses and deaths in the developing countries are caused by foodborne diseases and these cost billions of dollars in medical care and social expenditures. Globally, *Salmonell*a spp. are one of the pathogens that lead millions of cases of enteric diseases, thousands of hospitalizations and deaths each year [11,12]. CDC [13] reported that *Salmonella* spp. causes an estimated 1.4 million cases of foodborne illnesses and more than 500 deaths per year in the US. This pathogen is also responsible for collective food poisoning with approximately 65% of cases in France [14].

For past twenty years, the occurrence of antibiotic resistant *Salmonella* spp. has become a serious health challenge in the globe. The frequent use of antimicrobial agents in food animal production and the routine practice of giving antimicrobial agents to domestic livestock as a means of preventing and treating diseases has contributed to the occurrence of *Salmonella* spp. with decreased susceptibility to drugs [15]. The objectives of this study were to determine the prevalence and antibiotic resistance of *Salmonella* spp. isolated from raw meat and liver of cattle in the Wa Municipality of Ghana. The study also aimed at determining the presence of antibiotic residues in the meat and liver samples.

2. Materials and methods

2.1. Sample collection and preparation

A total of 100 swab samples (50 beeves and 50 livers) were randomly collected from the Wa, Abattoir and examined for the presence of *Salmonella* spp. Additionally, 35 samples (20 beeves and 15 livers) were randomly collected for antibiotic residue test. The swabs and beef/liver samples kept in an ice chest containing ice block and transported to the Spanish Laboratory of University for Development Studies, Nyankpala Campus. Microbial analysis was carried out immediately on arrival.

2.2. Isolation and confirmation of Salmonella spp.

2.2.1. Non selective pre-enrichment step

To allow recovery and growth of any stressed organism, each swab was placed in 10ml buffered peptone water and incubated at a temperature of 37°C for 18-24 hours.

2.2.2. Selective enrichment step

Two types of media were used for selective enrichment, Rappaport-Vassiliadis and Selenite broths to encourage growth of *Salmonella* spp. while inhibiting other microorganisms. After non selective pre-enrichment, 0.1ml of aliquant was transferred into Rappaport Vassiliadis (RV) broth and inoculated at 42°C for 18-24 hours while 1ml was transferred to Selenite broth and incubated at 37 °C for 18 to 24 hours.

2.2.3. Isolation step

Selective enrichment media were streaked unto two selective solid agars (Xylose Lysine Deoxycholate and Brilliant Green) containing one or more agents that inhibit non-Salmonella organisms. Xylose Lysine Deoxycholate and Brilliant Green Agar plates were incubated at 37°C for 18-24 hours. Colonies having a slightly transparent zone of reddish/ pink color with or without a black center on Xylose Lysine Deoxycholate plates and grey to reddish/pink and slightly convex colonies that caused the color of the medium to be red/pink in Brilliant Green Agar plates were suspected as Salmonella. The suspected colonies from both Xylose Lysine Deoxycholate and Brilliant Green Agar agars were plated onto nutrient agar plates for biochemical confirmation.

IOP Conf. Series: Earth and Environmental Science **287** (2019) 012006 doi:10.1088/1755-1315/287/1/012006

2.2.4 Confirmation step

Presumptive Salmonella spp. were purified on Trypticase Soya Agar and confirmed using Gram Stain, growth in Lysine Iron Agar, Triple sugar iron agar and Salmonella latex agglutination test kit. All media and reagents used were purchased from Oxoid, UK.

2.3 Antimicrobial susceptibility test

Antimicrobial susceptibility of *Salmonella* isolates was checked against some commonly used antibiotics such as Amoxycillin/clavulanic acid (30µg), Azithromycin (15µg), Ceftriaxone (30ug), Chloramphenicol (30ug), Ciprofloxacin (5ug), Gentamicin (10ug), Teicoplanin (30 µg), Tetracycline (30ug) and Suphamethoxazole/trimethoprim using Kirby-Bauer antibiotic discs diffusion method. The results interpreted as sensitive, intermediate, or resistance according to Clinical and Laboratory Standards Institute guidelines (CLSI) [16].

2.4 Procedure for antibiotic residues testing

Premi®Test Kit for determination of drug residues in meat was used for qualitative screening of antibiotics by following the manufacturer's instructions. Approximately $2cm^2$ of raw meat and kidney were used. The samples were frozen and thawed to obtain their extract/juice. 100µl extract/juice were pipetted onto the agar in the ampoule of the test kit and incubated at 37°C for 20 minutes for a pre-diffusion. The extract/juice was carefully flushed out by washing the test tubes twice with demineralized water. The test ampoule was closed with foil to avoid evaporation during incubation. The test ampoules were incubated in water bath at 64.0 °C until negative control change colour from purple to yellow. Color change in the negative control was observed within 3 hours. After the negative control has changed color the results were read using color chart. If there is no antibiotic residue in the extracted solution, the bacteria grow and the acidic pH causes the medium to change from purple to yellow color and this is marked as "negative. If antibiotics are present in the extracted solution, they would kill the bacteria; the whole or part of the tube remains purple and this is marked as "positive for antibiotic residue".

3. Result and discussion

3.1 The prevalence of Salmonella spp. *in meat and kidney of cattle*

The prevalence of *Salmonella* spp. in meat and liver of cattle in the Wa, Municipality of Ghana is shown in Table 1.

Table 1. Distribution of *Salmonella* spp. in liver and meat of cattle in the Wa, Municipality.

Source of meat	Number of samples examined	Number of positive samples	% Prevalence
Liver	50	16	32
Muscle	50	15	30
Total	100	31	24

Salmonella spp. were isolated from liver 32% (16/50) and beef 30%(15/50) samples. This means that the liver and beef samples were contaminated with *Salmonella* spp. Contamination of meat samples happens when maximum care is not taken during slaughtering and dressing of animals. The contamination of various meat types such as beef [5,6,7, 8], guinea fowl meat [9], and pork [8] by *Salmonella* spp. in Ghana have been reported. Adzitey [6] reported an overall prevalence of *Salmonella* spp. in beef samples to be 31% (22/70) in the Tamale Metropolis. The prevalence of *Salmonella* spp. was significantly higher in beef 75% (45/60) than that of table (60%), knife (60%) and apron (33%) samples collected from the Techiman Municipality [9]. Anachinaba et al. [8] found *Salmonella* spp. in beef samples sold at the Bolgatanga

Municipality. In Nigeria, Adesiji et al. [17] did not find *Salmonella* spp. in retail raw chicken, beef and goat meat. They found that 8% (6) of pork was positive for *Salmonella* spp.

3.2 Antibiotic resistance of Salmonella spp. *in meat and liver of cattle*

The antimicrobial susceptibility of the *Salmonella* spp. is shown in Table 2. The *Salmonella* spp. were mostly resistant to teicoplanin (96.77%). They were susceptible to chloramphenicol (100%), ciprofloxacin (100%), tetracycline (100%), suphamethoxazole/trimethoprim (100%), amoxycillin/clavulanic acid (93.55%), ceftriaxone (93.55%) and gentamicin (83.87%). Danikuu (2004) reported that *Salmonella* spp. isolated from farm animals in the Kumasi Metropolis, Ghana were all resistant to tetracycline but susceptible to ciprofloxacin, which agrees with the current study. An intermediate resistance of 12.90% was found for azithromycin. Intermediate resistance means that those isolates are not clearly resistant or susceptible, and such isolates have the tendency to easily become resistant [9].

Table 2. Percentage antibiotic resistance of Salmonella spp. isolated from cattle liver and meat.

Antimicrobial	R (%)	I (%)	S (%)
Amoxycillin/clavulanic acid 30µg (AMC)	3.23	3.23	93.55
Azithromycin 15µg (AZM)	29.03	12.90	58.06
Ceftriaxone 30ug (CRO)	0.00	6.45	93.55
Chloramphenicol 30ug (C)	0.00	0.00	100.00
Ciprofloxacin 5ug (CIP)	0.00	0.00	100.00
Gentamicin10ug (CN)	12.90	3.23	83.87
Teicoplanin 30 µg (TEC)	96.77	0.00	3.23
Tetracycline 30ug TE	0.00	0.00	100.00
Suphamethoxazole/trimethoprim (SXT)	0.00	0.00	100.00
Overall	17.77	2.87	81.36

S, susceptible; I, Intermediate; R, resistant

The antimicrobial resistance profile and MAR index of the *Salmonella* spp. isolates is showed in Table 3. The *Salmonella* spp. exhibited 5 antibiotic resistant patterns and the resistant pattern Tec (teicoplanin, MAR index = 0.11) was the most common. The resistant to 3 different antibiotics AmcAzmTec (MAR index= 0.33), and AzmTecCn (MAR index= 0.33) were observed in 1 and 3 isolates respectively. Thus 12.9% (4/31) of the *Salmonella* spp. exhibited multidrug resistant.

Table 3. Antibiotic resistance profile and multiple antibiotic resistance index of individual *Salmonella* species.

Code	Source	No. of Antibiotics	Antibiotic resistant profile	MAR index
L18	Liver	3	AmcAzmTec	0.33
M8	Muscle	2	AzmCn	0.22
eL15	Liver	2	AzmTec	0.22
M9	Muscle	2	AzmTec	0.22
M36-2	Muscle	2	AzmTec	0.22
M36-3	Muscle	2	AzmTec	0.22

L14	Liver	3	AzmTecCn	0.33
M11	Muscle	3	AzmTecCn	0.33
M28-2	Muscle	3	AzmTecCn	0.33
L1	Liver	1	Tec	0.11
L2	Liver	1	Tec	0.11
L7	Liver	1	Tec	0.11
L16	Liver	1	Tec	0.11
L20	Liver	1	Tec	0.11
L35	Liver	1	Tec	0.11
L36-1	Liver	1	Tec	0.11
L36-2	Liver	1	Tec	0.11
L42-1	Liver	1	Tec	0.11
L42-2	Liver	1	Tec	0.11
L43	Liver	1	Tec	0.11
L49-1	Liver	1	Tec	0.11
L49-2	Liver	1	Tec	0.11
M4	Muscle	1	Tec	0.11
M6	Muscle	1	Tec	0.11
M21	Muscle	1	Tec	0.11
M23	Muscle	1	Tec	0.11
M27	Muscle	1	Tec	0.11
M28-1	Muscle	1	Tec	0.11
M36-1	Muscle	1	Tec	0.11
M48-1	Muscle	1	Tec	0.11
M48-2	Muscle	1	Tec	0.11

In this study, 16.13% (4/31) of the *Salmonella* spp. were resistant to two antibiotics, while 70.97% were resistant to one. [9] found that the resistant pattern EVa was the commonest and was exhibited by 9 different *Salmonella* spp. isolated from beef in Techiman. They also found that some of the *Salmonella* spp. were resistant to 3 or more antibiotics and exhibited multidrug resistance. The increasing multi-drug resistance constitutes a potential source of transmission of resistant strains to humans and of public health concern.

3.3 Antibiotic residues in meat and liver of cattle

Table 4 shows the prevalence of antibiotic residues in cattle liver and meat samples. From Table 4, the prevalence of antibiotic residues in liver 17% (6/15) was higher than muscle 3% (1/35). In [19] reported that beef samples collected from selected markets in Kumasi, Ghana contained drug residues. Babapour et al. [20] screened 500 samples of beef and mutton collected from Iran for drug residues and reported a prevalence rate of 22.8% and 14% for beef and mutton, respectively. In Pakistan, 38.33% of beef samples were observed to be contaminated with antibiotic residues [16]. They observed that the presence of antibiotic residues in beef samples was higher at Karachi (48.33%) followed by Sukkur (41.67%), Hyderabad (36.67%), Mirpurkhas (33.34%) and Larkana (31.67%).

Table 4. Prevalence of antibiotic residues in cattle liver and meat.

Source of meat	Samples	Positive	% Prevalence
Liver	15	6	17
Muscle	20	1	3
Total	35	7	20

The use of antimicrobials in animal production is mainly responsible for the dissemination of antibiotic resistance bacteria isolates. To minimise antimicrobial use, a multifaceted integrated approach is recommended [21].

4. Conclusion

It is concluded that some liver and meat of cattle from the Wa, Municipality of Ghana are contaminated with *Salmonella* spp. which are resistant to some antibiotics. Kidney and meat samples also do harbor antibiotic residues.

Reference

[1] Maripandi A, Al-Salamah A. 2010. Multiple antibiotic resistance and plasmid profiles of S. enteritidis isolated from retail chicken meats. *Am J Food Technol.* 5:260-268.
[2] Bhandare S, Sherikarv A, Paturkar A, Waskar V, Zende, RJ. 2007. A comparison of microbial contamination of sheep/goat carcasses in a modern Indian abattoir and traditional meat shops. *Food Control.* 18:854—68.
[3] Huda N, Shen YL, Huey R, Ahmad, A. Mardiah 2010. Evaluation of physico-chemical properties of Malaysian commercial beef meatballs. *Am J Food Technol,.* 5:13-21.
[4] Adeyemo OK 2002. Unhygienic operation of a city abattoir in South Western Nigeria: Environmental Implication. *African J Environ Appl Microbiol.* 4(1):23-28.
[5] Soyiri IN, Agbogli HK, Dongdem JT. 2008. A Pilot microbial assessment of beef in the Ashaima Market, a suburb of Accra Ghana. *African J Food Agric Nutr Dev.* 8(1):91-103.
[6] Adzitey F. 2015. Prevalence of Escherichia coli and Salmonella spp. in beef samples sold at Tamale Metropolis, Ghana. *Int J Meat Sci.* 5(1):8-13.
[7] Adzitey F, Teye GA, Anachinaba I. 2015. Microbial quality of fresh and smoked guinea fowl meat sold in the Bolgatanga Municipality, Ghana. Asian. *J Poult Sci.* 9:165-171.
[8] Anachinaba IA, Adzitey F, Teye GA. 2015. Assessment of the microbial quality of locally produced meat (beef and pork) in Bolgatanga Municipal of Ghana. *Internet J Food Safety, 17: 1-6.*
[9] Adzitey F, Nsoah JK, Teye G.2015. Prevalence and antibiotic susceptibility of Salmonella species isolated from beef and its related samples in Techiman Municipality of Ghana. *Turkish J Agric - Food Sci Technol.* 3(8):644-650.
[10] Fratamico P, Bhunia A, Smith J. 2005. Food borne pathogens in microbiology and molecular biology. Wymondham,. *Norfolk, UK Caister Acad Press.* 270—5.
[11] CDC. 2009. Third Annual 2009 National Conference on Health Communication, Marketing, and Media. In: *Third Annual 2009 National Conference on Health Communication, Marketing, and Media.*
[12] Hur J, Jawale C, Lee JH. Antimicrobial resistance of Salmonella isolated from food animals: A review: *Food Res Int.* 2011;45(2):819 – 830.
[13] CDC. 2006. *Surveillance for Foodborne Disease Outbreaks - United States,.* Vol 58.
[14] Haeghebaert S, Sulem P, Deroudille L, et al. Two outbreaks of Salmonella Enteritidis phage type 8

linked to the consumption of Cantal cheese made with raw milk, France, 2001. *Eurosurveillance*. 2003;8:151–156.

[15] Angulo F, Johnson K, Tauxe R, Cohen M. 2000. Origins and consequences of antimicrobial-resistant nontyphoidal Salmonella: implications for the use of fluoroquinolones in food animals. *Microb Drug Resist*. 6(1):77-83.

[16] Mangsi As, Khaskheli M, Soomro A, Muhammad H, Shah G. 2014. Antibiotic Residues Detection In Raw Beef Meat Sold For Human Consumption In Sindh,. *Pakistan Int J Res Applied, Nat Soc Sci*. 2(7):15-20.

[17] Adesiji YO, Alli OT, Adekanle MA, Jolayemi JB. 2011. Prevalence of Arcobacter, Escherichia coli, Staphylococcus aureus and Salmonella species in retail raw chicken, pork, beef and goat meat in Osogbo, Nigeria. *Sierra Leone J Biomed*. 3:8-12.

[18] Danikuu M. 2004. Isolation and serotyping of Salmonella from slaughtered food animals in the Kumasi Metropolis. Available at: http://ir.knust.edu.gh/handle/123456789/1936 [Accessed 25 December, 2014].

[19] Abavelim DA 2014. Determination of Antibiotics Residues in Beef and Mutton from some Selected Markets in Kumasi -Ghana..

[20] Babapour J, Fartashmehr L, Azami A. 2012. Overview of antibiotic residues in beef and mutton in Ardebil, North West of Iran. World. *Appl Sci J*. 19(10):1495-1500.

[21] EFSA. European Food Safety Authority and European Centre for Disease Prevention and Control (ECDC) 2016. The European Union summary report on trends and sources of zoonoses, zoonotic agents and food-borne outbreaks in 2015. *EFSA J*. 14:4634-4865.

The effect of addition sweet orange essential oil and penicillin in tris yolk extender to Simmental liquid semen against percentage motility, viability and abnormalities of spermatozoa

S A Sitepu[1] and J Marisa[2]

[1] Department of Animal Husbandry, Faculty of Sains and Technology, Pembangunan Panca Budi University, Indonesia.
[2] Department of Agroecotechnology, Faculty of Sains and Technology, Pembangunan Panca Budi University, Indonesia.

Abstract. One of the causes a poor quality of liquid semen is increasing bacterial growth. The addition of sweet orange essential oil and penicillin in Simmental liquid semen is expected to improve the quality of liquid semen because it contains antibacterial. The purpose of this research is to know the value of the percentage of motility, viability and spermatozoa abnormalities in Simmental liquid semen with an addition of combination penicillin with sweet orange essential oil on tris yolk extender. The research hypothesis is the addition of combination penicillin with sweet orange essential oil to tris yolk extender can increase the value of the percentage of motility, viability and spermatozoa abnormalities in Simmental liquid semen. Materials used in this study is the Simmental fresh semen, tris yolk extender, penicillin, and sweet orange essential oil. The experimental design used randomized design with 5 treatments and 5 replications. Treatment on the research is the addition of sweet orange essential oil 0%, 0.25%, 0.5%, 0.75% and 1%. The parameters observed were the evaluation of motility, viability and abnormality spermatozoa before equilibration and after equilibration. The results showed the best result was the addition of 1% sweet orange essential oil and the lowest on the addition of 0% sweet orange essential oil before and after equilibration.
Keywords – sweet orange essential oil, penicillin, tris yolk extender, Simmental, percentage motility, viability, abnormalities, spermatozoa.

1. Introduction

Population and productivity Simmental cattle can be done through reproductive technology, such as Artificial Insemination. Mating with Artificial Insemination can optimize the use of a superior male to mating many females. Artificial Insemination can improve genetic quality and livestock population [12]. The benefits of Artificial Insemination are to minimize the risk of disease transmission through natural mating, not to require many bull for mating programs and the semen used from superior bulls. Superior bull to mating many cows. The benefits of Artificial Insemination are to minimize the risk of disease transmission through natural mating, not to require many bulls for mating programs and the cement used to come from superior bull. Artificial Insemination can increase the profits of farmers [8].

The success of Artificial Insemination is strongly influenced by sperm quality. The quality of sperm after the shelter will experience a decrease if it is not used immediately. Fresh undiluted semen will reduce spermatozoa fertility. Therefore, to maintain the quality of sperm during storage can be done by adding extenders. The addition of extenders aims to provide a source of energy for sperm so as to ensure the survival of spermatozoa during storage. An important requirement for sperm diluents is to provide protection against cold shock that occurs during freezing and as a buffer to maintain pH stability [11]. Besides being able to provide food as an energy source, maintain the same osmotic pressure as sperm and contain antibiotics to inhibit bacterial growth [15].

One of the factors of the low survival of spermatozoa during liquid semen storage is the development of bacteria. Generally, the addition of antibiotics in liquid semen extenders is done to minimize bacterial growth. Penicillin is one antibiotic that can be used as an antibacterial. The administration of penicillin to the extenders has been carried out by combining several other antibiotics such as GTLS (Gentamicin, Tylosin, and Linco-Spectin) [6] and Andromed® (gentamicin sulfate, spectinomycin, and lincomycin) [10] Essential oils of sweet orange peel containing limonene and linalool are toxic to bacteria [4].

The use of essential oils of sweet orange on Trish egg yolk extender in the liquid semen of Simmental Cattle was carried out and proved to be able to maintain the quality of semen of Simmental Cattle. By combining penicillin and essential oils of sweet orange peel, it is expected that the best antibacterial composition can be obtained that can inhibit bacterial growth but does not reduce the quality of Simmental liquid semen. This study aims to determine the effect of the use of a combination of penicillin and essential oils of sweet orange peel on the quality of Simmental liquid semen.

2. Research method

This study used a completely randomized design with 5 treatments and 5 replications. The treatments are:

P_0 = Penicillin + sweet orange Essential oil 0%
P_1 = Penicillin + sweet orange Essential oil 0.25%
P_2 = Penicillin + sweet orange Essential 0.5%
P_3 = Penicillin + sweet orange Essential 0.75%
P_4 = Penicillin + sweet orange Essential 1%

2.1. Evaluation of microscopic sperm parameters

2.1.1. Motility

The percentage of spermatozoa moving progressively forward. The evaluation was done by observing spermatozoa in eight different visual field with a light microscope magnification 400 times.

2.1.2. Viability

Percentage of dead spermatozoa after thawing was evaluated with eosin staining (eosin-Y 1.67g and sodium citrate 2.9 g dissolved in 100 ml distilled water) as described by [2]. The evaluation was done at a minimum of 200 spermatozoa was observed using a light microscope magnification of 400 times. Live spermatozoa characterized by a head that do not absorb the dye, while the die is characterized by a red head.

2.2. Abnormalitas Spermatozoa

Deviations in the form of sperm morphology that can reduce the fertility of spermatozoa. Calculated abnormalities are too large head abnormalities, head too small, double head (duplicate head), circular tail and double tail. Abnormalities are calculated by the formula:

$$\frac{\text{Number of spermatozoa abnormalities}}{\text{Total sperm counted}} x100\%$$

3. Results and discussion

The results of motility, viability, and abnormality of spermatozoa in liquid semen of Simmental Cattle before and after equlibration can be seen in the recapitulation table below.

Table 1. Recapitulation of motility, viability, and abnormalities of Simmental Cattle semen before equlibration and after equlibration.

Parameter	Treatment	Observation	
		Before Equlibration	After Equlibration
Motility	P_0	70$\pm$0.00	69$\pm$2.74
	P_1	72$\pm$0.00	70$\pm$2.24
	P_2	73$\pm$2.24	72$\pm$2.24
	P_3	75$\pm$2.74	74$\pm$2.74
	P_4	77$\pm$2.24	75$\pm$0.00
Viability	P_0	77$\pm$2.50	75$\pm$1.68
	P_1	78$\pm$2.50	77$\pm$2.50
	P_2	80$\pm$1.68	79$\pm$0.87
	P_3	82$\pm$2.72	80$\pm$2.72
	P_4	84$\pm$1.68	82$\pm$2.50
Abnormality	P_0	7$\pm$1.37	8$\pm$1.37
	P_1	6$\pm$1.24	7$\pm$1.45
	P_2	4$\pm$1.15	6$\pm$0.72
	P_3	3$\pm$0.75	4$\pm$1.24
	P_4	3$\pm$1.24	3$\pm$1.42

Description: Different superscripts in columns show very real differences (P <0.01).

3.1. Motility of Simmental spermatozoa

The results of research on observations of semen motility of Simmental Cattle before dilution using trish yolk extender and sweet orange essential oil and observations after equilibration showed different results for each treatment. According to [5], sperm motility or sperm motility is one of the determinants of the success of spermatozoa to reach the ovum in the fallopian tubes and the simplest way to assess sperm for Artificial Insemination. The results of variance analysis showed that the effect of essential oils as extenders had a very significant effect (P <0.01) on the motility of spermatozoa both after dilution and after equilibration. This result is also the same as the treatment of Boer Goats using gentamicin and sweet orange essential oil [13]. Judging from the average observation of each treatment that the motility of all treatments meets the standards to be used as cement for Artificial Insemination due to motility above 69% and that number has met the requirements for Artificial Insemination. Where the conditions for cement that are suitable for Artificial Insemination have motility of not less than 40% [3].

[7] states that hydroxinonenal is a lipid peroxidation that can inhibit glycolysis and sperm motility. In addition to damage caused by lipid peroxidation, decreased motility can also occur due to several factors. According to [15] factors that can reduce motility are changes in medium pH, osmotic pressure and electrolyte and non-electrolyte effects. The best sperm motility that can be used and meet the standards in this study is the addition of 1% sweet orange essential oil. The results showed that the sperm motility of liquid semen increased with the addition of sweet orange essential oil both before dilution and after dilution. Addition of sweet orange essential oil to liquid semen Simmental cattle is appropriate because it increases the percentage of spermatozoa motility Simmental Cattle. This is consistent with the statement of [11] that the quality of semen will decrease if the liquid cement extender is not appropriate.

3.2. Viability of Simmental spermatozoa

[14] states that immovable spermatozoa do not necessarily die so they do not absorb color, while in interpretation with a moving and immovable basis are considered immobile. In spermatozoa that are alive and moving but there is a defect in the cell wall, it can absorb colors considered dead, while other interpretations are considered not dead. The results showed that the average value of all treatments for the percentage of live spermatozoa before equlibration was 84% and after equlibration 82%. The results of the analysis of various percentages of dead life spermatozoa are very significantly different (P <0.01) both on the observation before equlibration and after equlibration.

This difference in average viability can be due to physical influence at the time of treatment so that it can cause death. Friction between spermatozoa can cause abnormalities as well as death. The occurrence of a decrease in spermatozoa viability after the cooling and freezing process can be caused due to physical influence during the treatment that caused death. The physical effect is caused by friction between spermatozoa, between spermatozoa and tube walls, or between fat globules of egg yolks so that there is a tendency to decrease viability along with different levels of dilution. The National Standardization Agency determines the quality of a cow's liquid cement after the freezing process must show a minimum viability percentage value of 40% [1]. Quality degradation after the cooling and freezing process is caused by spermatozoa experiencing cold shock. Another factor that can cause a decrease in sperm quality is because during the freezing process the formation of ice crystals occurs, so the concentration of electrolytes in the cell increases and dissolves the lipoprotein sheath of the spermatozoa wall [15]. [9] stated that during freezing and storage of cement there is a membrane imbalance, which can reduce sperm resistance so that after thawing the quality of cement becomes low.

3.3. Abnormalities of Simmental spermatozoa

Spermatozoa abnormalities are a condition in which spermatozoa experience disability in one or all parts of the body of the spermatozoa. Spermatozoa abnormalities consist of primary, secondary and tertiary abnormalities. Primary abnormalities occur during the process of spermatogenesis and the presence of testicular disorders. Secondary abnormalities occur after spermatozoa leave the seminal tube to the male reproductive tract, while tertiary abnormalities occur after ejaculation to the handling process. The results of variance analysis showed that the effect of the addition of streptomycin combination with sweet orange essential oil as a diluent had a very significant effect (P <0.01) on abnormalities of spermatozoa both after freezing and after freezing. The results of further BNT tests show that the best abnormalities are in P4 treatment, which is 3% before equlibration and 3% after equlibration.

Increased abnormalities of spermatozoa after cooling and freezing are caused by the physical influence of spermatozoa which can cause abnormal spermatozoa. temperature changes can cause changes in the permeability of the sperm cell wall membrane and result in disharmonism, membrane breakdown, and enzyme secretion. This condition can cause an increase in sperm abnormalities. The number of abnormal spermatozoa increases will cause low semen fertility of the animal. Spermatozoa cells are deformed, although they can fertilize an egg but usually end up with the death of a child before birth. Another factor that influences the increase in abnormalities is the inadvertent action at the time of treatment, liquefying the cement with the same isotonic fluid, cold shock, heat, and nutritional disorders. Spermatozoa who experience abnormality are caused by physical influence at the time of treatment, where spermatozoa rub against each other causing abnormalities as well as death. Factors that affect abnormal percentages are the act of not carefully diluting semen with fluid that is not the same as isotonic, cold shock, heat, nutritional disorders or endocrine disorders that affect normal spermatogenesis [16].

4. Conclusion

Addition of essential oils of sweet orange and penicillin in liquid semen extender Simmental cattle can increase the percentage value of motility, viability and spermatozoa abnormality and are suitable for use in Artificial Insemination. The best results obtained are the addition of essential oils as much as 1% and the lowest is without the addition of essential oils. Fertility testing is needed to see the success rate of Artificial Insemination using Simmental Cattle liquid semen.

5. Reference

[1] Anonimus, 2005. Persyaratan Mutu. Badan Standarisasi Nasional.
[2] Baqir, M., M.R. Fakhrildin, dan B.K. Kouty. 2009. Outcomes of Sperm Parameters, Hypo-Osmotic Swelling Test and Intra-Uterine Insemination for Varicocelic and Non-Varicocelic Infertile Patients. Journal Dohuk University, Vol. 12. No. 1.
[3] Evans G. and W. M. C. Maxwell 1987. Salamon's Artificial Insemination of Sheeps and Goats. Butterworths. London.
[4] Fisher, K., and C. A. Phillips. 2006. The effect of lemon, orange and bergamot essential oils and their components on the survival of Campylobacter jejuni, Escherichia coli O157, Listeria monocytogenes, Bacillus cereus and Staphylococcus aureus in vitro and in food systems. J Appl Microbiol. 2006 Dec;101(6): 1232-40.
[5] Hafez, E.S.E. 1987. Semen evaluation. In: Reproduction in Farm Animals. E.S.E. Hafez (Ed). 5th Edition. Lea and Febiger. Philadelphia. pp. 22: 466.
[6] Hasan, S., S.M.H. Andrabi, R. Munir, M. Jehangir, P. Shafique, M. Anzar and N. Ahmad. 2000. Effect Of New Antibiotic Combination On Post-Thaw Semen Quality Of Buffalo And Sahiwal Bulls. 33rd Annual Meet. Soc. Study Reprod., 62: 157. Abstract.
[7] Lodhi, L., A., M. Zubair, Z.I. Qereshi, I. Ahmad dan H. Jamil. 2008. Correlation Between Hypo-Osmotic Swelling Test and Various Conventional Semen Evaluation Parameters in Fresh Nili-Ravi Buffalo and Sahiwal Cow Bull Semen. Pakistan Veteriner Journal, Vol. 28. No. 4.
[8] Marisa, J. dan Sitepu, SA. 2018. Peningkatan Pemahaman Dan Pengetahuan Peternak Tentang Inseminasi Buatan Dengan Menggunakan Semen Beku Pada Kambing Di Desa Tanjung Selamat. Prosiding Seminar Nasional Kebangkitan Peternakan III, 3 Mei 2018, Undip Semarang, Indonesia.
[9] Maxwell W. M. C. and P. F. Watson. 1996. Recent Progress in the Preservasion of Ram Semen. Animal reproduction.
[10] Minitub. 2001. Certificate Andromed. Minitub Abfullund Labortechnik GmbH andCo KG. Germany.
[11] Mumu, M.I. 2009. Viabilitas Semen Sapi Simental Yang Dibekukan Menggunakan Kriopektan Gliserol. Journal Agroland 16 (2): 172-179.
[12] Sitepu, S.A. and Zaituni, U., 2018. Improved quality of frozen boer goat semen with the addition of sweet orange essential oil on tris yolk and gentamicin extender. In *IOP Conference Series: Earth and Environmental Science* (Vol. 122, No. 1, p. 012125).
[13] Sitepu, S.A., Udin, Z., Jaswandi, J. and Hendri, H., 2018. Quality Differences of Boer Liquid Semen During Storage with Addition Sweet Orange Essential Oil In Tris Yolk And Gentamicin Extender. *Journal of Community Research and Service*, *1*(2), pp. 78-82.
[14] Partodihardjo, S. 1982. Ilmu Reproduksi Hewan. Mutiara Sumber Widya. Jakarta.
[15] Toelihere, M. R. 1993. Fisiologi Reproduksi pada Ternak IPB Press, Bogor.
[16] Yulianti, E. R. 2006. Pengaruh Beberapa Pengencer Dengan Waktu Equilibrasi Yang Berbeda Terhadap Kualitas Semen Kambing Boer Sebelum Pembekuan. Fakultas Peternakan UB. Malang.

Egg production and quality of local female chicken by dietary self-selection reared under semi-scavenging system in the tropics

Syafwan* and Noferdiman

Faculty of Animal Sciences, University of Jambi, Jambi, Indonesia.
*Corresponding author e-mail: syafwan@unja.ac.id.

Abstract. The feeding standard for local chicken breeds in Indonesia has not been published yet. The present study were aimed to measure egg production and quality while estimating the ME and CP needs during production for local female chickens reared under semi-scavenging system in the tropics (Indonesia) through dietary self-selection. A total of 138 twenty two week-old chicks were randomly distributed into 12 sheltered pens, 10-14 birds each. Two feeding methods (control and self-selection) were assigned to pens, so each treatment consisted of 6 replicates. The control group received a control diet complying with the Hy-line Brown Nutrient Requirements Standard, whereas the self-selection group had access to the control and four other diets (high energy-high protein, low protein-high energy, high protein-low energy, and low protein -low energy diet). Feeds and drinking water were provided ad libitum to 53 weeks of age. Feed consumption (FC), CP intake, ME intake, concentration of dietary CP and ME and egg production were recorded weekly. Egg quality was measured three times. Daily temperature and relative humidity in the morning (07:00), noon (12:00), and afternoon (17:00) were 21.8 to 28.1°C and 46 to 88%; 24.7 to 34.5°C and 35 to 72%; and 23.5 to 34.5°C and 36 to 80%, respectively. Data were analyzed using Proc Mixed of SAS, but egg quality was analyzed by ANOVA. The results showed that feeding method had apparent effect on CP intake (P=0.018) and ME intake, dietary concentration of CP and ME, egg production (P<0.001) but not on FC. The effect of week and feeding method by week interaction were also very significant for all performance, except egg production was not affected with feeding method by week interaction. Weekly CP and ME intake of the self-selection group were greater than those of the control (105.8 vs. 101.3 g/bird; P=0.018 and 1,709 vs. 1,575 kcal/bird; (P<0.001), respectively). Dietary concentrations of CP and ME in the self-selection group was higher (P<0.001) than those in the control group (189.3 vs. 180.7 g/kg and 3,053 vs. 2,810 kcal/kg, respectively). Egg production of the self-selection group was higher (P<0.001) than those of the control (60.5% vs. 55%). Over all, egg quality (eggshell thickness, shell strength, shell weight, egg weight, Hmm, color, compare, HU, and yolk weight) was similar, except color and compare were higher in choice-fed birds when they grew older. Local female chickens were able to select diets to adjust their nutrient requirements by selecting more of an energy-rich diet and less of a protein-rich diet. CP and ME requirement by local female chickens were more likely greater than the current formulated for them in order to increase egg production.
Keyboard – egg production, local female chicken, dietary self-selection, semi-scavenging system, tropics.

1. Introduction

Local chickens have a potential to produce a considerable amount of meat and eggs. In 1997, the contribution of local chickens to total poultry meat production was about 32% [1] and to total egg production was about 17% [2]. In addition, the price of meat and eggs of local chickens are relatively stable and the taste of these products have a high preference of local consumers. However, Department of Agriculture of Indonesia [3] reported that the contribution to total poultry meat and egg production were decline to 18% and 12.8% respectively. The reduction in these contributions during the last 10 year depict the lower productive performance of local chickens. These conditions probably due to lack of improvement in nutrient requirements.

In Indonesia, there is no establish the nutrient requirements for any purpose production of local chickens yet. Diet formulated for these chickens are still using non tropical feeding standards such as the NRC [4] or standard for Hy-line Brown Laying hen. This is unlikely suitable to meet nutrient requirements of local chickens in the tropic because high ambient temperature would influence stress level [5], heat production, energy utilization, and metabolic activities of the chickens [6] [7]. The improvement an egg production was achieved by dietary of 2,900 kcal ME /kg with 15% crude protein [8]. The increasing CP from 13.6% to 16% in local chicken was increase egg production [9]. The improvement in egg production more than 58% was also found by diet containing 18.4% CP and 2,750 kcal ME/kg [10]. These all information provided by using intensive rearing systems and improving energy and CP, and therefore the chickens did not have an opportunity to choose the nutrient requirements. The present study were aimed to calculate egg production and quality while estimating the ME and CP needs during production for local female chickens reared under semi-scavenging system in the tropics (Indonesia) through self-selection feeding method.

2. Materials and methods

A total of 138 of 22 week-day-old local female chickens, the reminder chickens from the previous trial, were distributed to 12 sheltered pens connected with netted scavenging yard (pen dimension: 3 m long × 1.75 m wide × 2 m high; yard dimension: 3 m long × 1.75 m × wide × 2 m high). Sand was used as litter in roofed pen and it was regularly added to each pen to maintain a good litter condition. There was no material added as litter in yard. The birds had free access to go in and out to pen and yard.

Five feed troughs were placed in each pen in order to enable the birds an easy and equal access to feed. Birds had access to feed and water ad libitum. In order to avoid any confounding in place and choice of feed from feeding troughs, the site of each feeding trough in a pen was changed every day according to a predetermined random schedule. Each pen had one bell shaped drinker with a 3 liter capacity and the wooden slat of a perch measured 175 cm long x 4 cm wide x 3 cm high with rounded angles about 1 m above the floor.

The temperature and humidity cycle were recorded daily by using a wireless weather center (La Crose Technology, type WS2-550). The light of each pen was maintained 23-h light and 1-h dark per day to ensure the birds having an opportunity to choose the feed and to increase the feed intake during night time with probably low temperature. The light was provided by 75 Watt bulb which was positioned under a bluish white metal with Chinese shaped-head.

The experiment was conducted as a completely randomized design with repeated measures. The factor was a dietary treatment (control, single diet fed birds versus choice-fed birds). Each dietary treatment was assigned to six replicated pen with 10 and 14 birds each. Thus, the pen was the experimental unit.

The control groups received a control diet which was formulated to contain ME, CP, and other nutrients for laying phase [18-53 wk of age]) as recommended by The Hy-line Brown Commercial Management Guide [11]. The self-selection groups were given access to the control diet [2,814 kcal/kg and 18.4% CP] and four other diets (high protein -high energy, HPHE [23.0% CP and 3,101 kcal/kg], low protein-high energy, LPHE [14.3% CP and 3,133 kcal/kg], high protein-low energy, HPLE [23.4% CP and 2,638 kcal/kg], and low protein -low energy diet, LPLE [14.6% CP and 2,677

IOP Conf. Series: Earth and Environmental Science **287** (2019) 012008 doi:10.1088/1755-1315/287/1/012008

kcal/kg]). This gave the birds the opportunity to eat from all the diet and, thus, compose their own diet. Feeds and drinking water were provided ad libitum to 53 weeks of age.

Before meal time, all feeders were taken out of the pen and weighed, and new feeders were placed back. The protein sources were soybean meal and fish meal. The energy sources were maize, rice bran and vegetable oil.

Data of feed consumption (FC), ME intake, CP intake, concentration of dietary ME and CP and egg production were recorded weekly. All performance data were taken on the same experimental units, repeated in time: 31 weekly observations were available for each individual experimental unit. Repeated measurement on the same animal cannot be regarded as independent units of observation and mixed model can be used to account for the covariance structure among repeated observations [12]. Data were subjected to Proc Mixed, and differences (P≤0.05) among treatment means were distinguished by PDIFF with SAXTON macro [13].

The egg quality (Eggshell thickness (mm), Shell strength (lb/cm^2), egg weight, hmm, color, compare, HU, shell weight and yolk weight) were measured three times (29, 44 and 52 week). Egg quality were analyzed by ANOVA according to completely randomized design [13].

3. Results and discussions

3.1. Temperature and humidity

Temperature (T) and humidity (RH) are given as the average, minimum and maximum ± SD for each period time recorded. During the period of this experiment, average T and RH in the morning (07:00) were 24.2 ±0.8°C and 78±3.1%. The minimum T and RH were 22.5°C and 68% and the maximum T and RH were 27.0°C dan 81%. In the day (12:00), the average T and RH were 30.8±2.2°C dan 53±6.3%. The minimum T and RH were 25.6°C and 42% and the maximum T and RH were 33.5°C and 72%. In the afternoon (17:00), the average T and RH were 28.7±2.1°C and 62±8.4%. The minimum T and RH were 23.5°C and 46% and the maximum T and RH were 32.5°C and 80%.

Diurnal T and RH cycling in the house during the experiment was depend on the natural condition because no effort have been done to control the T and RH. The increasing and decreasing of T in the house was followed by the decreasing and increasing of RH. These conditions were very helpful for the birds to release the body heat to the environment, especially during the high temperature which generally occur in the period of the day (12:00) until afternoon (17:00). The high T and RH are the detrimental combination to the bird due to the difficulty in releasing the body heat. That humidity is consider having the key role due to the growth of broiler was decline when RH more than 60-65% either at 28 or 30°C [14].

3.2. Bird performance

Probability values for every parameter are presented in Table 1. Differences in performance of the local female chickens in each period at different dietary treatments are presented in Figure.

Table 1. Probability values of main effects and interaction between dietary treatment[1] (F) and week for different traits.

Main Effect	Feed Intake (g/bird/week)	CP Intake (g/bird /week)	ME Intake (kcal/kg/bird /week)	CP Concentration (g/kg)	ME Concentration (Kcal/kg)	Egg Production (%)
F	0,9609	0,0185	<.0001	<.0001	<.0001	<0.001
Week	<.0001	<.0001	<.0001	<.0001	<.0001	<0.001
F*Week[2]	<.0001	<.0001	<.0001	<.0001	<.0001	0,9997

[1]Control diet (CP: 18.4%; ME: 2,814 kcal/kg); high protein -high energy, HPHE [23.0% CP and 3,101 kcal/kg], low protein-high energy, LPHE [14.3% CP and 3,133 kcal/kg], high protein-low energy, HPLE [23.4% CP and 2,638 kcal/kg], and low protein -low energy diet, LPLE [14.6% CP and 2,677 kcal/kg]).
[2]F x week=interaction between dietary treatment and week.

The results showed that there was no effect of feeding methods on FC of the chickens, although fluctuations were observed during 37 to 41 weeks of age (Figure). However, average CP and ME intake indicated that the chickens given a choice to feed consumed more CP and ME than those offered the control diet (105.8 vs. 101.3 g/bird/wk and 1,708.7 vs. 1,574.8 kcal/bird/wk, respectively; P<0.018 and P<0.001) suggesting self-adjustment capability of the chickens to nutrients' need. Capability in adjusting protein and energy are always continue from the beginning and than the protein and energy intake of self-selected fed birds were higher than control [15]. Final body weight of broiler chickens given dietary low CP feedstuff were emulated with 20% CP-fed chickens if they have free access in yard [16].

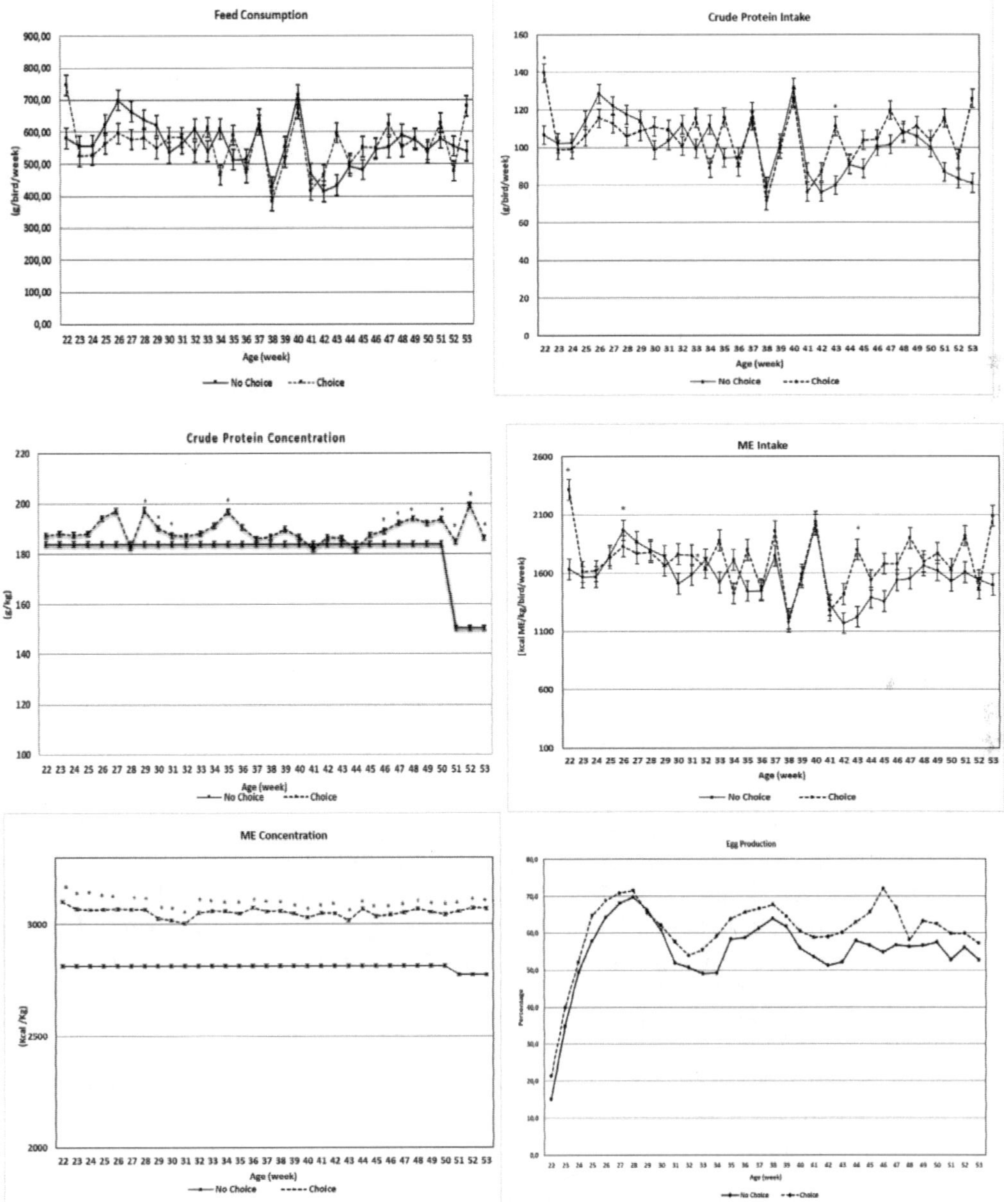

Figure 1. Least square means of performance parameter in local female chickens from 22 to 53 week (31 weeks) as affected by dietary treatments.

Feeding methods did not affect total FC, but free choice fed birds preferred more to consume HPHE diet (44.47%) and LPHE diet (36.26%) than the other diet (control, 17.13%; HPLE, 1.13% and LPLE, 1.0%). These preferences confirm the previous study that broiler chickens reared in high temperature [5] and unsex indigenous chickens in low land tropical region [17] chose more high energy diet than high protein diet. These preferences determined the concentration of CP and ME of the diet consumed by the chickens (Figure). The higher CP concentration consumed by the choice fed birds could indicate that the choice feeding method gives more chance to maintain its CP need especially from 22 to 53 weeks of age (in the average: 189.3 vs. 180.7 g/kg; P<0.001). The graph also clearly illustrated a greater ME concentration in the diet consumed by choice fed birds than control fed birds (3,053 vs. 2,810 kcal/kg; P<0.001). The greater CP and ME concentration in the diet consumed by choice fed birds indicated that chickens were able to select diets containing nutrients of their needs. The ability of chickens to meet their protein and energy requirement by composing several diet in broiler chickens [17] [18]. This finding demonstrates that local chicken likely requires more CP and ME than they used to be fed.

Feeding method had a very significant effect on egg production (P<0.001). The graph showed that egg production of choice-fed birds was higher than of control-fed birds (Figure). The egg production in choice-fed group was more from the beginning of production. The higher egg production could be due to higher CP concentration consumed by the choice-fed birds. Level of protein diet influence egg production [19] and 10% more protein diet in commercial layer chickens produced higher hen-day egg production compared with10% lower protein diet for every phase production [20].

Eggshell thickness, shell strength, shell weight, egg weight, Hmm, color, compare, HU, and yolk weight were similar (P>0.05), but color and compare were higher (P<0.001) in choice-fed birds when they grew older. The higher CP concentration in the diet consumed by the choice-fed birds had no strong effect on yolk weight. The weight of yolk increase significantly by increasing protein level of diet [20].

4. Conclusions

Local female chickens were able to select diets to adjust their nutrient requirements by selecting more of an energy-rich diet and less of a protein-rich diet. The results of this study suggested that CP and ME requirement by local female chickens were more likely greater than the current formulated for them in order to increase egg production. Footnotes

Footnotes should be avoided whenever possible. If required they should be used only for brief notes that do not fit conveniently into the text.

5. References

[1] Riethmuller, P. 1999. Livestock Industries of Indonesia Prior to the Asian Financial Crisis. FAO Regional Office for Asia and the Pacific, Bangkok 10200, Thailand.

[2] Diwyanto, K., and S. Iskandar. 1999. Livestock Industries of Indonesia Prior to the Asian Financial Crisis. FAO Regional Office for Asia and the Pacific, Bangkok 10200, Thailand.

[3] Department of Agriculture. 2011. Statistik Peternakan dan Kesehatan Hewan. Agriculture Department, Jakarta.

[4] NRC. 1994. Nutrient Requirements of Poultry. 9th rev. ed. National Academy Press, Washington, DC.

[5] Syafwan, S., G. J. D. Wermink, R. P. Kwakkel, and M. W. A. Verstegen. 2012b. Dietary self-selection by broilers at normal and high temperature changes feed intake behavior, nutrient intake, and performance. Poultry Science. 91: 537-549.

[6] Gous, R. M., and T. R. Morris. 2005. Nutritional interventions in alleviating the effects of high temperatures in broiler production. World's Poultry Science Journal 61: 463-475.

[7] Sakomura, N. K., F. A. Longo, E. O. Oviedo-Rondon, C. Boa-Viagem, and A. Ferraudo. 2005. Modeling energy utilization and growth parameter description for broiler chickens. Poultry Science. 84: 1363-1369.

[8] Argono, R. S., and S. Iskandar. 2005. Review hasil-hasil penelitian dan dukungan teknologi dalam pengembangan ayam lokal. Prosiding Lokakarya Nasional Inovasi Teknologi Pengembangan Ayam Lokal. Semarang, Indonesia. 10-19 .
[9] Muryanto. 2005. Evaluasi hasil-hasil penelitian dan pengembangan pada ayam buras. Prosiding Lokakarya Nasional Inovasi Teknologi Pengembangan Ayam Lokal. Semarang, Indonesia. 238-251.
[10] Adrizal, A., Y. Yusrizal, S. Fakhri, W. Haris, E. Ali, and C. R. Angel. 2011. Feeding native laying hens diets containing palm kernel meal with or without enzyme supplementations: 1. Feed conversion ratio and egg production. The Journal of Applied Poultry Research. 40-49.
[11] Hy-line, I. 2011. Hy-line Brown Commercial Management Guide. Hy-Line, Australia.
[12] Littell, R. C., P. R. Henry, and C. B. Ammerman. 1998. Statistical analysis of repeated measures data using SAS procedures. Journal of Animal Science. 76: 1216-1231.
[13] SAS. 2003. Statistical Analysis Institute (Version 9.1.3, service pack 4). SAS Institute Inc., Cary, NC, USA.
[14] Yahav, S. 2000. Domestic fowl—Strategies to confront environ-mental conditions. Poult. Avian Biol. Rev. 11: 81–95.
[15] Adrizal, A., S. Syafwan, N. Noferdiman, S. Zubaida, and T. M. Pasaribu. 2017. Energy and protein consumption and utilization of local female chicken reared under semi-scavenging system in the tropics. 21st European Symposium on Poultry Nutrition. Spain.
[16] Fanatico, A. C., V. B. Brewer, C. M. Owens-Hanning, D. J. Donoghue, and A. M. Donoghue. 2013. Free choice feeding of free-range meat chickens. The Journal of Applied Poultry Research. 22: 750-758.
[17] Syafwan, S., R. P. Kwakkel, and M. W. A. Verstegen. 2012a. Feed intake behaviour, nutrient intake and performance of indigenous chicken fed a choice diet under tropical climate condition in Jambi. First International Poultry Science Seminar. West Sumatera, Indonesia. 111-122.
[18] Cruz VC, Pezzato AC, Pinheiro DF, GonÁalves JC, and Sartori JR. 2005. Effect of Free-Choice Feeding on the Performance and Ileal Digestibility of Nutrients in Broilers. Brazilian Journal of Poultry Science. 7(3): 143 – 150.
[19] Parsons, C. M., K. W. Koelkebeck, Y. Zhang, X. Wang, and R. W. Leeper. 1993. Effect of dietary protein and added fat level on performance of young laying hens. The Journal of Applied Poultry Research. 2: 214-220.
[20] Shim, M. Y., E. Song, L. Billard, S. E. Aggrey, G. M. Pesti, and P. Sodsee. 2013. Effects of balanced dietary protein levels on egg production and egg quality parameters of individual commercial layers. Poultry Science. 92: 2687-2696.

Polymorphism of prolactine genes and its association with body weight in Bayang ducks, local duck from West Sumatera, Indonesia

Yurnalis[1, 2], **Arnim**[1], **Zulkarnaini Kamsa**[1], **and Dino Eka Putra**[1]

[1] Faculty of Animal Husbandry, Andalas University, Padang, West Sumatera, Indonesia.
[2] Author corresponding: yurnalisunand@yahoo.com.

Abstract. This study aims to determine the polymorphism of Prolactin (PRL I XbaI and PRL I DraI) genes in Bayang ducks using the PCR-RFLP (Polymerase Chain Reaction-Restriction Fragment Length Polymorphism) technique and its association with weight of ducks aged 1-10 weeks. This study used 200 Bayang duck blood samples consisting of 102 male ducks (♂) and 98 female ducks (♀). DNA from blood samples was isolated using the Genomic DNA Purifification Kit (Promega) using protocol from the manucfacture. The DNA was then amplified using two primers with F: 5'-AAA TTC CCT CTC ACA GTT ACA-3 'and R: 5'-GAT GCA GAG ACA AGT TTC ACC-3' and F: 5'- GAATAGAACACTTGACCCTG -3 'and R: 5'- TAGAGGAGGCAAGCATAG -3 'which produces fragments with a length of 416 bp and 566 bp. Restriction with *XbaI* enzymes that recognized the binding site (5 -TT CTAGA-3 ') resulted 3 genotypes: homozygote (+/+), heterozygote (+/-) and homozygote (-/-) with frequencies 0.455, 0.495 and 0.050 respectively and with frequency allel (+) 0.702 and frequency allel (-) 0.297. While the results of the restriction with enzyme DraI found three types of genotypes, namely (+/+), (+/-) and (-/-) with frequencies of 0.64, 0.35 and 0.01 respectively with frequency allel (+) 0.82 and allele frequencies (-) 0.18. From the results of the analysis, it was found that there was no relationship between these two diversity and weight ducks of duck.
Keywords – Bayang duck, PRL-XbaI, PRL-DraI, body weight, PCR-RFLP.

1. Introduction

West Sumatera, Indonesia has many species of local duck such as Pitalah, Bayang, Kumbang Jonti, and Kamang. All of this duck are layer with small body size and lower meat production. Bayang duck is the local duck found in Pesisir Selatan Distric, West Sumatra Province, Indonesia. Bayang duck is dual-purpose duck type with color of the feathers is predominantly white and brown, the body looks large and erect, short neck, black beak and Legs. Adult weight about 1.4 - 2.0 Kg. The eggs are light blue with a production of about 180 - 200 per year. Recently, the demand of duck meat in Indonesia are increasing together with the increasing of product diversification. To supply requirement and demand of duck meat, meat performance of native duck should be improved by selection.

The application of molecular genetics in identification of polymorphism in growth candidate genes that show association with specific economically relevant traits provide useful information to enhance genetic improvement programme in livestock and validation of genetic markers of growth traits is the initial and crucial step to establish a Marker Assisted Selection system (MAS).

PRL gene is one of the growth hormone gene family and is synthesised mainly in the anterior pituitary of all vertebrates [1]. The prolactin gene is a candidate gene that specifically controls the variation in the amount of egg production through reducing egg biosynthesis during the brood period [2]. The prolactin gene in ducks consists of five exon regions separated by four introns and encoding 229 amino acids [3]. Research on the prolactin gene been widely carried out by [1] [4] [5]. [1] found variations in exon two, four and five exons in local Chinese ducks. Other mutation points were identified in the non-coding part of the prolactin gene in Tsaiya and Gaoyou ducks [4] [5]. Research conducted by [4] in Chinese local ducks found T/C mutations at position 1326 bp in intron 1 region than can be identified using *Dra*I enzyme.

Prolactin is a single-chain polypeptide hormone synthesized in the anterior pituitary gland. The prolactin hormone in ducks is composed of 229 amino acids encoded by the prolactin gene [3]. Research on intron 1 duck prolactin gene polymorphism is associated with egg weight [4] while exon 5 shows the relationship with annual egg production [1]. Research on the polymorphism of intron one prolactin gene was associated with egg weight [4] while exson five showed a relationship with annual egg production [1]. Haplotype analysis shows that each mutation is related to egg production and reproductive traits [5]. The mutation of duck prolactin gene occurs in introns. Research conducted by [6] in Gaoyou ducks (Chinese local ducks) found prolactin gene SNP at intron 1, ie T/C mutation at position 1326 bp after sequencing was carried out three genotypes AA, AB and BB. And research conducted by [1] on several types of Chinese local ducks found SNP of prolactin gene in intron 1. They found C→A mutation at position 386 at introns which can be detected by *Xba*I enzyme. In Mulard ducks, *PRL/Xba*I polymorphism had an effect ($p < 0.05$) on body weight aged 10 and 12 weeks [7].

The present study was carried out to study the polymorphism in Prolactine gene and its association with weight in Bayang duck and to analyse the potential of this gene to act as markers for selection tools for duck breeding.

2. Materials and method

2.1. Research material

The material used in this study were 200 Bayang duck blood samples consisting of 102 male ducks and 98 female ducks. The fertile duck egg were collect from famer and hatched. Then the Bayang duck were kept intensively for 10 weeks and blood samples were taken through a brachial vein ± 1 ml. DNA isolation is carried out from a blood sample using the Genomic DNA Purification Kit from Promega based on the procedure from the manufacturer. PRL gene amplification was performed using primer 5'-GAATAGAACACTTGACCCTG F -3 'and 5'-TAGAGGAGGCAAGCATAG R -3' and primers F 5 'AAATTCCCT CTCACAGTTACA'3 and R 5' GATGCAGAGACAAGTTTCACC '3 which produce fragment length 566 bp and 416 bp. PCR amplification using Master Mix (Thermo SCIENTIFIC®) with the following composition: DNA sample 2 ul, Master Mix 12.5 ul, each primer 2 ul and, Nuclease-Free Water (REF P1193) 7.5 ul done using PCR machine (eppendorf® Mastercycler gradient) with pradenaturasi at a temperature of 95°C for 5 minutes, denaturation at a temperature of 95°C for 45 seconds, annealing at temperatures of 55°C for 45 seconds and extension 72°C for 1 minute, and a final extension of 72°C for 5 minutes with 35 cycles. Results of PRL gene amplification, visualisation on 1% agarose (Agarose TOPVISION ThermoSCIENTIFIC®) and stained by Ethidium Bromide Solution (MP Blomedicals®) and the results observed by using a UV trans-illuminator (SynGENE® G: BOX). The results of the subsequent amplification restricted to use 5 units of restriction enzyme *Dra*I (5 'TTTAAA↓ 3') and the enzyme *Xba*I (5'-T↓CTAGA-3').

2.2. Data analisis

Statistical analysis: The genotype and allele frequencies were calculated in each group of ducks. The genetic effects of the PRL|*Dra*I and PRL|*Xba*I polymorphisms on body weight were analyzed using a General Linear Model (GLM) procedure. The model used the Eq. 1:

IOP Conf. Series: Earth and Environmental Science **287** (2019) 012009 doi:10.1088/1755-1315/287/1/012009

$$Yijk = \mu + Gi + Sk + gijk \quad (1)$$

where, Yijkl is the observed value of the dependent variable, μ is the overall mean, Gi is the fixed effect due to genotype (i =1,2,3), Sk is the fixed effect due to sex (k = males or females) and gijkl is the random residual error. The Hardy-Weinberg equilibrium was assessed with the Chi-square test. The statistical significance of differences among the means was calculated in accordance with the SAS/STAT Software, Release 6.12 (SAS Institute Inc., Cary, NC, USA).

3. Result and disccussion

The amplification of the PRL gene fragments in Bayang ducks was carried out using the Polymerase Chain Reaction (PCR) method with two pairs of primers 5'- GAA TAG AAC ACT TGA CCC TG-3 (forward) and 5'-TAG AGG AGG CAA GCA TAG-3 ' (reverse) [4] and forward: 5'-AAA TTC CCT CTC ACA GTT ACA-3 'and reverse: 5'-GAT GCA GAG ACA AGT TTC ACC-3' which results in fragments of length 566 bp and 416 bp The results of gene amplification are presented in Figure 1.

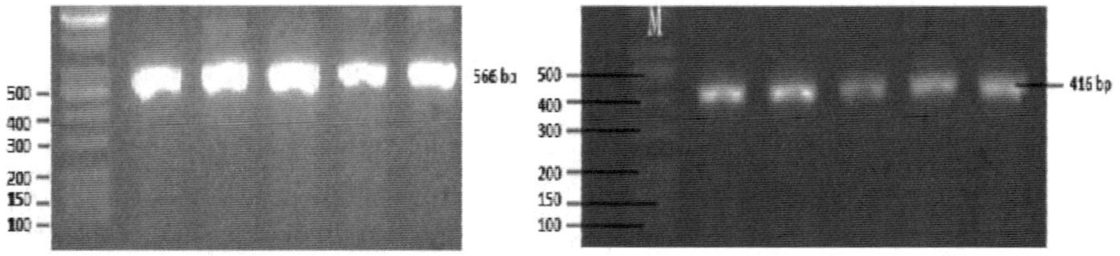

Figure 1. Electrophoresis of amplification of PRL|*Dra*I and PRL|*Xba*I genes on Agarose1% Gel (M = Marker (Kapa Universal 100 bp).

4. Genotyping of PRL|*Dra*I and PRL|XbaI

The PRL|*Dra*I gene genotype can be detected by the PCR-RFLP technique using the *Dra*I restriction enzyme. *Dra*I restriction enzymes cut DNA by recognizing TTT↓AAA located in the intron 1. The following DNA restriction fragments were obtained for PRL-*Dra* I polymorphism: 519bp/47bp for the TT genotype, 519/310bp/209bp/47bp for TC genotype, and 310bp/209bp/47bp for the CC genotype (Figure 2).

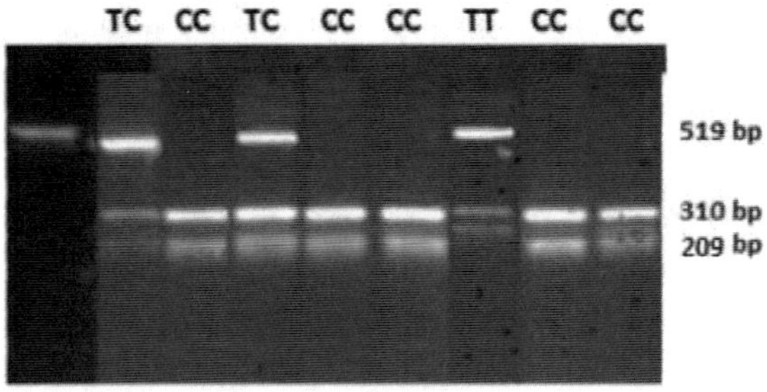

Figure 2. PCR-DraI PCR-RLP result

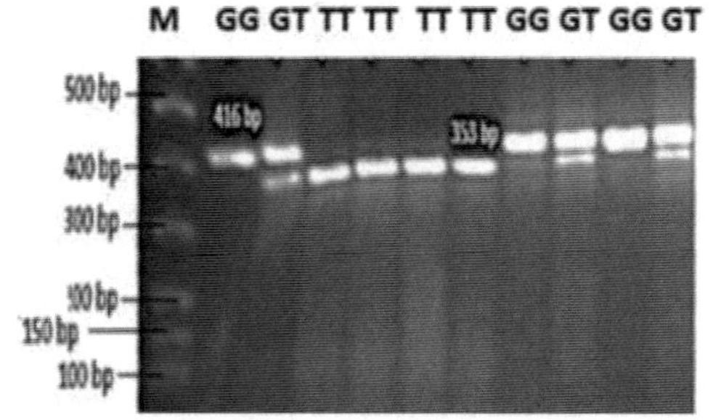

Figur 3 : PCR-*XbaI* RFLP result

Frequencies genotypic and allelic frequencies of of PRL|*Dra*I and PRL|*Xba*I were presented in Table 1

Tabel 1. Frequencies genotypic and allelic frequencies of of PRL|*Dra*I and PRL|*Xba*I.

*Dra*I	Frequencies Genotipy			Frekuensi Alel		n
	CC	CT	TT	C	T	
Male	0,684 (63)	0,293 (27)	0,021 (2)	0,831	0,168	92
Female	0,602 (53)	0,397 (35)	0	0,801	0,198	88
Sum	0,64 (116)	0,34 (62)	0,01 (2)	0,816	0,183	180
*Xba*I	GG	GT	TT	G	T	
Male	0.48 (49)	0.48 (49)	0.02 (4)	0.728	0.272	102
Female	0.42 (42)	0.50 (50)	0.06 (6)	0.673	0.327	98
Sum	0.455 (91)	0.495 (99)	0.05 (10)	0.702	0.297	200

The results of the analysis in Table 1 show that PRL/*Dra*I in male Bayang ducks obtained three genotypes consisting of homozygotes (CC), heterozygotes (CT) and homozygotes (TT) with frequencies of 0.684, 0.293 and 0.21, respectively. While in female Bayang ducks, only obtained two genotypes consisting of homozygotes (CC) and heterozygotes (CT) with frequencies of 0.602 and 0.397. Male Bayang ducks have the most homozygous (CC) genotypes, 63 samples with a frequency of 0.684 compared to Bayang ducks, females have genotypes (CC) only 53 samples with a frequency of 0.602, while for the most heterozygous (CT) genotypes obtained by Bayang ducks females, which were 35 samples with a frequency of 0.397 compared to male Bayang ducks, only got 27 samples with a frequency of 0.293 and for homozygotes (TT) it was only owned by male Bayang ducks which were 2 samples 0.021. so that the allele frequency (C) is 0.801 higher than the frequency frequency (T) which is 0.183.

The results of RFLP characterization analysis using the *Xba*I restriction enzyme on the exson 1 intron 1 PRL gene fragment obtained three kinds of fragments, namely non-cut fragments (416 bp) known as homozygous genotypes (TT), fragments that can be cut (353 bp and 63 bp) which is known as genotype (GG), but the band with a base length of 63 pb is not visible because it is too short, and the combined fragments (416 bp, 353 bp, and 63 bp) or heterozygotes are known as genotypes (GT).

The proportion of the Prolactin (PRL|*Xba*I) gene fragment alleles in the Bayang duck studied showed that the frequency of the heterozygote (GT) genotype was slightly greater than the homozygous (GG) genotype frequency, ie the heterozygous (GT) genotype frequency of 0.495 and frequency genotype (GG) of 0.455. However, the frequency of the homozygous (TT) genotype is quite low, there is only a genotype frequency of 0.050. This causes the allele (G) frequency of 0.702 to be more dominant than the allele (T) frequency of 0.297. The proportion of alleles obtained in the Prolactin (PRL|*Xba*I) gene fragments in the Bayang duck studied has a frequency of genotype and allele not much different from the genotype and allele frequency values in the study of [1] in Shaoxing local Chinese ducks obtained frequency values of heterozygous (GT) genotypes of 0.500, homozygous (GG) genotypes of 0.416 and homozygotes (TT) of 0.024 with allele (G) frequency of 0.738 more dominant from the allele frequency (T) of 0.262. From the analysis no relation between this polymorphism with body weight Bayang duck.

5. References

[1] Wang C, Liang Z, Yu W, Feng Y, Peng X, Gong Y, Li S. 2011. Polymorphism of the prolactin gene and its association with egg production traits in native Chinese ducks. S Afr J Anim Sci. 41:64–69.

[2] Chen, C. F., Y. L. Shiue, C. J. Yen, P. C. Tang, H. C. Chang and Y. P. Lee. 2007. Laying traits and underlying transcripts, expressed in the hypothalamus and pituitary gland, that were associated with egg production variability in chickens. Theriogenol 68: 1305-1315.

[3] Kansaku, N. Ohkubo, T. Okabayashi, H. Guémené, D. Kuhnlein, U. Zadworny, and D., K. Shimada. 2005. Cloning of duck PRL cDNA and genomic DNA. Gener Compar Endocrinol: 39-47.

[4] Li, H., F. Zhu, W., Q. Chen, K., W. Zhang, and T., J., W. Song. 2009. Association of polymorphisms in the intron 1 of duck prolactin with egg performance. Turky Journal Veteriner Animals Science. 33, 193-197.

[5] Chang, M. T., Y. S. Cheng and M. C. Huang. 2012. Association of prolactin haplotypes with reproductive traits in Tsaiya ducks. Anim Rep Sci. 135: 91-96.

[6] Hui, F. L., Q. Z. We, W. C. Kuan, J. Z. Tang, and T. S. Wei. 2009. Association of polymorphisms in intron 1 of duck prolactin with egg performance. Turk J Vet Anim Sci. 33 (3): 193-197.

[7] Mazurowski, A, A. Frieske, A. Wilkanowska, D. Kokoszyński, S. Mroczkowski, Z. Bernacki and G. Maiorano. 2016.Polymorphism of prolactin gene and its association with growth and some biometrical traits in ducks. Italian Journal of Animal Science , 15, P 200-206.

Effect of using GnRH plus P4 and estradiol plus P4 on ovarian response of embryo production of local Pesisir Selatan cows in West Sumatera

Zaituni Udin, Hendri, Masrizal, and Syafri Nanda

Faculty of Animal Science, Andalas University, Padang, West Sumatera, Indonesia.
Email: zaituniudin@yahoo.co.id.

Abstract. The objective of this study to evaluate the effect of GnRH plus P4 and estradiol plus P4 on superovulation responses and to know the relation between ovarian activity of local Pesisir Selatan cows. The data were obtained from 4 local Pesisir cows superovulated on breeding centre in west Sumatera. The treatment in superovulation was using GbRH plus P4 and estradiol plus P4 and data were analyzed using SPSS16 to determine theeffect of treatment on ovarian structure and the the regression correlation. The result showed that using GnRH plus P4 was greater ovulate rate, transferable embryos and lower the percentage of degerate and unfertize ($P<0.05$) than using estradiol plus P4. Linear regression between dominant follicle, thae number of CL and the number of embryos in coefficient correlation Rfcl= 0.950; Rfe=0.829 and Rcle= 0.980, respectively. In conclution the using of GnRH plus P4 for FTAI is greater the response in embryos production of local Pesisir Selatan cows.
Keywords – GnRH, estradiol P4, embryo production, FTAI.

1. Introduction

Embryo transfer can be used to enhance genetic improvement and to increase marketing opportunities with purebred cattle. Embryo transfer is use full for cattle, is caused low productivity rate and long generation interval. The primary use of embryo transfer in cattle has been to amplify reproductive rate of valuable cows. In local cattle transfer embryo technique is potential to applicate in order to increase the offspring and population as genetically resources of beef cattle in west Sumatera. However, the local cattle has been correlated to the lower fertility of beef herds. The advantages of using TE are acceleration and dissemination of local cows to increase the population as a beef cattle. In previous study showed ovulating follicle size has a greater effect on fertility of local Pesisir cow ranged from 6 mm – 11mm and largest follicle diameter has the highest ovulation rate and conception rate [1]. ET technique has been introduce as an efficient tool to improve fertility for repeat breeding dairy cows.

Multiple ovulation and embryo transfer (MOET) is a reproductive technology used to increase embryo production of cows, this technology has been used to increase the number of embryos for embryo transfer program. Currently, there are several protocols for fixed time artificial insemination(FTAI) and embryo transfer with different combination. Ovulation synchronization protocols such us ovsynch allow timed ET without the need for estrus detection in donor and recipient. Protocol for SOV without estrus detection are especially important when working with *Bos indicus* donor and high-yielding dairy *Bos Taurus* cows, due to the inherent difficulties with estrus with these animals [2] [3]. According [4] that delaying the timing of an ovulatory stimulus (GnRH or pLH) resulted in an increased superstimulatory response and embryo production following FTAI of lactating Holstein cows. Expression of estrus during protocols for TAI or TET is associated with an increase in fertility and reduction in pregnancy loss. During TAI programs, optimazing follicle diameter and increasing circulating P4 on day 7 after AI were also associated with increased fertility [5].

The timed-AI protocol commonly refered to as Ovsynch consists of three treatments; two injections of GnRH given 9nd apart, with an intervening injection of PGF2α given 7 d after the 1st GnRH treatment. According to [6] that the application timed – AI and timed-ET can improve reproductive efficiency in dairy cows. Furthermore, various mechanism for improvement fertility due increase estradiol concentration associated with estrus expression exist. The improve embryo quality and advancement in stage of development among heifer that exhibited estrus could also br a result of improved environment for embryo development. Exhibiting estrus before TAI resulted in improved embryo quality and advanced embryo stage on d 6 [7]. Arecent embryo transfer experiment that examined cows that did not exhibit estrus before TAI demonstrated that donor cows with increased concentration of estradiol at AI were more likely to yield a fertized embryo and increased estradiol tened to improve embryo quality and viability [8].

Protocols have now been developed that allow AI to be perform at a known time in relation to expected ovulation without the need for detection of estrus even in anovular cows [9] [10] [11]. Fixed time artificial insemination are important tools to improve reproductive efficiency in cattle. Therefore, this present study was designed to compare the protocol of time embryo transfer using GnRH plus P4 and using estradiol plus`P4 to increase the fertility of local Pesisir Selatan cows in West Sumatera. The objective of this study were to evaluate the ovulatory response of local Pesisir Selatan cows and to compare two protocol of time embryo transfer.

2. Material and methods

2.1. Location, animal and management

This study was conducted in breeding centre (BPTU-HPT) Padang Mengatas during June – August 2018 using multiparous cows local Pesisir Selatan (n=5). Then were housed during the treatment and then maintained in intensive pasture that allowed for continuous grazing, and salt and water was available all the time. The cows were selected on the basis of ovarium function (transrectal ultrasonography) and postpartum period, normal estrus cycls, and normal health status.

2.2. Experimental design and FTAI

The donors were divided into two groups: Treatments A). Time embryo transfer program using GnRH plus progesterone combination in local cows Pesisir Selatan cows. The protocol consists of the administration PGF2α (lutalyse) concurrent with P4 device insertion (Day0), followed by the administration of GnRH (fertagyl) on Day 7 AM. Treatment with gonadotropins is then initiated on Day 8 PM 936 h after GnRH, with twice daily administration of FSH until Day 12 AM. Donors are given GnRH on Day 13 AM, with FTAI 12h and 24 later. All donors was used Ovsynch protocol (GnRH; PGF2α) to synchronize ovulation and FTAI without estrus detection. Finally, embryo are collected on Day 20 (Figure 1) (Treatment B). Timed embryo transfer program using estradiol plus progesterone, the treatment of P4 implanted concurrent estradiol on Day 0. Treatment of FSH on Day 4 twice daily until Day 7 PM, and the GnRH treatment should be given 24 h after the P4 device removal (Day 8 PM), followed by FTAI both 12 h (Day 9 AM) later. Embryo are then collected (flushing) on Day 15 PM (Figure 2). The embryos were recovered by standard nonsurgical uterine flushing 7 days after AI. The number of ovulatory follicles were assessed by Transrectal Ultrasonography immediately before GnRH and corpus luteum (CL) immediately before embryo collection. The cows with 3 or more CL were considered to have responded to the superovulation treatment (SOV).

After collection, the embryos were classified according to the criteria defined by the international embryo transfer society [16]. The quality grade was assessed as good (grade1), fair (grade 2, poor (grade 3), or degenerated (grade 4). Only good and fair quality embryos at the morula and blastocyst stage were used for transfer and freezing. The number of ovulatory follicles were assessed by transrectal ultrasonography examination performed before GnRH injection and 7 days later immediately before embryo collection to assess the number of CL. Percentage of transferable embryos was calculated by dividing total number of grade 1, 2, and 3 embryos by the total numbers of embryos. The percentage of degenerated embryos was calculated by dividing the total number of generated embryos by the total number embryo. Percentage freezable embryos was calculated by dividing total number of grade 1 and 2 embryos by the total number of embryos. Ovulation rate was calculated by dividing the number of CL by the number of ovulatory follicle.

Table 1. Timed embryo production programs using GnRH plus progesterone combination in local Pesisir Selatan cows (A).

Treatment day	AM	PM
0	P4 device + PGF2α	-
7	GnRH	
8		FSH (20%)
9	FSH (20%)	FSH (15%)
10	FSH (15%)	FSH (10%)
11	FSH(10%)	FSH (5%) +PGF2α
12	P4 device removal+FSH (5%)	
13	GnRH	FTAI
14	FTAI	
20	Flushing and evaluation	Freezing

Table 2. Timed embryo production programs using estradiol plus progesterone combination in local Pesisir cows (B).

Treatment day	AM	PM
0	P4 device + EB (2 mg)	-
4	FSH (20%)	FSH (20%)
5	FSH (15%)	FSH (15%)
6	FSH (10%) + PGF2α	FSH (10%)
7	FSH (5%)	FSH (5%)+P4 device removal
8	GnRH	FTAI
9	FTAI	
15	Flushing+evaluation	Freezing

The freezing medium consisted of 5 % ethylene glycol and 6% propylene glycol in Dulbecco's phosphate-buffered saline (PBS) and 20 % fetal bovine serum. Embryo at the morula and blastocyst stage were transferred directly into the freezing medium, and each embryo was then loaded into a 0.25 ml plastic straw and allowed to equilibrate for 15 min. the straws were then placed in an alcohol bath of programmable freezer and cooled at -6.5 ºC for 5 min. subsequently, seeding was induced by touching the straws with forceps pre-cooled in liquid nitrogen. Five min later, the straws were cooled to - 30ºC at the rate of – 8.3 ºC min^1, and then the straws were plunged into liquid nitrogen and stored in liquid nitrogen.

To compare the two timed embryo transfer program was analyzed by using SPSS 16 program. The results in table are expressed as a mean ±standard deviation, graphs and regression correlation between the parameter was used.

3. Result and discussion

3.1. Effect of treatment on ovulation rate and quality of embryos of Pesisir Selatan cows

The ovulation rate was higher in treatment A is 70.37 % that in treatment B is 50 % (P.<0.05), with the number follicular ovulation was 19 from total follicular ovulation was 27 in treatment A and 9 follicle was ovulate in total follicular follicle 18. This present study showed that treatment A using GnRH plus P4 is more effective to superstimulation ovary than the estradiol plus P4. Ovulation rate was increase by using GnRH plus progesterone, this association with the function of GnRH induce the follicle emergence wave and ovulation. FTAI is used of GnRH 2x and PGF2 (ovsynch protocols) to stimulate of follicle emergence wave ovulation. Administration of GnRH protocol depended to ovarium status (medium anestrus is better response than deep anstrus). Ovsynch plus CIDR protocol can induce synchronized ovulation and luteal formation with a normal life span in early postpartum beef cows even though they are in non- cycling status [12] [13]. Meanwhile, estradiol and progesterone- based synchronization protocols have been successfully used to control follicular and luteal dynamics and to synchronize ovulation enabling AI without estrous detection [14] [15]. Many factors may influence how donors respond to superovulation and generate a high number of fertilized good to excellent quality embryos. According to [16] founded that the ovulation in different protocols superovulation range from 38.7±10.6 % to 88.0±4.9%. According to [17] and [18] that treatment with GnRH on day 6 and 7 of the estrus cycle is the optimal time for ovulation and fertility, however, treatment GnRH at random reproductive stage generally results in ovulation in 50 % or less. [9] reported that regression of dominant follicle does not occur in all dairy cows following the standard treatment with 2 mg EB ant P4 implant and 3 mg EB does not improve follicle wave synchronization. Furthermore, treatment GnRH that can couse ovulation of CL and thus increase circulating P4 during growth of the ovulating follicular wave. Treatment with E2 esters can induce CL regression, potentially decreasing circulating P4 during ovulatory follicle growth. This present study is supported by [19] that the increased P4 was due to greater ovulation to the GnRH than the Eb at the star of the protocol and reduced premature CL regression for GnRH compare with EB.

The percentage of transferable embryo is higer in treatment A (61.11%) than treatment B (0%) transferable embryo (P<0.05) and the number of transferable embryo in GnRH plus P4 is 11 embryos and did not have the transferable in using estradiol plus P4. [20] that the number of transferable embryo dairy cows in different postpartum period range from 1.7±0.74 – 4.4±0.93. the quality and stage of the embryo development influenced the CR. [21] [22] founded the number of transferable embryo was 2.01±0.39. In treatment A greater the percentage of transferable embryo caused GnRH protocol can induce an LH surge, ovulate a dominant follicle and thus initiated a synchronized new follicular wave, although the magnitude of these responses depend on stage of the estrus cycle when GnRH treatment is given. That percentage of embryos classified as grade 1 and 2 were 21.4 % vs 32.8 % (P<0.05) and lower than the percentage s of embryos with a quality code excelent and /or good were 81.5 % to 83.8 %. In addition, that on day 6 of estrus cycle a reduction of the dose of GnRH to synchronize follicular wave emergence as a pretreatment for superstimulation promote transferable embryos [23].

Table 3. Effect of treatment on ovulation rate and embryo quality of local Pesisir cows.

Superovulation Response	Time Embryo Transfer Program	
	GnRH +P4 (treatment A)	Estradiol+P4(treatment B)
Ovulation rate (%)	70.37	50
Transferable embryo (%)	61.11	0
Degenerated embryo (%)	22.22	66.66
Unfertilized (%)	16.66	33.33

The degenerate embryo percentage in using GnRH plus P4 22.22 %) was lower (P<0.05) than using estradiol plus P4 (66.66). This present study indicated that GnRH plus P4 is also effect in quality of embryo associated with FSH. The number of degenerate of embyo in using GnRH plus P4 was 4 and in using estradiol plus P4 was 2 (P<0.05). The number of generated embryo in the present study was higher than reported by [22] that 0.93±0.25 and [24] was founded the transferable embryo in Holstein 12.46 % and higher in inbreeding group. According to [25] that a lower number of transferable embyos in cows with Fx = 9.0%-30.0% compare to those with Fx – 0%-8.9% (P<0.05).

The unfertilized in treatment A (16.66%) is low (P<0.05) than treatment B (33.33%). The ferlized embryo is lower in treatment A than treatment B (P<,0.05). The number of unfertilized embryo in GnRH plus P4 was 3 and in using estradiol plu P4 was 2. This result was higher than reported by [22] that the the average unfertized ova was 1.31±0.37 and [24] was 15.77 %. According to [26] that E2 or P4 treatments are used at the initiation of protocols suppression circulating LH and FSH concentrations and regression the follicles in the current follicular wave, resultimg in a synchronized follicular wave emergence 3 to 5 d later, depending on the dose type of estradiol. [13] founded that higher synchronization and transfer rates, and shorte postpartum interval to ET, can be achieved with timed ET following the Ovsynch plus CIDR protocol than after estrus with the single PGF2α treatment.

3.2. Regression correlation of ovarian activity in two treatment of timed embryo transfer program

The average number of follicle dominant is 27, the number of corpus luteum is 19 and the number of embryo is 118 in GnRH plus progesterone combination. In estradiol plus progesterone combination the number of follicle dominant is 18, the number of CL is 9 and the number of embryo is 6. The average number of embryos in treatment A is 6 .0 embryo and treatment B is 3.0 embryos /donor. This present study the number of embryos is lower than reported by [27] was 14,3 ±2.1 90.5 % (in hyaluronan 0.5%) and 14.4±2.0 (in hyaluronan 10%) and 10.2±1.8 (in FSH). This present study the transferable embryos is lower than earlier study was range from 47.8±8.4 to 62.5±7.5 %. The largest number of cows showing estrus in GnRH group may be could be explained due to the induction of an earlier follicular wave. Thus, the greater size of follicle has more time to grow and therefore it was able to produce more levels of estradiol at the time of CIDR removal [28].

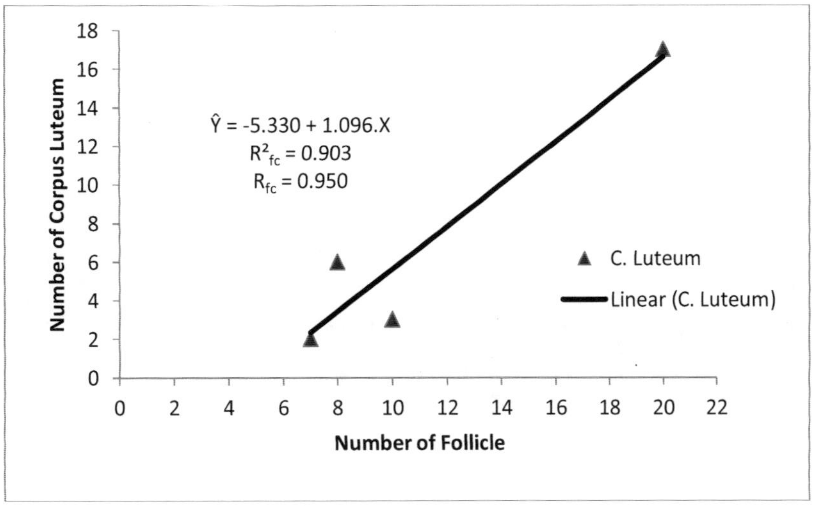

Figure 1. Linear regression lines for the number of follicle dominant with the number of CL.of local Pesisir Selatan cows.

The linear regression for the number of follicle dominant in relation to the number of CL (Figure 1), with high coefficient correlation is R = 0.950 and linear regssion (Figure 2.) with the embryos is R= 0.879. Furthermore, the linear regression for the number of CL in relation to the number of embryo (Figure 3), with the high coefficient correlation is R= 0.980. At the time of flushing not all the donor have the equal number of follicle with the CL and embryo is tend to decrease the number, even the linear regression. In treatment GnRH plus P4 from 27 follicle dominant produced 19 CL and and produced 18 embryos. In treatment B at the time of flushing the number of embryos was 6 from 9 and 10 follicle dominant. This present study showed that the ovarian activity response, tend to decrease at the step of cyclic status of reproduction of donor. This means the number of embryo production depended to the number of follicular ovulation and yhe number of CL at the flushing.

This precense study supported by [16] that the number of ovulatory follicles increased, the number of CL, also increased (r= 0.84) and the number of CL increased, also increased the number of embryo (r= 0.38). Thus more total embryos collection because more total structure ovarian were recovered. According to [29] compared twice daily im FSH injections to once daily sc FSH injection to superovulate beef heifers, had the greater superovulatory response, more recovered structures, and more embryos grade 1and (twise daily is greater than once daily. In addition, that no effect of the number ovulateion on the percentage of transferable embryos.

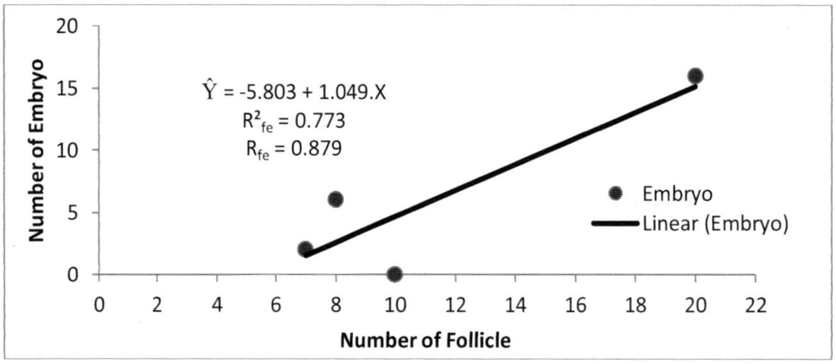

Figure 2. Linear regression for the number of follicle dominat on the number of embryo of local Pesisir Selatan cow.

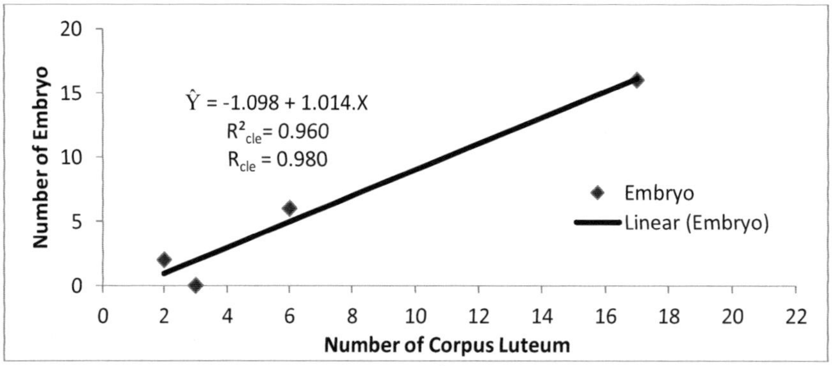

Figure 3. Linear regression for follicle dominant, the number of CL and the number of embryos local Pesisir Selatan cows.

In this result showed that the greater number of follicle, tend to greater number of CL and embryos production and treatment GnRH plus P4 more embryo production than treatment estradiol plus P4. According to [30] that large ovulatory follicles have associated with the formation of the larger CL with greater capacity of synthesizing P4. The manifestation of estrus behavior prior to timed-AI was associated with large ovarian follicle, greater plasma P4 concentrations during the luteal phase and greater conception rate [31]. When animal are bred following a TAI protocol, where ovulation is induced by GnRH, estrus response e and increased preovulatory concentration of estradiol before ovulation are critical determinants of subsequent embryo quality and potential pregnancy loss. According to [24] that inbred animals with a lower coefficient Fx had a very small and insignificant difference in quality and proportion of transferable embryos compared with group Fx = 0. At greater inbreeding coefficients (3.1%- 25%), the reduction in embryo quality increased.

The donor of Pesisir Selatan cows had the inbreeding without breeding management control, caused of limited bull to bred the cow at the time breeding seasons. According to [25] that a decrease in the count of transferable embryos in the inbreeding group Fx= 3 %- 5.9%. The effect of inbreeding depression on ovarian activity in superovulated cows. The regression equation was significant for transferable embyos was`R^2= 0.91 [24], founded that increased inbreeding of embryos had a negative impact on reproduction.

4. Conclusion

The ovulation rate, and the quality of embryos were greater in using GnRH plus P4 than Estradiol plus P4. The response of ovarium activity have the linear regression between number of follicular ovulation, the number of CL and the number of embryo of local Pesisir Selatan cows in west Sumatera. linear regression of ovarian avtivity was high in donor local Pesisir Selatan cows.

5. References

[1] Zaituni, U, Hendri, M. Masrizal. 2017. Fertility in south Pesisir cows following ovsynch and cosynch protocols of estrus synchronization in west Sumatera. IJASEIT. 7: 2100- 2107.

[2] Lopez, H. 2005. Reproductive hormones and follicular growth during development of one or multiple dominant follicle in cattle. Biol. Reprod. 72: 788-795.

[3] Baruselli, P. S., M. F. Sa Filho, C. M. Martin, L. F. Naser, M. F. G. Nogueira, C. M. Barros. 2006. Superovulation and embryo transfer in *Bos indicus* cattle. Theriogenology. 65: 77-88.

[4] Martin, C. M., C. A. Rodrigues, L. M. Vieira, R. J. Mapletoft, G. A. Bo, M. F. Sa Filho, P. S. Baruselli. 2012. The effect of timing of the induction of ovulation on embryo production insuperstimulated lactating Holstein cows undergoing fixed- timed artificial insemination. Theriogenology 78: 974-980.

[5] Pereira, M. H. C., M. C. Wiltbank, and J. L. M. Vasconcelos. 2016. Expression of estrus improve fertility and decreases pregnancy losses in lactating dairy cows that receive artificial insemination and embryo transfer. J. Dairy Sci. 99: 2237-2247.

[6] Ambrose, D. J., M. G. Colazo, J. P. Kastelic. 2010. The applications of timed artificial insemination and timed embryo transfer in reproductive management of dairy cattle. R. Bras. Zootec. 29: 383-392.

[7] Larimore, E. L., O. L. Amundson, S. L. Bird, B. J. Funnell, S. G. Kruse, G. A. Bridges and G. A. Perry. 2015. Influence of estrus at fixed – timed artificial insemination on early embryonic development in beef cattle. J. Anim. Sci. 93: 2806-2812.

[8] Jinks, E. M., M. F. Smith, J. A. Atkins, K. G. Pohler, G. A. Perry, M. D. Macnell, A. J. Robert, R. C. Waterman, L. J. Alexander, and T. W. Geary. 2013. Preovulatory estradiol and the establishment and maintenance of pregnancy in suckled beef cows. J. Anim. Sci. 9: 1176- 118.

[9] Souza, A. H., H. Ayres, R. M. Ferreira, and M. C. Wiltbank. 2008. Anew presynchronization system (double ovsynch) increases fertility at first postpartum timed AI in lactating dairy cows. Theriogenology 70: 208-215.

[10] Pereira, M. H. C., A. D. P. Rodrigues, R. J. De Carvalho, M. C. Wiltbank, and J. L. M. Vasconcelos. 2014. Increasing length of an estradiol and progesterone timed artificial insemination protocol decreases pregnancy losses in lactating dairy cows. J.Dairy Sci. 97: 1454-1464.

[11] Stevenson, J. S., J. R. Pursley, H. A. Garverick, P. M. Fricke, D. J. Kesler, J. S. Ottobre, and M. C. Wiltbank. 2006. Treatment of cycling and noncycling lactating dairy cows with progesterone during Ovsynch. J.Dairy Sci. 89: 2567-2578.

[12] Sakase, M., Y. Seo, M. Fukushima, M. Noda, K. Takeda, S. Ueno, T. Inaba, H. Tamada, T. Sawada, N. Kawate. 2005. Effect of CIDR-based protocols for timed-AI on the conception rate and ovarian functions of Japanese Black beef cows in the early postpartum period. Theriogenology. 64:1197-1211.

[13] Sakase, M., N. Kawate, C. Nakagawa, M. Fukushima, M. Noda, K. Takeda, S. Ueno, T. Inaba. K. Kida, H. Tamada, T. Sawada. 2006. Inhibitory effects of CIDR-BASED ovulation- synchronization protocols on uterine PGF2α secretion at the following luteal phase in early postpartum non- cycling beef cows. J. Reprod. Dev. 52: 497-502.

[14] Bo, G. A., L. Cutaia, L. C. Peres, D. Pincinato, D. Marafia, P. S. Baruselli. 2007. Technologies fixed-time artificial inseminationand their influence on reproductive performance of *Bos indicus* cattle. Soc. Reprod. Fertil. Suppl. 64: 223-236.

[15] Sa Filho, O. J., M. Meneghetti, R. Peres, G. Lamb, J. L. M. Vasconcelos. 2009. Fixed–timed artificial insemination with estradiol and progesterone for *Bos Indicus* cowsII: strategies and factor affecting fertility. Theriogenology. 72: 210-218.

[16] Carvalho, P. D., K. S. Hackbart, R. W. Bender, G. M. Baez, A. R. Dresch, J. N. Guenther, A. H. Souza, P. M. Fricke. 2014. Use of a single injection of long- acting recombinant bovine FSH to superovulate Holstein heifer: A preliminary study. Theriogenology. 82: 481-489.

[17] Lopez, G., J. O. Giordano, A. Valenza, M. M. Herlihi, J. N. Guenther, M. C. Wiltbank, and P. M. Fricke. 2013. Effect of timing of initiation of resynchronization and presynchonization wiyh gonadotropin-releasing hormone on fertility of resynchronized inseminations in lactating dairy cows. J.Dairy Sci. 96: 3788-3798.

[18] Bisinoto, R. S., E. S. Ribeiro, F. S. Lima, N. Martinez, L. F. Green, L. Barbosa, P. P. Bueno, L. F. S. Scagion, W. W. Thatcher, and J. E. P. Santos. 2013. Targeted progesterone supplementation improve fertility in lactating dairy cows without a corpus luteum at the initiation of the timed artificial insemination protocol. J.Dairy Sci. 96: 2214-2225.

[19] Melo, L. F., P. L. J. Monteiro Jr. R. S. Surjus, J. N. Drum, M. C. Wiltbank, and R. Sartoni. 2016. Progesterone – based fixed- timed artificial insemination protocols for dairy cccowsw: gonadotropin- releasing hormone versus estradiol benzoate at initiation and estradiol cypionate versus estradiol benzoate at the end. J. Dairy Sci. 99: 1-11.

[20] Stadnik, L., J. Bezdicek, A. Makarevich, E. Kubovicova, F. Louda, I. Fellnerova, Z. Hegedusova, R. Holasek. 2017. Ovarian activity and embryo yield in relation to the postpartum period in superovulation dairy cows. Acta, Vet. BRNO. 86:51-57.

[21] Ongubo, M. N., H. A. Rachuenyo, F. N. Lusweti, D. K. Kios, J. K. Kitilir, K. Muser, W. K.Tonui, L. K. Lokwaleput and G. O. Oliech. 2015. Factors affecting conception rates in cattle following embyo tramfer. Uganda Journal of Agricultural Sciences. 16:19-27.

[22] Lamb, G. C., D. R. Brown, R. C. Wasson, J. E. Larson, C. R. Dahlen, and A. DiCostanzo. 2004. Effect of Trace Mineral Nutrition on Folicular Response, Ovulation, and Embryo Production in Superovulated Angus Heifers. J. Reprod. Fertil. 16(1,2): 289 (Abstr.

[23] Sato, T., K. Akada, Y. Uchiyama, Y. Kimura, N. Ujiwara, Y. Sato, M. Umeda, T. Furukawa. 2005. The effect of pretreatment with different doses of GnRH to synchronize follicular wave on superovulation of follicular growth in dairy cattle. J. of Repro.duction and Development. 5: 573-578.

[24] Bezdicek, J., L. Standnik, A. Makarevich, E. Kubovicova, F. Louda, Z. Hegedosova, R. Holasek, J. Beran, M. Nejdloka. 2014. Effect of inbreeding on yield and quality of embryos recovered from superovulated Holstein cows. Turkish Journal of Veterinary and Animal Science. 38: 681-685.

[25] Alvarez, R. H., M. V. Gualberto, J. B. P. Carvalho, M. Bunelli, 2005. Effect of inbreeding on ovarian responses and embryo production from superovulated Mantqueira breed cows. Theriogenology 64: 1669-1676.

[26] Bo, G. A., L. C. Peres, L. E. Cutaia, D. Pincinato, P. S. Baruselli and R. J. Mapletoit. 2012. Reproduction, Fertility and Development, 24: 272-277.

[27] Bo, G. A., D. R. Rogan, R. J. Mapletoft. 2018. Pursuit of a method for single administration of pFSH for superstimulation in cattle: what we have learned. Theriogenology 112: 20-33.

[28] Kim, U. H., G. H. Suh, H. W. Nam, H. G. Kang and L. H. Kim. 2005. Follicular wave emergence, luteal function and synchronization of ovulation following GnRH or estradiol bemzoat in a CIDR – treated lactating Holstein cows. Theriogenology. 63:260-268.

[29] Walsh, J. H., R. Mantovani, R. T. Duby, E. W. Overstrom, J. R. Dobrinsky, W. J. Enright. 1993. The effects of once or twice daily injection of pFSH on superovulatory response in heifers. theriogenology. 40: 313-321.

[30] Frade, M. C., C. Frade, M. B. Cordeiro, M. F. de Sa Filho, F. S. Mesquita, G. de P. Nogueira, M. Binelli, C. M. B. Membrive. 2014. Anim. Repro. Sci. 151: 85-90.

[31] Sa Filho, M. F., J. E. P. Santos, R. M. Ferreira, J. N. S. Sales, P. S. Baruselli. 2011. Importance of estrus on pregnancy per insemination in suckled bos indicus cows submitted to estradiol /progesterone – based timed insemination protocols. Theriogenology 76: 455-463.

Acknowledgements

The author want to thank the Andalas University, BOPTN and LPPM Andalas University for financial suppot and facilities for doing experiment.

The prevalence and risk factor of gastrointestinal nematode infestation in cattle based on the environmental conditions of the farming locations in Aceh Province

Zulfikar [1], S Umar[2], T R Ferasyi[3] and M Tafsin[4]

[1] Ph.D. Student of Natural Resources and Environmental Management of the University of North Sumatera, Medan, Indonesia.
[2, 4] Department of Agriculture of the University of North Sumatera, Medan, Indonesia.
[3] Department of Veterinary Medicine of Syiah Kuala University, Banda Aceh, Indonesia.
E-mail: drh.zulfikar68@yahoo.com.

Abstract. This study was conducted to determine the prevalence and risk factor of gastrointestinal nematode infestation (GIN) in cattle based on environmental conditions, i.e., elevated altitude locations in Aceh Province. Cross-sectional surveys were conducted in the Districts of Gayo Lues and East Aceh, representing high altitude location (HAE) and low altitude location (LAE), respectively. A total of 64 samples (HAE) and 93 (LAE) samples were randomly collected from selected cattle. Fecal samples were examined with the centrifuge technique, and McMaster-counting chamber was used for egg count per gram feces (EPGF). Data obtained were analyzed for GIN infestation prevalence and risk factors based on the calculation of the Odds Ratio (OR). The results revealed that 14 HAE samples (21.9%, CI 13.5-33.43) and 35 LAE samples (37.6%: CI 28.46-47.79) were positive. Then, the number of EPGF in the GIN was significantly lower in the HAE (in average at 567 eggs/grams feces) compared to the LAE (an average at 669 eggs/grams feces) ($P<0.005$). The results of the OR calculation showed that the factor of raising cattle in LAE location significantly increased the risk of GIN infestation which was about 2.16 times higher than that of HAE location (($P<0.05$). In conclusion, the LAE location for raising cattle is a risk factor for higher prevalence of GIN infestation in Aceh Province.
Keywords – cattle, altitude, environment, GI nematodes, risk factor.

1. Introduction

Cattle is one of the most important agricultural commodities as it is considered to be livestock that serve as a source of protein for the life of living things, especially humans, in fulfilling the body's needs in carrying out their lives. According to [1] [2], most sources of animal protein to build human life generally come from livestock and its products.

Cattle are kept and grown by farmers on a small or large scale, both traditionally and intensively. Moreover, they are raised for meat production. However, some of the world's cattle production faces constraints in meeting the mentioned needs, especially health problem leading to low production and productivity. One of the health problems is caused by the presence of worm parasites such as nematode worm attacking the livestock. This problem is common in the farms of developing countries with low health management. Indeed, poor animal health condition will impact on production.

This aspect was pointed out by [3]. In fact, most of the developing countries in the growing of livestock showed low limits after the investigation demonstrating low production due to high cases of diseases caused by worm parasites, primarily gastrointestinal (GI) nematode that attacks the digestif system of livestock causing large economic losses.

Worm parasitic infections, especially nematode classes found in the digestive (gastrointestinal) system of livestock, are one of the main and biggest causes of production losses [4]. As explained by [5] [6] [7], health problems repeatedly occurred because the GI nematode infestations will cause huge losses for both the government and other actors such as farmers and livestock breeders.

In another study by [8], it was reported that cattle farms in Sri Lanka had an economic loss of up to 230 million rupees/year due to nematode infestations, especially the gastrointestinal types. These losses were also observed in Indonesia. According to reports from [9] [10], the costs incurred for the prevention/ treatment of parasite worms can reach 240 billion rupiahs each year.

According to [11], gastrointestinal nematode (NGI) infection can cause a continuous decrease in livestock production, and if ignored, it can cause tremendous losses in livestock production as a whole. Generally, an increase in gastrointestinal nematode parasites in cattle can be associated with poor maintenance system. For instance, in Indonesia, farmers raise the cattle in traditional or semi-intensive settings by letting the livestock feed themselves, although they are not kept in the shed, leading to a great opportunity for parasite breeding.

Nematodes need an appropriate environment before interacting with the host (animal or human) as explained in some study results in parasitology. Some authors, [12] [13], explained that nematodes are persistent in the environment, and the survival in a host body will last for a long time between 1 and ten years, even though the host immune system continues to react. Because the nematodes can trick the host defense system and repeat infections in the host which may lead to chronic infection with high morbidity in adult animals and contamination of young animals. Furthermore, gastrointestinal nematodes need an environment to survive outside the body of animals, such as topography, temperature, humidity, rainfall, soil conditions or soil pH, because after living the animal or human body, this nematode will continue to live and react with the other host.

[14] [15] explained that an increase in the spread of diseases, especially worm parasites attacking livestock in recent years, could be associated with climate changes, including altitude. This theory is also explained by [16]. Indeed, several main factors increase the spread of diseases, especially those caused by parasites, including the altitude, topography, geographical area, unsuitable climate, different age, unsuitable population, lack of health handling and level of knowledge. According to [17], the relationship between the survival process and the parasite-free life stage of nematode worms in the environment is strongly influenced by temperature, humidity, soil pH, topography and other conditions, including staying alive in extreme conditions such as winter and soil chemical conditions contaminated with disinfectant.

In Aceh Province, the cattle are raised in both high (HAE) and low (LAE) altitudes. The HAE area represents the districts in this province covering an area of 166.06 km^2 with hills and mountainous terrain and an average height between 800-2.000 m asl (above sea level). It includes Mount Leuser which is the highest mountain in Aceh Province (4.446 m asl). An example of HAE is the District of Gayo Lues [18].

LAE is the district with a total area of 6,040.60 Km^2. Generally, a low topographic area is hilly, partly marshy with mangrove forests at an altitude of 0-308 masl. Livestock commodities in the low topographic area, at present, include cattle, buffalo, goats, sheep, and poultry. For example, the districts on the eastern coast of Aceh Province, such as District of East Aceh. Based on reports, the total population of cattle was 64.169 heads, and LAE is one of the regencies in Aceh Province becoming cattle hulls [19]. Therefore, the purpose of this study was to determine the prevalence and risk factor of gastrointestinal nematode infestation (GIN) in cattle farming in HAE and LAE in Aceh Province.

2. Materials and methods

2.1. Time and location of the study

Cattle fecal collection was conducted from April to June 2018 in two different areas, i.e., HAE in Gayo Lues (>800 mdpl) and LAE in East Aceh (<100 mdpl). A precise position of each location is shown in the map presented in Figure 1.

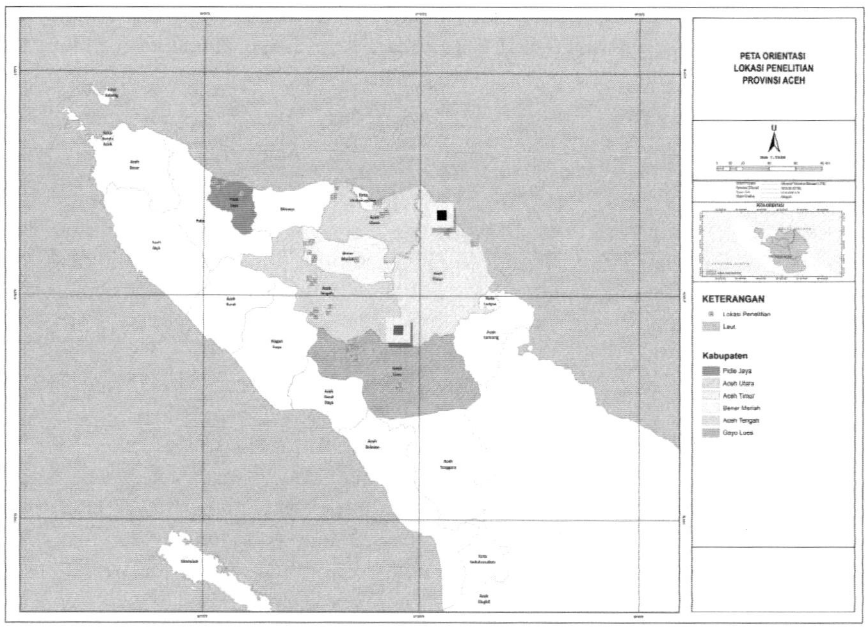

Figure 1. Map of the study locations (black dot: East Aceh District; red dot: Gayo Lues District).

2.2. Materials

The materials used in this study were as follows: spoons, plastic basins, refrigerators, scales, test tubes, Petri dishes, measuring cups, scissors, glass objects, centrifuges, Becker glasses, Pasteur pipes, microscopes, ice boxes, and digital cameras. The feces samples were collected with plastic, connective rubber, cardboard, pens, distilled water, 0.4% formalin, and saturated salt or NaCl Physiology.

2.3. Sample examination method

Fecal sample examination to identify positive and negative gastrointestinal nematode worm eggs were conducted through the centrifuge method. Initially, the samples were added with the saturated salt solution as much as 40 ml so that the total volume becomes 60 ml. The fecal solution was taken/sucked while stirring until even/homogeneous using a special pipette which at the end of the filter has a size of 250 ~ L according to the size of the largest worm eggs with a transverse diameter of 130-200 p [20].

The solution was taken using a pipette and rapidly incorporated into the glass room count tool (Universal Whitlock) with four count rooms. Each room has a volume of 0.5 ml. The number of eggs per gram of feces (EPG) was then counted [21] [22].

2.4. Data analysis

The prevalence was calculated as the number of positive samples from HAE and LAE that were examined. Data analysis was performed as per the EPGF of GI nematodes in each location. The obtained data were analyzed for GIN infestation prevalence and risk factors based on the calculation of the Odds Ratio (OR).

The formula is as shown below:

a. Prevalence:

$$\mathbf{P = \frac{X}{N} \; X\,100\%}$$

P = Prevalence, X = exposed, N = Total Sample

b. Faecal egg count (EPGF) test:

$$\mathbf{TTGT = n/Bt \times Vt/Vh}$$

N = the number of worm eggs in the room,
Vt = total volume of stool samples,
Vh = volume of calculated rooms,
Bt = stool weight.

c. Calculation of odd ratio:

$$\mathbf{OR = ad/bc}$$

Notes:

- OR = odd ratio;
- a = cell a positif;
- b = cell b negatif;
- c = cell c positif : and,
- d = cell d negatif

3. Results and discussion

3.1. The percentage of Gastrointestinal Nematodes (NGI) in HAE and LAE

The percentages of gastrointestinal nematodes (NGI) in cattle maintained in both high and low topographic areas are presented in Table 1. High topographic areas showed a smaller percentage gastrointestinal nematodes (NGI) (<21.9%: CI 13.5-33.43) compared to a low topographic area (>37.6%: CI 28.46-47.79). There were significant differences (P<0.005) in the percentages or prevalence of GI nematodes between areas at different altitudes.

Table 1. The percentages of gastrointestinal nematodes (NGI) infection in cattle.

Areas	District	Gastrointestinal Nematodes				
		Total Sample	Sample Positive	Sample Negative	Prev (%)	CI
HAE	Gayo Lues	64	14	50	21,9	13,5-33,43 [a)]
LAE	East Aceh	93	35	58	37,6	28,46-47,79 [b)]

Note: [a) and b)]different superscripts show significant differences (P<0.005).

The results showed that microenvironmental conditions such as altitude (area differences), LAE, high temperatures, high humidity, and low rainfall, are likely to affect the prevalence of gastrointestinal nematodes (NGI) in cattle. These results follow those of [15], explaining that the intensity and distribution of worm parasites, especially gastrointestinal nematodes (NGI) types, have been strongly influenced by changes in environmental conditions such as temperature, climate, and altitude (topography).

[23] stated the high prevalence of gastrointestinal nematodes (NGI) in LAE areas could be caused by the number of adult females producing thousands of eggs per day. Thus, causing rapid larval pastoral contamination associated with the outbreak of hemosiosis due to the influence of hotter temperatures. This result was explained by [24], stating that the climate with high temperature and humidity can be the cause of over hatching in the farming area which occurred as these parasites have high biotic potential increasing their pathogenicity in a hot climate.

3.2. Fecal Egg Counts (FEC) of gastrointestinal nematodes infestation in HAE and LAE.

Fecal Egg Counts (FEC) of gastrointestinal nematodes infestation that appeared in cattle in HAE and LAE are presented in Table 2.

Table 2. Fecal Egg Counts (FEC) of gastrointestinal nematodes infestation in cattle in the different areas.

Area Category	District	Gastrointestinal Nematodes			
		Total Sample	Sample Positive	Sample Negative	Average FEC/gram feces
HAE	Gayo Lues	64	14	50	567[a)]
LAE	East Aceh	93	35	58	669[b)]

Note [a) and b)]different superscripts show significant differences (P<0.005)

Table 2 above showed the number of eggs per gram of feces. Cattle kept in HAE area had lower egg infestation compared to those kept in LAE area. Fourteen positive samples from the HAE obtained GI nematode eggs had an average of 567 eggs/gram feces. Meanwhile, the LAE of 35 positive samples turned out to produce an average of 669 eggs/gram feces. These results showed significant differences in various altitudes (P<0.005). The GI nematode egg infestations were lower in HAE environment compared to the LAE ones. However, the results of the average number of eggs, fecal egg counts (FEC) of gastrointestinal nematode infestation in HAE were categorized as the low category (<500 eggs), they were categorized as the medium category (between 500-1000 eggs) in LAE. The types of gastrointestinal nematode eggs can be seen in Figure 2.

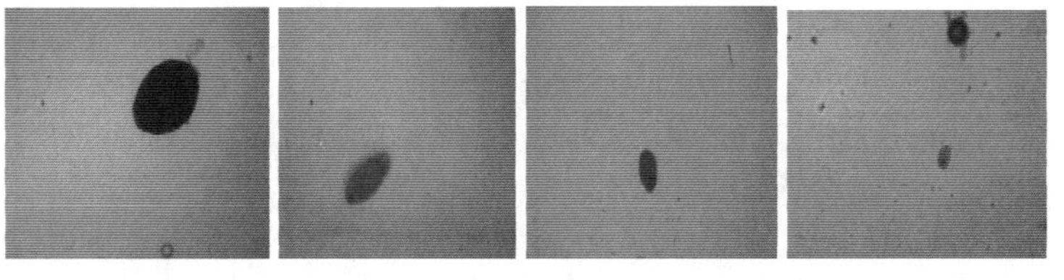

Figure 2. Types of gastrointestinal nematode eggs found in fecal samples.

The number of eggs per gram fecal (EPGF) causing high gastrointestinal nematode (NGI) infections was found in cattle kept in HAE and LAE but was still considered low. This result follows that of [25] [26], stating that the number of egg (1-1000 eggs) is still considered light and medium category. However, the number of eggs, ≥ 1000 eggs, is considered a high category.

3.3. Risk Factor

Risk factors of gastrointestinal nematodes in cattle kept in different environments can be seen in Table 3. The results of OR calculation showed that the factor of raising cattle in LAE locations significantly increased the risk of gastrointestinal nematodes infestation about 2.16 times higher than that of HAE locations (P<0.05).

Table 3. Analysis of gastrointestinal nematodes infestation risk factors in cattle based on the farming areas.

Areas Category	Nematoda GI					
	Total Sample	**Sample Positive**	**Sample Negative**	**Prev (%)**	**OR**	**P (Value)**
HAE	64	14	50	21.9 (95% CI: 13.5-33.43)	2.16 (95% CI: 1.04-4.45)	<0,05
LAE	93	35	58	37.6 (95% CI: 28.46-47.79)		

Note : Prev= prevalence, OR = Odds Ratio, CI = Confidence Interval.

The high level of prevalence and transmission risks of gastrointestinal nematode (NGI) in cattle farming environment in LAE area can be associated with environmental factors such as temperature, humidity and pH, population density, poor house conditions, and poor treatment. The gastrointestinal nematode egg can survive in the farming environment. According to [27], the high infestation in LAE areas can be associated with poor environmental conditions, maintenance management and housing located in densely populated areas with limited land, while cattle are constantly being carried to the same location during both dry and rainy seasons. Furthermore, [28] explained that the high prevalence of gastrointestinal nematode infestation in the environment is caused by exposure occurring due to environmental conditions such as unsuitable humidity and temperature.

4. Conclusion

In conclusion, these results indicated that the number of EPGF of GIN was significantly lower in HAE compared to LAE location. Also, the LAE location for raising cattle is a risk factor for higher GIN infestation prevalence in Aceh Province.

5. References

[1] Domke, A V., Chartier C, Gjerde B, Leine N, Vatn S, Osteras O and Stuen S 2011 Worm control practice against gastro-intestinal parasites in norwegian sheep and goat flocks *Journal. Acta Vet Scand,* **53** 29.

[2] Williams, E. B 2013 Gastrointestinal parasites in ruminants at selected abattoirs in the greater accra region, *thesis* legonin partial fulfillment of the requirement for the award of mphil microbiology degree. University of Ghana.

[3] Pathak, A. K. 2013 Potential of using condensed tannins to control gastrointestinal nematodes and improve small ruminant performance *Int J of Molecular Veterinary Research* **3** 8.

[4] Sibbald A M,, Erhard H W, MCleod J E and Hooper R J 2009 individual personality and the spatial distribution of groups of grazing animals: an example with sheep *Journal Behav Processes*. **82**.

[5] Perry, Perry G A, Perry B L 2009 GNRH treatment at artificial insemination in beef cattle fails to increase plasma progesterone concentrations or pregnancy rate *Journal Theriogenology*. **71**.

[6] Williams, E.B 2013 Gastrointestinal parasites in ruminants at selected abattoirs in the greater accra region, ghana. thesis.university of ghana legonin partial fulfillment of the requirement for the award of mphil microbiology *degree* University of Ghana.

[7] Kothalawala A., K.H.M., Fernando, G.K.C.N. and Kothalawala, H. 2007 Estimation of economic losses on nematode infestation in goats in Sri Lanka *Procedding.ISAH Tartu Estonia.*

[8] Dirjen PKH 2014 Pedoman pelaksanaan penanggulangan penyakit hewan kegiatan Direktorat Kesehatan Hewan Pusat dan Daerah *Laporan Tahunan* Direktorat Jenderal Peternakan dan Kesehatan Hewan Kementerian Pertanian.

[9] Zulfikar, Sayed U, T.Reza F and M. Tafsin 2018 Association of infestation of gastrointestinal nematode with altitudes of location of cattle raising in two different districts in Aceh Province *Int. J. Trop. Vet. Biomed. Res.* **3** 1.

[10] Johannes, C., Johan höglund, Georgvon Samson-Himmelstjern, Pierre Dorny' and Jozef Vercruysse. 2009 Gastrointestinal nematode infections in adult dairy cattle: impact on production, diagnosis and control *Journal Veterinary Parasitology*. **164** 1.

[11] Oliveira, M.C.S. M.M. Alencar, A.C.S. Chagas, R. Giglioti and H.N. Oliveira 2009 Gastrointestinal nematode infection in beef cattle of different genetic groups in Brazil *Journal Veterinary Parasitology* **166**.

[12] Erwin, N., K. Mustaka dan R. Indah. 2010 Identitas jenis telur cacing parasit usus pada ternak sapi (Bos Sp) dan kerbau (Bubalus Sp) di rumah potong hewan Palembang *Jurnal Edisi Khusus Juni (D). Jurnal Penelitian Sains.* **10**.

[13] Zulfikar, Sayed U, T.Reza F and M. Tafsin 2017 Hubungan lingkungan dengan tingkat infestasi nematoda gastrointestinal pada sapi di Aceh *Jurnal Serambi Engineering* **2**.3.

[14] Mitchell, G.B.B and Somerville, D.K. 2015 Effects of climate change on helminth diseases in scotland". *Journal. SAC Publication*. **1**.

[15] Fox Naomi J. Glenn Marion, Ross S. Davidson, Piran C. L. White and Michael R. Hutchings 2012 Livestock helminths in a changing climate: approaches and restrictions to meaningful predictions *Journal Animals* **2** 1.

[16] Bhattachryya, D. K and Ahmed K 2005 Prevalence of helmintic Infection in Cattle and Buffaloes". *Indian Vet. J.* **82.**

[17] O'Connor, L. J., Kahn, L. P., and Walkden-Brown, S. W. 2007 Moisture requirements for the free-living Development of Haemonchus contortus: quantitative and temporal effects under conditions of low evaporation. *Journal Veterinary Parasitology*, **150**.

[18] BPS [Badan Pusat Statistik] 2017 Gayo Lues Dalam Angka. *Badan Pusat Statistik Kabupaten Gayo Lues.*

[19] BPS [Badan Pusat Statistik] 2017 Aceh Timur Dalam Angka. *Badan pusat statistik Kabupaten Aceh Timur.*

[20] Thienpont, D. Rochette, F and Vanparijs, O.F.J. 1986 Diagnosing helminthiasis through coprological examination *Edition 2* Publisher Janssen Research Foundation.

[21] Whitlock, H. V. 1948 Some modifications of the McMaster helminth egg-counting technique and apparatus *Journal of the Council for Scientific and Industrial Research.* **21** 3.

[22] Jaroslav V, Miloslav P, Igor Z, Zuzana Č, Ivana J, Iva L and Milan M. 2010 Which mcmaster egg counting technique is the most reliable *Journal Parasitol Res.* **109**.

[23] Roeber F, Jex A R, and Gasser R. B. 2013 Impact of gastrointestinal parasitic nematodes of sheep, and the role of advanced molecular tools for exploring epidemiology and drug resistance-an Australian Perspective *Journal. Parasites Vectors.* **6**.

[24] Waller P, and Chandrawathani P. 2005 Haemonchus contortus: parasite problem number 1 from tropics-polar circle. problems and prospects for control based on epidemiolog *Journal Trop. Biomed.*.**22** 2.

[25] Soulsby E J L 1986 Helminths, Arthopods and Protozoa of domesticated animal. *7th (edn),* ELBS and Bailliere Tindall, London.

[26] Zulfikar 2013 Kajian epidemiologi parasit nematoda gastrointestinal pada sapi di Dataran Tinggi dan Dataran Rendah Provinsi Aceh *Tesis* Program Studi Kesehatan Masyarakat Veteriner Program Pascasarjana University Syiah Kuala-Banda Aceh.

[27] Zvinorova P I T E, Halimani F C, Muchadeyi O, Matika V, Riggio and K, Dzama 2016 Prevalence and risk factors of gastrointestinal parasitic infections in goats in low-input low-output farming systems in Zimbabwe *Journal Small Ruminant Research* **143**.

[28] Moussouni, L., Mokhtar, B., Mokhtar.S., and Abdelhanine A.2018 Prevalence of gastrointestinal parasitism infections in cattle of bass kabylie area: case of Bejaia Province Algeria *Journal Vet Rev.* **1**.

Acknowledgment

The author would like to thank Gayo Lues District Food Crops Agricultural Service and the Animal Husbandry Agriculture in East Aceh, Aceh Province, Indonesia, and the promoter team who provided the opportunity to complete the present research. This paper is part of the Doctoral Dissertation Program at the University of North Sumatra, and the Natural and Environmental Resources Study Program, Medan, Indonesia. The research was funded by the Ministry of Research, Technology and Higher Education of the Republic of Indonesia.

Utilization of local feed to support new entrepreneur in poultry business

Adrizal Adrizal[1], James Hellyward[2], and Dwi Yuzaria[2]

[1] Department of Nutrition and Feed Technology, Andalas University, Padang, West Sumatera, Indonesia.
Corrsponding e-mail: adrizal@ansci.unand.ac.id.
[2] Department of Social Economic, Andalas University, Padang, West Sumatera, Indonesia.

Abstract. Nowadays, Indonesian poultry business dominated by big enterprises like Charoen Phokpand Indonesia, Japfa Comfeed, Gold Coin etc. The enterprises depend on imported feedstuffs, so the feed price is very volatile because influenced by international trade and dollar value change. The condition cause the new entreprises difficult to survive, so that it will be reasonable for them to utilize local feed. The objectives of the research were utilization of local feedstuffs in ration formulation in order to minimize feed cost and ensure availability of feed continuously.
Keywords – feed, price/cost, entrepreneur, poultry business.

1. Introduction

Nowadays the poultry business in Indonesia is dominated by large companies such as Charoen Phokpand Indonesia, Japfa Confeed, Gold Coin and others. The companies producing ration use feed stuffs mostly imported producks. This causes the price of ration became unstable and tends to rise in accordance with the exchange rate of the dollar against the rupiah. The condition causes a small scale enterprises especially that managed by new entrepereneurs being difficult to grow. On the other hand the local feed ingredients available in sufficient quantities to support ventures with small scales, but how is the optimal use of each feed stuff has not been studied in real. This research aims to formulate rations using local feed materials for three types of poultry i.e. broiler, local chicken and duck.

2. Materials and methods

The research material was used to formulate rations are corn, rice bran, coconut cake, soy beans, dried fish, bone meal and topmix. The rations were formulated for broiler, local chicken and duck in starter period. The data used were price, nutrient content, limitations, as well as nutritional needs each of poultry. The feed ingredient price data obtained through the survey to the source of raw materials in the city of Padang. The data content of nutrition and nutritional needs as well as limitations on the use of the feed materials obtained by the study of literature. Ration formulation is done with the algorithm of fuzzy linear programming using Microsoft excel software which comes with solver function. The following mathematical model was applicated to formulate the rations [1].

$$\text{Minimize} \quad Z = \sum_{i=1}^{n} C_i X_i$$

$$\text{Subject to}: \sum_{i=1}^{n} \sum_{j=1}^{m} a_{ij} X_{ij} \geq, \leq \text{ or } = L_j$$

Where:
Z =cost of complete ration
n = the number of feedstuffs
C_i = cost of ith feedstuff
X_i = Percent of ith feedstuff
m = number of constraints
a_{ij} = technical coefficient of ith feedstuff j^{th} constraint
L_j = j^{th} constraint limitation

3. Result and discussions

3.1. Identification of Local Feedstuffs

The result of identification of local identification presented on Table 1. On the table it looks that there are 10 kinds of potential feed ingredients used as a constituent of rations. As an energy source is corn and rice bran, while as a source of protein is the coconut cake, soybean and fish meal. On the table was also seen that the soybean in addition to serve as a source of protein, also serves as an energy source because of its high content of ME also. Bone meal, crab shells, limestone and topmix serves as a source of minerals.

3.2. Ration Formula

Ration formulation results for three kinds of poultry kept by new entrepreneurs are presented in Table 2. In the table was presented that the most expensive ration was broiler ration and followed by ducks and local chicken. The price of chicken broiler rations (IDR 4,103/kg) caused by high nutrient needs. The broiler chicken require minimum 22% protein and ME 3000 kcal/kg. Ducks just need protein by 17% and 2800 kcal/kg, so the ration price not too high i.e. IDR 3,470/kg. The cheapest ration was local chicken i.e. IDR3,185/kg because it requires only as many as 16% protein and 2800 kcal/kg of ME. Based on the results of the ration formulation It is evident that the cost of ration closely relate with nutrient content of the ration especially protein and ME.

Table 1. Price and nutrient content of local feed stuffs.

No	Feedstuff	Price (IDR/kg)	ME (Kcal/kg)	Nutrient content (%)							
				CP	Fat	CF	Ca	P	Lys	Meth	Trip
1	Corn	3000	3430	8,7	3,9	2	0,02	0,3	0,2	0,18	0,1
2	Rice brand	2000	1630	13,5	13	12	0,12	1,5	0,8	0,29	0,1
3	Coconut cake	2600	1540	21	1,8	15	0,2	0,6	0,6	0,29	0,2
4	soybean	8000	3510	38	18	5	0,25	0,6	2,4	0,51	0,55
5	Fish meal	5000	2970	50	8	1	5,5	2,8	5	1,8	0,8
6	Bone meal	4000					24	12			
7	Crab shells	4000	1870	31	2	11	15	1,6	1,4	0,5	0,3
8	Limestone	1000					40				
9	Topmix	10000									

On the table was also seen that a large part of energy source were supplied by corn. More half of the ration comes from corn. Other energy sources are soy bean, nevertheless because it was expensive, only the broiler using it, while local chickens and ducks did not use. Protein source was dominated by dried fish and added with coconut cake and rice bran. Soy beans in addition to serve as a source of energy also serves as a source of protein. Other materials such as crab shell, limestone and premix serves as a source of minerals.

Table 2. The results of the feed formulation using local materials for three kinds of poultry.

No	Feedstuffs	Formula (%)		
		Broiler	Local Chicken	Duck
1	Corn	51.7	57.0	55.3
2	Rice Bran	2.1	19.9	18.9
3	Coconut cake	14.5	7.7	8.1
4	Soy Beans	10.1	0.0	0.0
5	Dried fish	20.0	12.8	15.3
6	Crab shells	1.0	1.0	1.0
7	Limestone	0.0	1.0	1.0
8	Premix	0.5	0.5	0.5
Total		100	100	100
The price (IDR/kg)		4,103	3,185	3,470
No	Nutrient Content			
1	ME (Kcal/Kg)	3000	2800	2800
2	CP (%)	22.0	16.0	17.0
3	Fat (%)	6.0	6.0	6.0
4	CF (%)	4.3	4.9	4.8
5	Ca (%)	1.3	1.3	1.4
6	P (%)	0.9	0.9	0.9
7	Lysine (%)	1.5	1.0	1.1
8	Methionine (%)	0.6	0.4	0.5
9	Tryptophan (%)	0.3	0.2	0.2

4. Conclusion

Broiler ration, local chicken and duck can be formulate with feed local. The price of rations were IDR 4.103, 3.185 and 3.470 for broiler rations, local chicken and duck successively.

5. References

[1] Adrizal dan Marimin. 2004. Aplikasi Fuzzy Linear Programming untuk Optimasi Formulasi Ransum Unggas. Jurnal Keteknikan Pertanian 18: 77-85.

Acknowledgments

Thanks presented to the Direktorat Penelitian dan Pengabdian Masyarakat, Kementerian Riset, Teknologi, dan Pendidikan Tinggi has financed these activities in accordance with the contract of No. 001/SP2H/PPM/DRPM/2018.

The effect of duration of giving Cinnamononi Extract as a growth promoter feed additive and antimicrobial to broiler's performance on organic breeding

Ahadiyah Yuniza, Yose Rizal, Afriani Sandra

Lecturers of Animal Husbandry, Andalas University, Padang, West Sumatera, Indonesia.

Abstract. Cinnamononi Extract is a growth promoter feed additive and also plays role as antimicrobial that can substitute the use of antibiotic on the ration. This research was conducted to examine the effect of duration of giving the Cinnamononi Extract (EC) to the broiler's performance. This research used 160 broiler chicks on the age of 8 days which was caged on 20 boxes of cage. Each cage contained 8 chicks. There were 5 treatments concerning the duration of giving Cinnamononi Extract: A = no EC given (control), B = EC given every day for 5 weeks, C = EC given every day for the first 2 weeks and the last week, D = EC given every day for 3 weeks, and E = EC given every day for 4 weeks. The treatments were repeated 4 times. The result of this research showed that the treatment has a real impact ($p<0,05$) to the increasing of weight and ration conversion, but was not really effective to the ration consumption. All EC-giving treatments (treatment B, C, D, and E) can increase the weight gain and decrease the ration conversion ($p<0,05$). The treatment C (giving EC for the first 2 weeks, stop the giving, and give EC again on the last week) showed bad responses on weight gain and ration conversion ($p<0,05$) compare to treatment B, D, and E. the conclusion of this research is that the giving duration of Cinnamononi Extract every day for 5 weeks, 4 weeks, and 3 weeks (B, E, and D) gave a better response on weight gain and ration conversion compare to the giving for 2 weeks.
Keywords – Cinnamononi extract, feed additive, antimicrobial, organic breeding.

1. Introduction

The use of antibiotic on ration causes the chickens exposed by antibiotic continuously, because the ration will be consumed continuously until the chickens are ready to be harvested, so that the antibiotic residue will remain on the carcass produced. This matter will reduce consumers' interest to consume broiler chickens. As the applying of the Government Regulations that ban the giving of antibiotic to ration on the early 2018, it aroused the producers and broiler breeder to find the alternative to substitute antibiotic. This is considered as important because broiler chickens are susceptible to disease, so they cannot be freed from the giving of chemical drugs especially antibiotic, anthelmintic, and anticoccidial (Coccidiostat). Furthermore, the high temperature and humidity in Indonesia is a conducive weather for patogen microbial to reproduce. Therefore, the utilization of natural material that able to have a role as antimicrobial (antibacterial, anti-fungi, antivirus, etc.) is needed to substitute the use of antibiotic on the ration.

Cinnamononi Extract (EC) is an extract from the mixture of cinnamon leaves, noni fruit leaves, and noni fruit (ratio 1:2:2) with water as the solvent. This extract has been tested and it has been proved that this extract can stop the growth of *Escherichia coli* and *Salmonella Sp* bacterial [1], it also can stop the growth of *Ascaridiagalli worm* [2]. Biological test to broiler chickens showed that the right dose of the EC giving is 250 mg/kg weight [3]. The giving of Cinnamononi Extract with the aforementioned dose can decrease the feed conversion ratio (FCR), meat cholesterol, and abdominal fat content; it also can increase weight gain and broiler carcass weight [3].

Beside the antibacterial ability, Cinnamononi Extract also can push the growth and the production. This happens because the EC contains *proxeronine* which extracted from noni fruit. *Proxeroninase* enzyme will change *proxeronine* into *xeronine* [4]. *Xeronine* is an alkaloid that activates the work of synthesis protein enzymes in the body [5]. So, the giving of EC can be a solution on organic farming system.

That natural herbal plantations regarded safer and have smaller side effect. It happens because the constructive nature of herbal plants, but it also needs to be understood that not all natural things are safer if the utilization is excessive [6]. The side effect of the utilization may appear because of overdose and wrong utilization timing [7]. Continuous giving can also create side effects especially to liver and kidneys. The continuous giving of herbal extract can be avoided by putting duration of giving and stopping. Hence, it is important to examine the perfect duration of giving EC that positively affect the performance and the production, and how long the positive effect will last after the giving was stop. Therefore, we will know the effective duration of giving Cinnamononi Extract to broiler chickens.

2. Materials and method

This research use 160 broiler chicks in the age of 8 days from strain Cobb and caged in 20 boxes of cage which make each box will contain 8 chicks. The program that will be used is CRD. There are 5 treatments and 4 repetitions. The treatments are the varying durations of giving the Cinnamononi Extract (EC): A = no EC given (control), B = EC given every day for 5 weeks, C = EC given every day for the first 2 weeks and the last week, D = EC given every day for 3 weeks, and E = EC given every day for 4 weeks. Cinnamononi Extract is given at dose of 250 mg/kg weight per chicks per day through limiter drinking water. Before the giving of EC, the chicks are fasted for 1,5 hours. They will be feed again after all Cinnamononi Extracts are run out. This experiment is stopped after treatment for 5 weeks (42 days old bird). The parameters measured are: ration consumption, weight gain, and ration conversion. The ration is a mixture of refined yellow corn, fish flour, meat bone meal, soybean meal, palm oil, bone meal, and Neobro feed supplement.

Composition and content of the ration is showed on Table 1. The parameters measured are ration consumption, weight gain, and feed conversion ratio. All obtained data will be analyzed by analysis of variance [8]. For the significant result of the analysis, the Duncan Multiple Range Test analysis will also be conducted [8].

Table 1. Composition and nutrient content of the ration.

Ration Composition	Research Ration	
	Age 0 – 3 weeks	Age 3 – 6 weeks
Yellow Corn (%)	59	62,5
Soy meal (%)	20	18
Fish meal (%)	5	6
Meat bone meal (%)	12	7
Palm oil (%)	2,5	4
Neobro (%)	0,5	0,5
Bone meal (%)	1	2
Total (%)	100	100

ME* (kkal/kg)	3027,5	3004,04
Protein* (%)	23,10	20,30
Fat* (%)	6,14	7,08
Rough fiber* (%)	3,74	3,58
Ca* (%)	1,11	1,22
P* (%)	0,50	0,57
Lysin* (%)	1,10	1,13
Metionin* (%)	0,45	0,39

Note: *Calculation result.

3. Result and discussion

Responses observed are ration consumption and cumulative weight gain from the age of 8 days to 42 days, it followed by feed conversion ratio calculation. This data showed on Table 2 which is average of four repetitions for each treatment.

Table 2. Average ration consumption, weight gain (pbb), and feed conversion ratio during research.

Treatments	Ration Consumption^{ns}	Body Weight Gain	Feed Conversion Ratio
	----- (gram/bird/day) -----		
A = without EC	118,26	57.36 ^{b}	2.11 ^{a}
B = EC duration during research (5 weeks)	117,08	62,36 ^{a}	1.94 ^{b}
C = EC duration the first 2 weeks and the last 1 week	115,62	59,82 ^{ab}	1.93 ^{b}
D = EC duration 3 weeks	118,46	62,12 ^{a}	1.96 ^{b}
E = EC duration 4 weeks	114,81	61,97 ^{a}	1.89 ^{b}

Note: ns = *non significant*
Different superscript on the same column shows significant differences ($p<0.05$).

Analysis of variance result on data obtained shows that the giving of Cinnamononi Extract and the duration of giving are not significant effect ($p>0,05$) to the ration consumption, but give a significant effect ($p<0,05$) to the weight gain and feed conversion ratio. Further analysis (DMRT) on the weight gain variable shows that Treatment A (where the chicks were not given any EC) result the weight gain significantly lower ($p<0,05$) compare to the other birds that given EC (B, D, and E), except treatment C. The result of DMRT on feed conversion ratio variable also shows that the birds which was not get any Cinnamononi Extract (treatment A) resulted lower weight gain ($p<0,05$) compare to treatment B, C, D, and E.

This result shows that the giving of Cinnamononi Extract could accelerate growth without affecting the ration consumption. In consequence, the feed conversion ratio became better. It happened because there is *proxeronine* substance gained from noni fruit. *Proxeroninase* enzyme will change *proxeronine* into *xeronine* [4]. *Xeronine* is an alkoloid that activates the work of synthesis protein enzymes in the body [5]. Therefore, by giving Cinnamononi Extract through drinking water, it can make ration giving more efficient.

A (without EC) and also difference, but not much, compare to treatment B, D, E. This result shows that the stoppage of giving EC on the early period of grower (3rd week from the treatment) for 2 weeks can affect the growth.

Table 3. Average body weight gain every treatment week.

Treatments	1st Week^{ns}	2nd Week^{ns}	3rd Week	4th Week	5 th Week^{ns}
	--------------- (gram/bird/day) -------------				
A = without EC	39,49	66,08	59.29 ^{b}	59,38^{b}	53,49
B = EC duration during research	38,96	66,70	66,40 ^{a}	72,78 ^{a}	61,97
C = EC durationfirst 2 weeks and the last week	36,69	64,11	67,17 ^{a}	63,60 ^{b}	67,49
D = EC duration 3 weeks	39,35	63,13	67,54 ^{a}	79,64 ^{a}	62,35
E = ECduration 4 weeks	36,71	68,09	69,01 ^{a}	72,75 ^{a}	64,51

Keterangan: ns = *non significant*
Different superscript font in the same column shows real differences ($p<0.05$).

To do further study on this matter, so the analysis of variance was done every treatment week on the weight gain data (Table 3). From Table 3, it shows that the effect of the stoppage of giving EC (treatment C) can only be seen on the 4th week from the treatment. On the 4th week from the treatment, it can be seen that birds that given treatment C (duration of giving EC for 2 weeks) have a body weight gain which difference non significantly ($p>0,05$), compare to the broiler with no EC give (treatment A); but it is a significantly difference compare to treatment B, D, and E. It indicates that after stop giving the EC after 2 weeks (treatment C) on the early grower period, the growth rate decreased. However, after the giving of EC for 1 week before harvested, the body weight gain was improved and able to be equal to the result of treatment B, D, and E.

DMRT result also shows that body weight gain and feed conversion ratio on the broiler that get treatment D and E differ non significantly ($p>0,05$) compare to the chicken that get treatment B. It shows that giving EC every day for 3 weeks (treatment D) and 4 weeks (treatment E) result the same performance compare to treatment B which gave the EC every day for 5 weeks during the research. Therefore, the most efficient duration to give EC is for 3 weeks (treatment D).

Based on those data, it can be conclude that the giving of Cinnamononi Extract can improve broiler chicken's performance, which increase body weight gain and decrease feed conversion ratio. The most efficient duration of giving EC is 3 weeks (treatment D).

4. References

[1] Yuniza, A., and Yuherman. 2015. Content of phytochemical compound and antibacterial activity of cinnamon leaf (cinnamomumburmanii) and noni fruit and leaf (Morindacitrifolia L) mixture extract to replace antibiotics. Pakistan Journal of Nutrition, Vol. 14 (8): 492-497

[2] Suhartiti. 2016. Penggunaan ekstrak cinnamononi sebagai agen anti bakteri dan anthelmentik terhadap penurunan total bakteri *Escherichia coli* dan total telur cacing *Ascaridia galli* pada ekskreta ayam broiler. Thesis draft. Faculty of Animal Husbandry. Padang (unpublished).

[3] Yuniza, A., and Y. Rizal. 2015. Pengaruh Cara Pemberian dan Dosis Ekstrak "Cinnamononi" Sebagai Feed Aditive Pengganti Antibiotik Bacitracin Terhadap Performans Ayam Broiler. Laporan penelitian Hibah Bersaing, Faculty of Animal Husbandry, Andalas University (unpublished).

[4] Heinicke, R.M. 1999. Xeronine. Morinda Inc. Hawai Hertrampt, J. 2001. Alternative antibacterial performance promoters. *Poult. Int.* 40: 50-55.

[5] Bangung AP. and B. Sarwono. 2002. Khasiat dan Manfaat Mengkudu. Agromedia Pustaka. Jakarta.

[6] Hermanto, N., and Subroto, MA. 2006. Herbal dan jamu (Pengaruh dan efek sampingnya). http://www.elexmedia.co.id.

[7] Primandini, Y., LD. Mahfudz, B. Sukamto. 2012. Interval waktu penambahan ampas mengkudu (*Morinda citrifolia L*) dalam ransum terhadap performans ayam broiler. Agripet: Vol (12) No. 1: 16-22.

[8] Steel and Torrie. 1991. Prinsip dan Prosedur Statistika. Translated by: B. Sumantri. PT. Gramedia Pustaka Utama. Jakarta.

Fermentation characteristic with addition of Jengkol (*Archidendron jiringa*) peel powder on in vitro

Desi Purnama Sari[1], **Sulaila**[1], **Ade Ahmad Almas Pabengsyah Putra Gustari**[1], **M Rofid Hibatullah**[1], **Nur Hidayah**[2*]

[1] Undergraduate Program of Animal Husbandry Department, Faculty of Agricultural, University of Muhammadiyah Bengkulu, Bengkulu, Indonesia.
[2] Department of Animal Husbandry, Faculty of Agricultural, University of Muhammadiyah Bengkulu, Bengkulu-38119, Indonesia.
*Corresponding author e-mail: nurhidayah@umb.ac.id.

Abstract. This study was aimed to utilize jengkol (*Archidendron jiringa*) peel as a natural feed additive for ruminant livestock. Jengkol (*A. jiringa*) peel has a good nutritional content and contains of bio-active compounds (tannin and saponin) that can be potential to increase the productivity of ruminant livestock. Tannin had the function to bind protein and decreased methanogen population, and saponin had used as defaunation agent to decreased the rumen protozoa population. The treatments were arranged in a randomized block design with 4 treatments (0%, 2%, 4%, 6%) and 4 replications. The variables observed included pH value, dry matter and organic matter digestibility, and NH3 concentration. Data were tested using Analysis of Variance (ANOVA) and the differences among treatments means were examined by Duncan Multiple Range Test. The results showed that the addition of Jengkol (*A. jiringa*) peel powder until 6% did not disturb (P> 0.05) pH value and digestibility (dry matter and organic matter digestibility), and strong significant (P <0.01) increased NH3 concentration. The research concluded that the addition of Jengkol (*A. jiringa*) peel powder had potency to increase livestock productivity because increased NH3 and did not disturb pH rumen, dry matter and organic matter digestibility.
Keywords – fermentation, jengkol pell, powder, in vitro.

1. Introduction

One of the most important aspect and determinant of success beef cattle business are cheap feed cost, good quality, and sufficient throughout the year. Good quality of feed is very important for livestock business because feed took 70% of the total production costs [1]. Another aspect, the feed that given for livestock must provide benefits to farmers and sufficient for livestock requirements. Feed supplementation is one of manipulation technique to increase beef cattle productivity. Tannin and saponin are bioactive compounds that can modified rumen microbe. These bioactive compounds can be beneficial or detrimental to ruminant livestock according to their structure and concentration [2]. Jengkol (*A. jiringa*) is one of plant that contain tannin and saponins with a total production in Bengkulu Province as much 2,822 tons in 2017 [3]. Another from production, Jengkol has potential from nutrition aspect. Jengkol peel contains of ash (3.39%), CP (8.83%), EE (0.65%), CF (27.50%), NFE (59.62%), tannins (7.82%), and saponins (56.92%) [4]. This study aimed to evaluate supplementation of Jengkol peel powder on digestibility (dry matter and organic matter) and pH in vitro.

2. Materials and methods

2.1. Preparing materials

Jengkol peel and forage was dried for 5-6 hours under the sun until got stable weigh. After that, the materials were milled with grinding machine to got materials into a fine powder form.

2.2. In vitro fermentation

In vitro fermentation was conducted according to the method of Tilley and Terry [5]. Into each 100 mL fermentation tube, 500 mg substrate, 40 mL McDougall buffer, and 10 mL rumen fluid were added at conducted at 39 °C. The rumen fluid for this experiment was collected after 3 h morning feeding from the 3 rumens fistulated Ongole crossbred beef cattle with Ethical Approval from Animal Care and Use Committee (AUAC) 01-2013b LIPI Cibinong. Samples from aliquol were taken after 4 h incubation for pH and after 48 h incubation for dry matter and organic matter digestibility analysis.

2.3. Sampling and measurement

The rumen's pH was measured with pH meter and the dry matter digestibility (DMD) and organic matter digestibility (OMD) were measured using Tilley and Terry method [5].

2.4. Statistical analysis

The experiment was conducted in a randomized block design with 4 treatments (0%, 2%, 4%, 6%) and 4 replications. The treatment tested was the ration:

A = Native grass (100%) + Jengkol peel powder (0%)
B = Native grass (98%) + Jengkol peel powder (2%)
C = Native grass (96%) + Jengkol peel powder (4%)
D = Native grass (94%) + Jengkol peel powder (6%)

Data were tested using Analysis of Variance (ANOVA) and the differences among treatments' means were examined by Duncan Multiple Range Test [6].

3. Results and discussions

3.1. Rumen pH value

Table 1. Rumen pH value with addition of jengkol peel powder.

Jengkol Peel Powder (%)	pH
0	6,95 ± 0,05
2	6,96 ± 0,07
4	6,90 ± 0,08
6	6,95 ± 0,09

The addition of 2-6% Jengkol peel powder did not affect (P>0.05) on rumen pH value (Table 1). The result indicated that tannin and saponin on Jengkol peel powder did not distrubted activity of rumen microbial fermentation. The rumen pH value for all treatments was 6.90-6.96, which is still the normal range of rumen pH value 6.5-7.0 [7]. This is indicator that feed degradation process goes well, so that selulolitic bacteria can live optimally in the rumen [8]. The fermentation pH value of in all treatments at 6.8 which is still in the normal range of normal rumen pH.

3.2. Dry Matter and Organic Matter Digestibility

Table 2. Dry Matter and Organic Matter Digestibility with Addition of Jengkol Peel Powder.

Jengkol Peel Powder (%)	DMD (%)	DMO (%)
0	45,83± 4,08	46,04± 3,30
2	44,61± 3,96	45,21± 3,07
4	44,32 ± 2,32	45,23± 2,09
6	45,15 ± 3,38	44,95 ± 3,33

The results showed that the addition of Jengkol peel powder at 2-6% did not affect ($P>0.05$) on dry matter and organic matter digestibility (Table 2). This indicated that the addition of Jengkol peel powder until 6% did not disturb rumen microbial activity to digested feed. Dry matter and organic matter digestibility arrange from 45.21% until 46.04%. Complete feed range was used of dry and organic matter digestibility at 48.26 - 53.75% [9]. The addition of lerak extract (81.5% saponins) at 0.6 and 0.8 mg / ml of rumen fluid did not affected of dry and organic digestibility [10]. The addition of pure tannins 0.5 mg / ml on the hay substrate significantly ($P<0.05$) decreased organic digestibility up to 4.4% - 5.9% [11]. The different result depend on concentration and structure of saponin and tannin.

3.3. Rumen Ammonia (NH3) Concentration

The results showed that the the addition of Jengkol peel powder had strong significant effect ($P<0.01$) on NH3 concentration (Table 3). The higher addition of Jengkol peel powder given higher of ammonia concentration. The lowest ammonia concentration was found in the control (0%) treatment (5.92 mM) and the highest that found with the highest (6%) addition of jengkol peel powder (11.78 mM). The increased of ammonia concentration caused to be due to saponin in Jengkol peel powder can reduced protozoa population so increased bacteria population. The optimum concentration of ammonia in rumen between 85 - 300 mg /l or 6-21 mM [12]. This indicated that ammonia concentration still in the ideal range, which did not affected and disturbted rumen metabolic activity.

Table 3. Rumen Ammonia (NH_3) Concentration

Jengkol Peel Powder (%)	NH_3 (mM)
0	$5{,}92 \pm 0{,}52^a$
2	$6{,}78 \pm 0{,}30^b$
4	$7{,}80 \pm 0{,}61^c$
6	$11{,}78 \pm 0{,}77^d$

Note: Means in the same coloumb with different superscript differ strong significantly ($P<0.01$)

The result showed that range of ammonia concentration from 5.92 to 11.78 mM in optimum conditions for microbial protein synthesis, and the normal ammonia concentration for rumen microbial growth at 4-12 mM [13]. The supplementation 0.1% extract lerak on high forage ration produced ammonia concentration at 4.11 - 9.86 mM [10]. Another result was reported, the addition of tannins and saponins in concentrated produced significant effect on increasing ammonia concentration [14]. The highest concentration of ammonia was produced with the addition of combination 1.5% tannin and 0.3% saponin (21.74 mM).

4. Conclusion

The research concluded that the addition of Jengkol (A. jiringa) peel powder had potency to increase livestock productivity because increased NH3 and did not disturb pH rumen, dry matter and organic matter digestibility.

5. References

[1] Rahmat and Bagus Harianto. 2017. Pakan sapi potong. Jakarta: Penebar swadaya.
[2] Suharti, S., Astuti, D.A. and Wina, E. 2009. Kecernaan nutrien dan performa produksi sapi potong Peranakan Ongole (PO) yang diberi tepung lerak (Sapindusrarak) dalam ransum. JITV 14: 200-207.
[3] BPS [Badan Pusat Statistik]. 2018. Statisitik Indonesia 2018. https://www.bps.go.id/publication/2018/07/03/5a963c1ea9b0fed6497d0845/statistik-indonesia-2018.html.

[4] Hidayah, N and Suliasih. 2017. Potential of Bioactive Compounds Archidendron jiringa by Product to be Natural Feed Additive for Sustainable Animal Production. Proceeding International Seminar SAADC "Fakultas Peternakan, Universitas Brawijaya Malang.

[5] Tilley, J.M.A. and Terry, R.A. 1963. A Two-Stage Technique for in Vitro Digestion of Forage Crops. Journal British Grassland Society, 18, 104-111.

[6] Steel, R.G.D. and Torrie, J.A. 1990. Principles and Procedures of Statistics. McGraw Hill. New York.

[7] Dehority, B. A. 2005. Effect of pH on viability of entodinium caudatum, entodinium exiguum, epidinium caudatum, and ophryoscolex purkynjei in vitro. J. of Eukaryotic Microbiol. 52: 39-342.

[8] Syahrir. 2009. Potensi Daun Murbei dalam Meningkatkan Nilai Guna Jerami Padi sebagai Pakan Sapi Potong. Sekolah Pascasarjana Institut Pertanian Bogor.

[9] Firsoni, J. Sulistyo., A.S. Tjakradidjaja, and Suharyono. 2008. In vitro fermantibility test of feed supplement in complete feed. Pusat Aplikasi Teknologi Isotop dan Radiasi BATAN. Seminar nasional Teknologi Peternakan dan Veteriner.

[10] Suharti S, Kurniawan A, Astuti DA, and Wina E. 2010. Microbial population and fermentation characteristic in response to sapindus rarak mineral block supplementation. MediaPeternakan 33(3): 150-154.

[11] Jayanegara A, H. P. S. Makkar and K. Becker. 2009. Emisi metana dan fermentasi rumen in vitro ransum hay yang mengandung tanin murni pada konsentrasi rendah. Media Peternakan 32 (3): 184-194.

[12] McDonald, P., Edwards, R.A and Greenhalgh, J. F. D. 2002. Animal Nutrition. 6th Edition. Longman, London and New York. 543 pp.

[13] Sutardi,T. 1979. Ketahanan Protein Bahan Makanan Terhadap Degradasi oleh Mikroba Rumen dan Manfaatnya bagi Peningkatan Produktivitas Ternak. Pros.Seminar Penelitian dan Penunjang Peternakan. LPP. Deptan 2: 91-103. Bogor.

[14] Wahyuni, I M. D., Anis, M., and Marry, C. 2014. Penentuan Dosis Tanin dan Saponin untuk Defaunasi dan Peningkatan Fermentabilitas Pakan. JITP 3 (3): 133-140.

Acknowledgments

This experiment was funded by Directorate General of Higher Education, Ministry of National Education of Indonesia through "Program Kreativitas Mahasiswa" grant 2018.

IOP Conf. Series: Earth and Environmental Science **287** (2019) 012015 doi:10.1088/1755-1315/287/1/012015

Macro minerals (Ca, P, Mg and K) concentration mixed of elephant grass (Pennisetum purpureum) CV. Taiwan and Legume *Indigofera zollingeriana*

Evitayani[1*], Khalil[1], Lili Warly[1], and Kensuke Kawamura[2]

[1] Faculty of Animal Husbandry, Andalas University, Padang, West Sumatra, Indonesia.
[2] Grassland Science, Hiroshima University, Japan.
[*] Corresponding author: evitayani.unand@gmail.com.

Abstract. This study aims to determine the content of macro minerals (Ca, P, Mg and K) of Elephant grass CV. Taiwan and Legumes of *Indigofera zollingeriana* are grown mixed with different doses of N fertilizer on land that was given Fungi Mycorrhizae of Arbuskula (FMA). This research was conducted by planting grass and *Indigofera* in the land by different N fertilizer: 0% N, 25% N, 50% N, 75% N. The measured parameters are macro mineral contents of Ca, P, Mg and K. Statistical Analysis calculated by the method of randomized block design with 4 treatments and 3 replications. The results showed that the treatment effect is not significant (P> 0.05) on the content of macro minerals (Ca, P, Mg and K) of elephant grass CV. Taiwan and Legume *Indigofera*. The mineral content of elephant grass CV. Taiwan ranged from of Ca: 5.6 to 6.6, P: 3.2 to 3.6, Mg: 2.9 to 3.5, K: 5.0 to 5.8 g/Kg DM, respectively. While, macro minerals of legume *Indigofera* was ranged from of Ca: 4.1 to 4.9, P: 4.0 to 4.3, Mg: 2.8 to 3.0, K: 4.4 to 5.4 g/Kg DM, respectively. From the results of this study concluded that different doses of fertilizer N up to a dose of 75% gives the same result by giving N 0% to macro mineral (Ca, P, Mg and K) of elephant grass CV. Taiwan and Legume *Indigofera*.
Keywords – Pennisetum purpureum, *Indigofera zollingeriana*, macro minerals, N fertilizers and Mycorrhizae.

1. Introduction

One important factor in the improvement of farm-related livestock productivity, which is necessary to improve in terms of quality, quantity and suistainability is feed. Lack of feed, could be contribute to low livestock production. Therefore, it is necessary to attempt to improve forage feed especially in terms of quality and quantity. Forage is required by ruminants, because 74-90% of food consumed comes from good forage in the form of fresh or in dried form [1]. Forage for livestock feed can be sourced from the type of grass and legumes. Forage used are elephant grass. Elephant grass (*Pennisetum purpureum*) is an annual plant (*perennial*) which can be grown on a wide range of soil types that exist in Indonesia. This plant is suitable developed by breeders who have limited land area for production of elephant grass plants are known to have high production, but the grass has the disadvantage of nutrient and low mineral and nutrient content.

The nutritional content of elephant grass is relatively low due to this plant content high in fiber and low protein, while the mineral content of elephant grass lower than livestock needs. Efforts to improve nutrient and mineral content of grass that can be planted in a mixture of grasses and legumes for *Indigofera*. *Indigofera* resistant to drought stress and could be fixation Nitrogen from the moisture. *Indigofera* used is *zollingeriana*. Crude protein and high mineral potential make this plant as a source of replacement protein concentrates as well as mineral resources. The weakness of elephant grass in addition to the content of nutrients and low mineral content that is greedy for nutrients, so as to meet the nutrient needs of the elephant grass needs to be done tillage. For optimum soil treatment it is necessary intensive fertilization. Therefore, grass needs nitrogen fertilizer. Nitrogen is the main nutrient required by all plants for optimum growth and production. However, excessive N fertilizer and in the long term can damage the quality of the land, so that the soil needs to be enriched with *Mycorrhizae*. Mycorrhizal fungi used type of Mycorrhizal Fungi of Arbuskula (FMA). FMA can stimulate rooting in the soil. FMA hyphaes association with the roots are able to absorb soil nutrients more so that fertilizer use can be reduced.

Indigofera and elephant grass are planted with mixed systems could be improved the nutritional value of plants and is expected to increase the mineral content. The mineral content should be available in sufficient quantities in the plant because of the animal's body can not make its own minerals. Minerals have an important role in the animal body. The early symptoms caused by deficiency of minerals in the form of reproductive decline of about 20-75%, retention of the placenta, the child born to be weak, and high child mortality [2]. Other diseases that arise are pneumonia, diarrhea, stomatitis, anorexia, and decreased milk production in dairy cows. Other symptoms are more severe fractures, dry and scaly skin, as well as a great emaciation. In the living body's own mineral is a constituent component of the body, 4-5% body weight consists of minerals, about 50% of the body's mineral composed of calcium, 25% phosphorus, and another 25% made up of other minerals. Therefore, minerals are needed for the survival of livestock due to the mineral content in forage fodder needs to be considered. The function of some minerals are calcium (Ca) as the formation of bones and teeth, blood clotting, activation of enzymes, and the contraction of tendons, phosphorus (P) as the formation of bones and teeth, part of many enzymes, the release of the body's energy, a part of DNA and RNA, magnesium (Mg) as an enzyme activator, part of a network framework and potassium (K) as the guardian of electrolyte balance, enzyme activators, the function of the tendon [3].

Based on the above a few information regarding the content of macro mineral (Ca, P, Mg and K) mixed by Elephant Grass (*Pennisetum purpureum*) CV. Taiwan and *Indigofera* which is grown contained of *Mycorrhizae* with different doses of N fertilizer in Padang, West Sumatra, Indonesia.

2. Materials and methods

2.1. Land preparation and planting

Land used for field trial research is the Technical Implementation Unit of Faculty of Animal Unand the top area of 300 m2. The land is divided into three plots. Each plot consists of four sub-plots. Each sub plot was given 18 holes then given FMA. Planting grass and *indigofera* is a mixture of the hole that has been enriched FMA. Elephant grass seeds that are used in the form of cuttings and seedlings *indigofera* used is *indigofera* who are 2 months old.

2.2. Fertilizing and plant maintenance

Elephant grass and planted *indigofera* given at a dose of N fertilizer 0% N, 25% N, 50% N, 75% N. Fertilization is done at age 15 HST (the day after grown) half-dose and half dose 30 days after planting. Plants maintained for 60 days for grass and legume *indigofera* for 120 days. During maintenance and cleaning watering crops from weeds regularly.

2.3. Harvesting and sample preparation

Harvesting is done at age 60 HST. Samples were separated between grass and *indigofera*, then weighed fresh weight, then chopped plant with a size of 2-3 cm and put into containers. Foliage along the container is dried in an oven 60 0 C for 24 hours. After drying weighed so that the data get dry weight. Forage dried and then ground into powder and prepared for analysis.

IOP Conf. Series: Earth and Environmental Science **287** (2019) 012015 doi:10.1088/1755-1315/287/1/012015

2.4. Parameters

The parameters measured were the content of macro minerals are calcium (Ca), phosphorus (P), Magnesium (Mg) and potassium (K) of elephant grass (*Pennisetum purpureum*) CV. Taiwan and *Indigofera zollingeriana.*

2.5. Data analysis

This study used a randomized design with 4 rations treatments and 3 replications. All data obtained were analyzed by one way analysis and the differences between treatments were tested by Duncan's Multiple Range Test (DMRT), with p<0.05 indicating a significant difference.

3. Results and discussion

In Table 1 are presented the data minerals and *Indigofera* elephant grass planted in a mixture and were given different doses of N fertilizer. This analysis results showed that the treatment effect is not significant (P> 0.05) on the mineral content of grass and legume *Indigofera.* While, significant effect between the forage species grass and legume (P>0.05).

Table 1. Macro minerals of Elephant Grass (EG) CV. Taiwan and Legume *Indigofera zollingeriana* (g/kg DM).

No	N Doses	Ca		P		Mg		K	
		EG	IZ	EG	IZ	EG	IZ	EG	IZ
1	0 %	5,6	4,9	3,6	4,3	2,9	2,8	5,0	4,5
2	25 %	6,6	4,4	3,4	4,0	3,5	3,0	5,1	4,4
3	50 %	5,6	4,8	3,2	4,2	3,3	2,7	5,5	5,4
4	75 %	6,3	4,1	3,5	4,2	3,0	2,9	5,8	4,9
	Rataan	6,0	4,6	3,4	4,2	3,2	2,9	5,3	4,8
	SE								
	EG vs IZ	*	*	*	*	*	*	*	*

Note: Between the N doses, macro minerals have not significantly different of (P>0,05).
SE = Standart of Error; EG vs IZ: Between species of the macro mineral have shown that significantly different (P>0.05).

Table 1 showed that the mineral content of macro minerals namely of Ca was ranged from 5.6 to 6.6 g/kg DM with the average of 6.0 g/kg DM, P content ranged from 3.2 to 3.6 g/kg DM with average 3.4 g/kg DM, Mg content ranged from 2.9 to 3.5 g/kg DM with average of 3.2 g/kg DM and K content ranged from 5.0 to 5.8 g/kg DM with average 5.3 g/kg DM, respectively. While, the mineral content *Indigofera* namely Ca ranged from 4.1 to 4.9 g/kg DM with the average 4.6 g/kg DM, P content ranged from 4.0 to 4.3 g/kg BK with the average of 4.2 g/kg DM, Mg ranged from 2,7 to 3,0 g/kg BK with the average of 2.9 g/kg DM and K content ranged from 4.4 to 5.4 g/BK kg with the average of 4.8 g/kg DM, respectively.

A comparison between the mineral content of elephant grass and *Indigofera. Indigofera* P content is higher than grass, whereas the content of Ca, Mg and K content of Indigofera lower than on grass. The high content of P in *Indigofera Zollingeriana* caused by the high content of P in the soil, while the low content of Ca, Mg and K *Indigofera* tend to caused by soil acidity and drought stress. On acid soils will decrease the ability of root nodules to fix N from the moisture. Also, it is could be seen planted crops where soil pH ranged from 4.81 to 4.98. Packaging of land does not affect the grass because grass is a plant that is resistant to soil acidity. Including tolerant crops on acid soils [4]. When compared with the research reported that the mineral content of legumes is higher than on grass this is in line with the content of P but is inversely proportional to the content of Ca, Mg and K [5].

The mineral content of elephant grass when compared with the results which reported mineral content (g/kg DM) for Ca ranged from 2.1 to 3.1, P 1.1 to 2.1, and K 3.6 to 5.3, respevtivelt [6]. While the Mg content compared with the elephant grass research at 2.8 g/kg DM [7]. The mineral content of elephant grass in this study is higher. The high content of Ca, P, Mg and K in this study due to the N, FMA and planting a mixture of grasses and legumes. N fertilization aimed at creating a high nutrient levels, so as to improve the quality of the crop. FMA aims to conserve the use of N fertilizer because FMA is able to extend hyphae - hyphae roots so that the absorption of nutrients in the soil can be improved. Planting a mixture of grasses and legumes can improve the nutritional content of the grass as the ability of legumes to fix N from the moisture and transfer it to the grass.

Indigofera mineral content compared with the results of research [8] reported *Indigofera* mineral content (g/kg DM) for Ca 3.4, P 1.0 and K 1.7. The content of Ca, P and K *Indigofera* in this study is higher, while the Mg content in research [8] 3.2 g/kg DM did not differ much from the results of 2.9 g/kg DM. The high mineral content of *Indigofera* in this study thought to be due to fertilizer N and FMA content. N fertilizer that serves to stimulate root growth and FMA that can help the absorption of nutrients in the soil.

If connected macro mineral contents of Elephant Grass and *Indigofera* in this study with the needs of cattle which has a weight of 300 kg [9], the cattle which consume 100% grass shortage (g/kg BS) Ca 20.7, P 33.1, 20.7 Mg and K 203,1. The giving of 100% *Indigofera* shortcomings (g/kg BS) Ca 8.3, P 5.5, 8.3 Mg and K 120.9. Giving 50% of elephant grass + 50% *Indigofera* (g/kg BS) Ca 13.4, P 19.2, 14.3 Mg and K 159.4, respectively.

4. Conclusion

From these results it could be concluded that the content of macro minerals (Ca, P, Mg and K) in the elephant grass CV. Taiwan and Legume *Indigofera zollingeriana* planted with mixed system in mycorrhizal soil with different N fertilizer dose to dose 75% gave the same results with control.

5. References

[1] Susetyo, S. 1980. Padang Pengembalaan. Departemen Ilmu Makanan Ternak. Fakultas Peternakan Institut Pertanian Bogor, Bogor.

[2] Gartenberg, P.K., L.R. McDowell, D. Rodriguez, N. Wilkiinson,J.H. Condrad, and F.G. Martin.(1990). *Evaliation of trace mineral status of ruminants innortheast Mexico*. Livestock Rest. For Rural Development 3(2): 1-6.

[3] Anggorodi, R. 1994. Ilmu Makanan Ternak Umum. Gramedia Pustaka Utama. Jakarta.

[4] Sumarsono, S. Anwar, D.W. Widjajanto, S. Budiyanto. 2009. Penerapan Pupuk Organik untuk Perbaikan Penampilan dan Produksi Hijauan Rumput Gajah pada Tanah Masam. Fakultas Peternakan Universitas Diponegoro. Semarang.

[5] Walzl,K,P., Soegaard,K., Jensen,H,H., Eriksen, J., Sanderson,M,A., Rasmussen, A. 2008. Forage herbs improve mineral composition of grassland herbage. USA.

[6] Snijders, P, J.M., Bram.P.W., John,N.K. 2011. Wageningen UR Livestock Research. Kenya Agricultural Research Institute (KARI-NAHRC). Kenya.

[7] Chew, W, Y., Joseph, K, T and Ramli, K. 1975. The Influence of Lime and Nitrogen Fertilization on the Yield and Nitrogen Uptake of Napier Grass on Malaysian Peat Soil., Mardi Res. Bull. 4,2: (43 - 50).

[8] Gafar, M, K., A,U,Itodo., F,A,Atiku., A,M, Hasan and I,J, Peni. 2011. Proximate and Mineral Composition of the Leaves of Hairy Indigo. Pakistan Journal of Nutrition 10(2): 168-175.

[9] NRC. 1996. Adapted from NRC.Nutrient Requirements of Beef Cattle, Sixth Edition.

Acknowledgment

The author is indebtedness to Directorate of Reserach and Community Service, Directorate General for Higher Education of the Ministry of Education and Culture of Indonesia, for the full financial support of our research in International Research and Scientific Publication with the contract number is 03/H.17/KLN-PI/LPPM/2017.

IOP Conf. Series: Earth and Environmental Science **287** (2019) 012016 doi:10.1088/1755-1315/287/1/012016

The effect of urea supplementation and incubation time in fermentation process of bagasse by using *Ganoderma lucidum* on the growth of *G. lucidum* and the nutritive value of bagasse

Fauzia Agustin[1], Elihasridas[1], and Indri Juliyarsi[2]

[1] Department of Nutrition and Feed Technology, Faculty of Animal Science, Andalas University, Padang, West Sumatera, Indonesia.
Corresponding author email: fauziaagustin@ansci.unand.ac.id.
[2] Department of Animal Production Technology, Faculty of Animal Science, Andalas University, Padang, West Sumatera, Indonesia.

Abstract. Fermentation of bagasse with fungi *Ganoderma lucidum* was designed to determine the effect of urea and optimum time to get the best growth of *G. lucidum* and the nutritive value of bagasse and the optimum laccase activity of *G. lucidum*. Treatments were combination of nitrogen dose (0% urea and 0.3% urea) and incubation time (0, 10, 20, 30, 40, 50, 60 days). The treatments were arranged in factorial 2x7 and allocated in completely randomized design with three replications. The result showed that there was no interaction of urea supplementation and incubation time (P>0.05) on laccase activity, but each factor was significant different. The laccase activity in bagasse which supplemented with 0.3% urea was higher than bagasse without supplemented. Dry matter content of the substrate decreased as much as 27.45% in urea treatment, was higher than non urea supplementation (15.45%). Organic matter content of fermented bagasse decreased as much as 31.64%, was higher than non urea supplementation (21.86%). It can be concluded that (1) urea can be used 0.3% as nitrogen source in fermentation process using *G. lucidum*, (2) the highest VFA content of fermented bagasse was 98.25 mM in the length of fermentation up to 60 days with 0.3% urea with the NH3 value was 15.99 mg%, (3) The highest dry matter and organic digestibility occurs in bagasse which is fermented with the addition of using *Ganoderma lucidum* which is at 40 days fermentation time with dry matter digestibility value 43.39% and organic matter digestibility value 40.97%.
Keywords – urea, laccase, Ganoderma lucidum, bagasse, fermentation.

1. Introduction

Bagasse is one of feedstuffs which have potential to be used as a source of energy for ruminant. Howeever, it contains a high-lignin (11-22%), low protein and very low digestibility (16-22%). Lignin is a limiting factor for digestibility of fibrous feed. Its nutritive value can be improved through fermentation process. Solid state fermentation is one methode of fermentation which can be applied to bagasse. Solid state fermentation is a process in which microorganism is grown on solid substrate.

Ganoderma lucidum is a white-rot fungi, a species of the class Basidiomycetes, which can degrade substances which contain lignin by producing ligninolytic enzyme, such as laccase [9] [27]. Previous studies have shown that *G. lucidum* can degrade raw material such as rice straw [26], palm by-product [11], bagasse [27]. The ability of *G. lucidum* to grow in fibrous feed are depended on the nutrients content of substrtate and incubation time.

Plant cells are mainly composed by lignocellulosic material, which includes cellulose, hemicellulose, pectin and lignin (lignocellulosic complex) [12]. In nature, microorganisms, especially fungi, are able to degrade the plant cell wall through a set of acting synergistically enzymes. This phenomenon leads to glucose being released in a free form, which can enter the metabolism of the microorganism, providing its energy [12]. The main groups of fungal ligninolytic enzymes (ligninases) are lignin peroxidase (LiP), manganese peroxidase (MnP) and laccase (Lac). Among the ligninases, Lac are the mostly studied [13] [14].

Laccase (EC1.10.3.2, p-diphenol oxidase) is one of enzyme that most widely distributed in white rot fungi [15] such as *G. Lucidum.* Fungal enzyme is responsible in mechanism for removing toxic phenols from the medium in which this fungi grow under natural conditions. Laccase only attack the phenolic subunits of lignin [16]. Laccases are oxidative enzymes related to the degradation of phenolic compounds, including lignin units [17]. Laccases are distributed among some plants [18], fungi [18] [19] [20], and bacteria [21] [22]. Laccase produced by some wood-rotting fungi from the genus Basidiomycetes play a major role in the biodegradation of lignin. Laccase produced by *G. lucium* can be used to improve feed quality of sugarcane bagasse.

There is still little information available concerning the use of urea in fermentation techniques for obtaining *Ganoderma lucidum* mycelium and its valuable components. Therefore, this study was conducted in order to determine the effect of urea and optimum time to get the best growth of *G. lucidum* and the nutritive value of bagasse. The ability of *G. lucidum* to degrade lignin is depended on the nutrients content of substrtate and the ability of *G. lucidum* to produce ligninolytic enzyme i.e laccase; and it is also affected by incubation time. Moreever, the effect of incubation time and nutrients content of the substrate on the development of *G. lucidum* and optimum condition need to increase enzyme activity, have not available yet. Dry matter content and organic matter content of the substrate supplemented with urea in different incubation time is also determine in order to evaluate the growth of *G. lucidum.*

2. Material and method

2.1. Microorganism and starter

Ganoderma lucidum was obtained from the collection of Indonesian Biotechnology Research Institute for Estate Crops (IBRI), Bogor. Starter was obtained by rejuvenating inoculums stock from IBRI at Ruminant Nutrition Laboratory, Faculty of Animal Science, Andalas University. The mycelia of *G. lucidum* was grown in sterile petri dishes containing potato dextrose agar (PDA).

2.2. Making process of Ganoderma lucidum inoculum

Inoculums of *G. lucidum* was made by using the medium of rice bran and corn that has been autoclaved and then was inoculated with starter of *G. lucidum* that has grown in petri dishes with PDA medium until the mycelium of *G.lucidum* meet medium of rice bran and corn.

2.3. Substrate preparation and inoculation

The main material used in this research was sugarcane bagasse. It was chopped to the size of 2-3 cm and was added with water up to 70% moisture content. Mixed substrate was inserted into the bottle 300 ml, covered with aluminum foil, and then was sterilized by using autoclave for 30 min with a pressure of 1.2 and a temperature of 121^0C. After cooling, sterilized bottle were inoculated with 4% inoculums aseptically in the laminar flow cabinet. The inoculated substrates was incubated at 27-28^0C for 0, 10, 30, 40, 50 and 60 days. Fermentation products were harvested according to the time of incubation.

2.4. Experimental design

The experiment was arranged in a completely randomized design (CRD) 2 x 7 factorial with 3 replications [28]. The first factor was urea supplementation (0%, 0.3%) and the second factor was incubation time (0, 10, 20, 30, 40, 50, 60 days) The combination between factors was decided as treatment. Parameter:

I. The growth of *G. lucidum*: 1. pH value of fermentation product of *G. lucidum*. 2. dry matter content of fermented bagasse. 3. organic matter content of fermented bagasse,
II. The nutritive value of fermented bagasse: 1. Rumen Characteristic: pH, VFA, NH3. 2. Invitro Nutrient Digestibility of Bagasse Fermentation Product Using *Ganoderma lucidum*.

2.5. Sample and data analysis

Percentage of dry matter (DM) content and organic matter (OM) content in each incubation time were calculated by drying it in oven. The data were statistically processed by using Analysis of Variance (ANOVA). The difference between treatments was tested by orthogonal contrasts [28].

3. Result and discussion

3.1. The growth of Ganoderma lucidum

3.1.1. Growing percentage of Ganoderma lucidum

Based on the results of investigation was known that *G. lucidum* can grow 100%. The growing of *G. lucidum* is determined by internal and external factors, namely the nutrient content in the substrate, the substrate moisture content, pH, temperature and humidity [8]. The water content of the substrat, sugarcane bagasse in this study was 70%, with a ± 5.5 pH. Fermentations with *G. lucidum* in this research have been done at 28 °C. The effect of temperature on growth and product formation has not been systematically studied.and a temperature of 28^0C is suitable for maximum growth for *G. lucidum*. in this research.

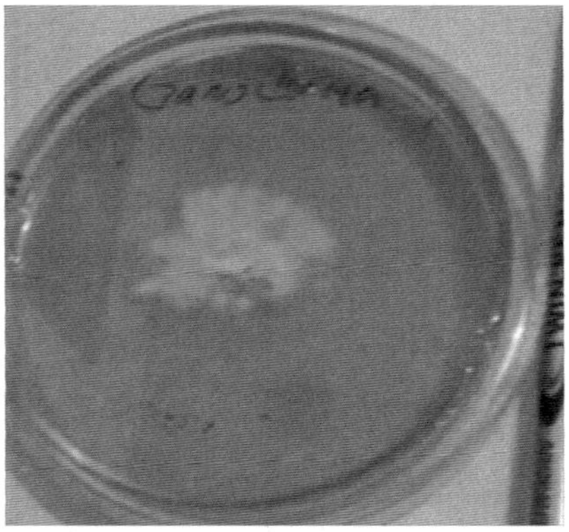

Figure 1. Starter of *Ganoderma lucidum.*

The substrate used for the growth of fungi *G. lucidum* is sugarcane bagasse. *Ganoderma lucidum* grows well on substrates with a moisture content of 70%. There were no contamination during the fermentation processs until the end of the incubation period of 60 days. Artificial cultivation of *G. lucidum* has been achieved using substrates such as grain, sawdust, wood logs [30, 31, 32], Lignocellulose is a common substrate for laccase and the laccase ability to break down nonphenolic ligno-cellulose is provided by certain phenolic compounds acting as mediators. The production of the inoculum is a crucial step that often receives only minor attention. The size, viability and homogeneity of the inoculum can affect the performance of the subsequent fermentation significantly [33] and therefore must be standardized in order to obtain a reproducible process. The optimum fermentation conditions depend on its biomass. Various factors interact to influence the relative productivity of these products.

3.1.2. pH of fermentation

Medium pH is an important parameter affecting growth and product formation in fermentation process. The fermentation process of bagasse supplemented with urea using *G. lucidum* was strongly influenced by the degree of acidity (pH). Prior to fermentation, pH of bagasse supplemented with urea showed a higher value than without urea. This is obviously due to urea is alkaline, so that with the addition of 0.3% urea in the substrate before fermentation, it increases the pH value of the bagasse is from 5:22 became 5:58. This initial substrate pH conditions also influents the growth of *G. lucidum*. Mycelial growth is good for *G. lucidum* is at a pH value 5.0-5.5 [8]; and based on these recommendations, the bagasse substrate, both plus urea or have no added urea, can be covered by *G. lucidum*; but based on observations during the study, it was noted that the faster growth of *G. lucidum* on the bagasse substrate is substrate with plus urea. It is proved that *G. lucidum* also needed a source of nitrogen for its growth. Initial pH of substrate also influence the growth of mycelial. *Ganoderma* adapts during the early stages of the fermentation. Different culture conditions and medium compositions have also been reported to strongly influence mycelial growth and the production of biopolymers [34]. Culture pH had a significant effect on exopolysaccharides yield, chemical composition and molecular weight, and mycelial morphology [35]. The pH and temperature optimal for laccases were found to be pH 5.0 and 30°C, respectively [36] Laccase was active in a wide range of pH values. At pH values above pH 5.0, the enzyme activity decreased gradually by 50% at pH 8 [36]. Laccase activity was relatively stable in the range of pH 3.0–7.0. Microorganisms grow best at their optimum growth pH. Growth occurs slowly or not at all below the minimum growth pH and above the maximum growth pH. The enzyme activity declined when the temperature was increased from 30 to 80°C [36].

Based on the results revealed that there is no interaction (P> 0.05) between the addition of urea with incubation time, as well as each factor was not significant different the pH value. At 10 and 20 days of fermentation, there is a decrease of pH value, this indicates that there has been a degradation of nutrients that produce organic acids that caused a decrease in the pH value. The pH value of the fermentation period of 30 days had no significant difference with fermentation time of 40 days (P> 0.05). The highest pH values after fermentation sugarcane bagasse is at long fermentation 40 days. In the fermentation period of 30 days is visible growth is also very good, and this can be seen from the amount of mycelium increasingly meet all of bottle fermentation. After 40 days of fermentation, the pH value declined. (Figure 2) and this also shows that the fermentation process is still continue until 60 days because they still form organic acids produced from fermentation by *G. lucidum*. During a fermentation the pH has a tendency to change for several reasons.

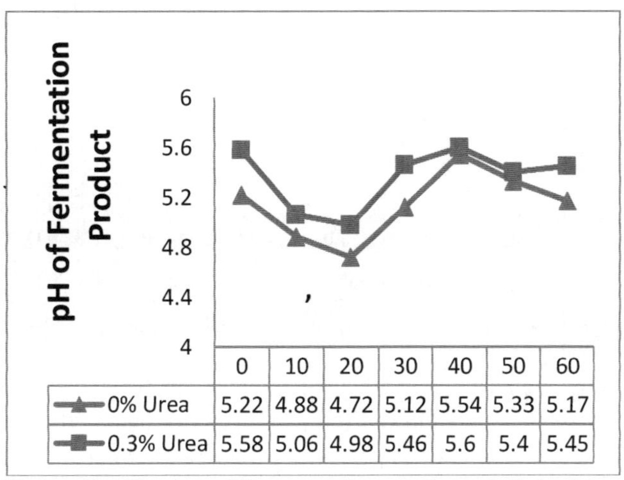

	0	10	20	30	40	50	60
0% Urea	5.22	4.88	4.72	5.12	5.54	5.33	5.17
0.3% Urea	5.58	5.06	4.98	5.46	5.6	5.4	5.45

Figure 2. pH value of fermentation product of *G. lucidum*.

3.1.3. Dry matter and organic matter content of fermentation product of G. lucidum

The effect of urea supplementation in different fermentation time on dry matter content was also evaluated at concentration 0.3% The content of dry matter (DM) of fermented sugarcane bagasse

IOP Conf. Series: Earth and Environmental Science **287** (2019) 012016 doi:10.1088/1755-1315/287/1/012016

(Figure 3) was highly significant difference (P <0.01) This result indicate that the nutrients contained in the sugarcane bagasse has been used by *G. lucidum*. The period during which the dry matter content decreased corresponded with the period in which the mycelial growth was increasing rapidly at day 20. Dry matter content continues to decrease with the length of fermentation up to day 60. Dry matter contained in the sugarcane bagasse is used by *G. lucidum.* We know that dry matter consist of inorganic and organic matter, in which minerals such as calcium, phosphor and magnesium required for the mycelial growth are component of inorganic matter. Urea can be used as nitrogen source for the mycelial growth of *G. lucidum.* Nitrogen is component of protein and carbon is component of carbohydrate and lipid are also required for the mycelial growth of *G. lucidum.* The addition of supplement in this research such as urea was found to accelerate mycelial growth The mycelial can be grown well by using dry matter of sugarcane bagasse, so that dry matter content decrease until day 60 fermentation. Dry matter content of the substrate decreased as much as 27.45% in urea treatment, was higher than non urea supplementation (15.45%). The mycelial growth and the production of bioactive component was ccelerate by addition another supplement such as fatty acids [37]. The increasing of mycelium number can be seen on the growth of *G. lucidum.* The main mechanism through which fungi and other microorganisms degrade plant biomass consists of production and secretion of enzymes acting synergistically in the plant cell wall, releasing monomers that can be used by the microorganism as chemical energy [33].

Ganoderma lucidum is one of the the saprophytic. Saprophytism, one of the most common lifestyle of microorganisms, involves living in dead or decaying organic matter, mainly composed by plant biomass. In this context, microorganisms developed cellular mechanisms in order to take energy from plant biomass, and one of this mechanisms involves the production and secretion of carbohydrate-active enzymes. These enzymes degrade the plant cell wall, releasing sugars monomers that can be used as substrates for the metabolism of the microorganism.

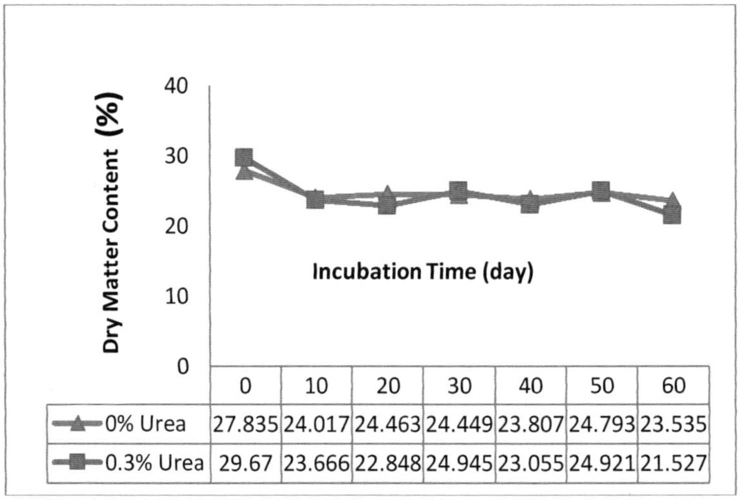

Figure 3. Dry matter content of fermented sugarcane bagasse.

3.1.4. Organic matter content of fermentation product of G. lucidum

Changes in organic matter content followed the pattern of dry matter content, which is influenced by the fermentation time (P <0:01); and also there is an interaction between the factors adding urea with the fermentation time (P <0:01). In Figure 4 it can be seen that the point of intersection of the fermentation time occurs in 50 days, where the value of organic matter content in sugarcane bagasse plus urea was continued to decline compared with bagasse without the addition of urea. The decline in organic matter content is a result of the *G lucidum* in the fermentation process. At that time, nutrient of substrate is reduced. At the urea treatment, organic matter content of the substrate decreased as much as 31.64%, was higher than non urea supplementation (21.86%). *G. lucidum* can utilize the fraction of fibre that are difficult to degrade by producing laccase enzyme that can degraded lignin. Laccase produced by *G. lucidum* will reduce the content of organic material. Laccase enzyme activity increases with increasing duration of fermentation as described below.

Plant cell wall [38] polysaccharides are the most abundant organic compounds found in nature. These compounds consist mainly of polysaccharides such as cellulose, hemicelluloses and pectin, as well as the phenolic polymer lignin. Together, the polysaccharides and lignin provide high complexity and rigidity to the plant cell wall. Fungi play a central role in the degradation of plant biomass, producing an extensive array of carbohydrate-active enzymes responsible for polysaccharide degradation [wagner shouza].

Concerning to lignin degradation, many white-rot basidiomycetes and some actinomycetes are able to produce lignin-degrading enzymes, especially peroxidases. For instance, *Phanerochaete chrysosporium* and *Phlebia radiata* are well known producers of extracellular peroxidases [39] as well as *Coriolus tersicolor*, which was shown to produce the intracellular haem peroxidase upon the induction by phenolic compounds [40]. A white-rot basidiomycete, *Rigidoporous lignosus*, is known to secrete two oxidative enzymes, laccase and Mn peroxidase, responsible for solubilizing the lignin in a synergistic way [41].

By using white rot fungi, lignin component can be degraded and among white rot fungi, *Ganoderma lucidum* is efficience producer of ligninolytic enzyme such as laccase.

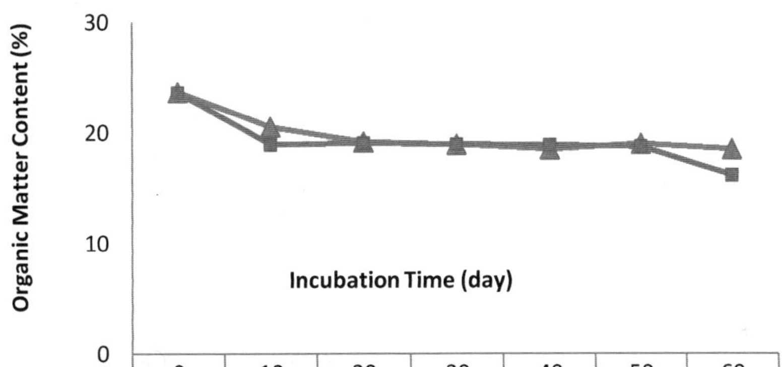

	0	10	20	30	40	50	60
0% Urea	23.6757	20.5325	19.1248	18.916	18.4737	18.9929	18.5013
0.3% Urea	23.6323	18.9328	19.0144	18.9455	18.8881	18.7213	16.1558

Figure 4. Organic matter content of Fermantaion Product of *G. Lucidum.*

3.1.5. Laccase activity

Laccase activity showed that the addition of urea and long fermentation gave a different effect on the enzyme activity of laccase produced by the *G. lucidum* (Figure 5). Based on Figure 5 appears that the addition of urea to produce laccase was very significantly higher than the fermented sugarcane bagasse without the addition of urea, at every length of fermentation 10 days to 60 days. The addition of 0.3% urea will increase laccase activity until fermentation 60 days, with the highest value of enzyme activity is 6.54 UmL^{-1}. Laccase activity decreased after fermentation 50 days, but the fermentation of 30 days, the activity of the enzyme laccase almost identical, both plus urea and non plus urea with a value of 6:13 U mL- and 5.91 U mL^{-1} respectively.

Ganoderma lucidum (Curt, Fr) P. Karst is a white-rot fungi, a species of the class Basidiomycetes, which can degrade substances which contain lignin by producing ligninolytic enzyme, such as laccase [9, 27]. Laccases produced by some wood-rotting fungi from the genus *Basidiomycete* play a major role in the biodegradation of lignin [42] Laccases can catalyze the oxidation of non-phenolic benzyl alcohols in the presence of a redox mediator, such as 2,2'-azinobis-[3-ethylthiazoline-6-sulfonate] (ABTS) [43] [44]. Aproximatelly 70% of lignocellulosic biomass is made of cellulose and hemicellulose. They are tigh linked to the ligninthrough hydrogenic and covalent bonds, making the structure resistant for any treatment.

In particular, the ability to decompose the aromatic lignin polymers in wood is mostly restricted to the white rot basidiomycetes. The white-rot decay of wood is possible due to secretion of organic acids, secondary metabolites, and oxidoreductive metalloenzymes, heme peroxidases and laccases, encoded by divergent gene families in these fungi [45].

Laccase is an enzyme involved in lignin degradation, beside lignin peroxidase a manganese peroxidase. Laccases were found in eukaryotes (fungi, higher plants, insects), [46]. The first gene and/or cDNA sequences were recorded for laccase from the *Ganoderma lucidum, Phlebia brevispora, Lentinula edodes* and *Lentinus tigrinus* [47].

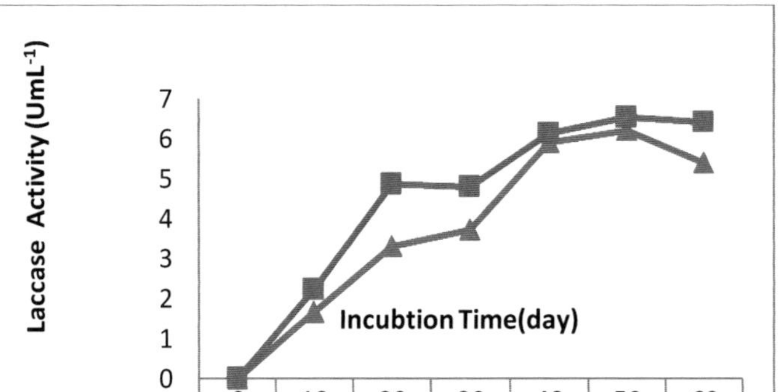

	0	10	20	30	40	50	60
0% Urea	0	1.66	3.31	3.73	5.91	6.21	5.4
0.3% Urea	0	2.22	4.87	4.8	6.13	6.54	6.42

Figure 5. Laccase activity of fermented bagasse using *G.lucidum.*

3.2. The nutritive value of bagasse

3.2.1. Rumen pH

The growth and the activity of rumen microbial are influended by many factors, one of them is rumen pH. The results showed that the pH value of rumen fluid in the in-vitro experiment of fermented bagasse with the addition of urea was highly significant different ($P < 0.01$) and the fermentation time was not significant ($P < 0.05$), although at 60 days fermentation period there was a decrease in pH to 6.8. The pH value of 6.8 is an ideal pH for the growth and microbial activity in the rumen. The resulting pH value is still in the normal range for rumen microbial growth with a pH value of 6.8-7.06 (Figure 6). Decreasing pH in 60 days fermentation is closely related to cellulose content and cellulose digestibility of fermented bagasse because cellulose is fermented into VFA and an increase in the amount of fermented cellulose also increases the VFA produced. Most VFAs are fermented products from carbohydrates, especially cellulose. VFA is a fermentation product in the rumen which is the main energy source for ruminants. If the pH value drops below 6 (<6) then the microbial activity to digest the fiber fraction will be disrupted and consequently can reduce the digestibility value of the fiber fraction, but in this study the pH value did not drop below 6 (<6), even the pH value was very support so that microbes are able to produce cellulolytic enzymes.

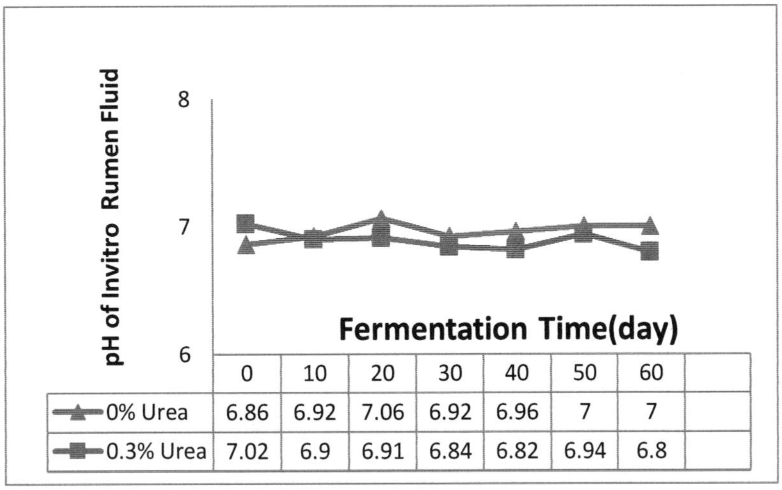

	0	10	20	30	40	50	60
0% Urea	6.86	6.92	7.06	6.92	6.96	7	7
0.3% Urea	7.02	6.9	6.91	6.84	6.82	6.94	6.8

Figure 6. Rumen pH of fermented bagasse using *G. lucidum* by adding urea at fermentation times of 0, 10, 20, 30, 40, 50 and 60 days.

3.2.2. Levels of VFA of rumen liquid from fementation products of sugar cane using Ganoderma lucidum

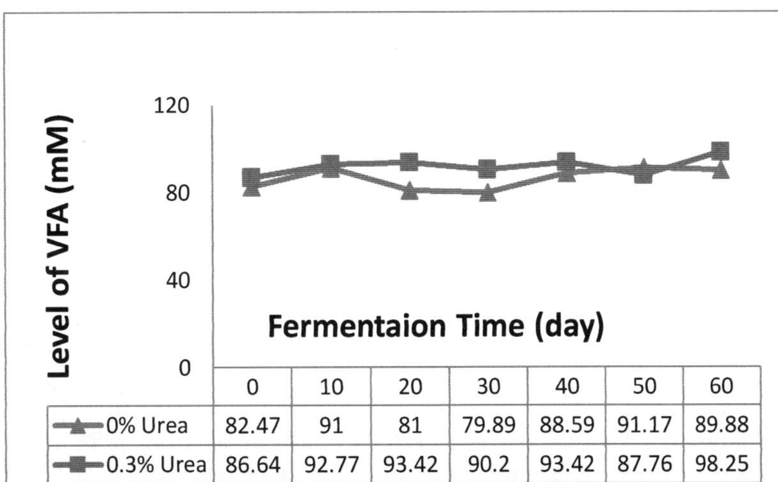

	0	10	20	30	40	50	60
0% Urea	82.47	91	81	79.89	88.59	91.17	89.88
0.3% Urea	86.64	92.77	93.42	90.2	93.42	87.76	98.25

Figure 7. Level of VFA in vitro from fermented bagasse using *G. lucidum* with the addition of urea at 0, 10, 20, 30, 40, 50 and 60 days.

Based on the results of the study it was found that the VFA levels in vitro from fermented bagasse using *G. lucidum* added with urea showed significantly different ($P < 0.05$). VFA levels from fermented bagasse added with urea were significantly higher than those without urea added during the fermentation process to 60 days. After 50 days of fermentation, it was seen that the VFA levels of fermented bagasse without added urea began to decline, while the bagasse added with urea, the VFA levels increased and even reached the highest value of 98.25 mM. This shows that the fiber fraction contained in bagasse pulp fermentation products is easily degraded in the rumen resulting in the highest fiber fraction fermentation product, VFA. The ease of degradation of this fiber fraction is the result of the work of an enzyme that begins with the enzyme laccase produced by *G. lucidum* in the process of bagasse fermentation outside the rumen, then continued by enzymes produced by rumen microbes.

3.2.3. Levels of NH3 in vitro rumen liquid

NH3 levels of rumen fluid showed significantly different values (P <0.01), higher in sugarcane ampas fermented by the addition of urea. The addition of urea in the bagasse fermentation process using Ganoderma lucidum will be utilized by *G. lucidum* as a source of nitrogen to increase the growth of *G. lucidum* mycelium, which in turn will increase the protein content of bagasse pulp fermentation products. This protein from bagasse is degraded in the rumen with the help of enzymes produced by rumen microbes. Fermentation product of protein in the rumen is NH3. NH3 produced in this rumen will be utilized by rumen microbes as a source of nitrogen for microbial protein synthesis. Increased microbial population will increase the amount of enzymes produced by microbes that will be used to digest feed. This increase in rumen microbial population is not only determined by the availability of a source of Nitrogen in the form of NH3, but is also determined by the availability of VFA which has a dual role, namely as an energy source and carbon framework for rumen microbial protein synthesis. So there must be synchronization between NH_3 produced with VFA which is formed from the degradation of carbohydrates in the rumen.

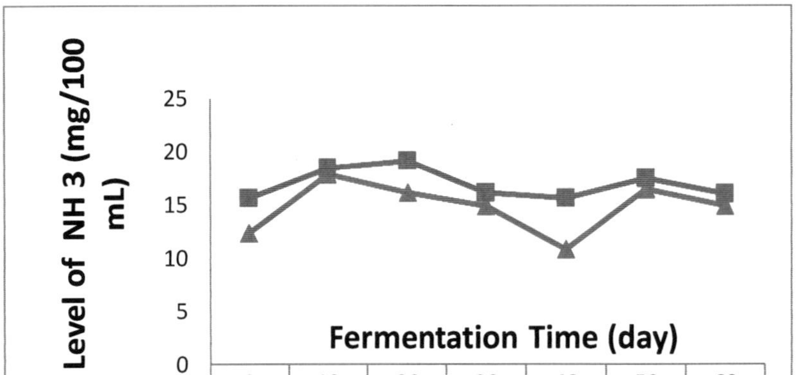

	0	10	20	30	40	50	60
0% Urea	12.33	17.88	16.1	14.86	10.78	16.44	14.85
0.3% Urea	15.62	18.41	19.08	16.1	15.58	17.49	15.99

Figure 8. Level of NH3 in vitro from fermented bagasse using *G.lucidum* with the addition of urea at 0, 10, 20, 30, 40, 50 and 60 days.

3.2.4. In vitro digestibility of dry matter of fermented bagasse with G. lucidum

The results showed that there was a very significant difference (P <0.01) between the addition of urea to bagasse and without urea in fermentation to the digestibility of dry matter. Digestibility of dry matter up to 10 days fermentation did not show any difference, but after 10 days of fermentation, the addition of urea gave a different response with no addition of urea. The highest value of dry matter digestibility occurred at 40 days fermentation period, namely in fermented bagasse added with urea with digestibility value of 43.39% (Figure 9); after that there was a decrease in digestibility of dry matter until fermentation of 60 days; while the fermented bagasse without urea showed results that decreased the dry matter digestibility until the fermentation time was 50 days, then the dry matter in fermentation was 60 days.

IOP Conf. Series: Earth and Environmental Science **287** (2019) 012016 doi:10.1088/1755-1315/287/1/012016

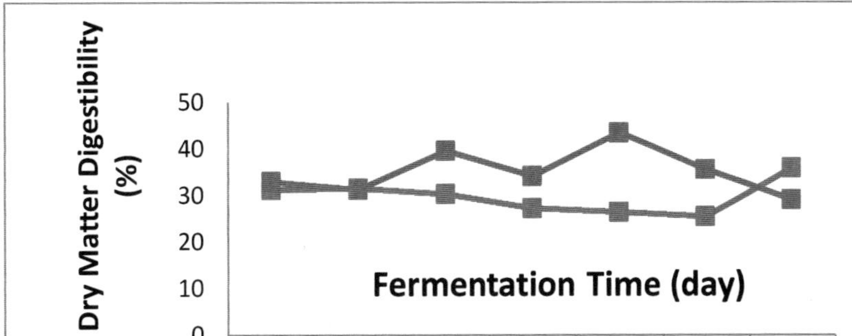

	0	10	20	30	40	50	60
0% Urea	31.23	31.46	30.18	27.16	26.37	25.35	35.71
0.3% Urea	32.8	31.23	39.46	34.13	43.39	35.51	28.98

Figure 9. In vitro digestibility of dry matter from fermented bagasse using *G. lucidum* with the addition of urea at 0, 10, 20, 30, 40, 50 and 60 days.

3.2.5. In vitro digestibility of organic matter of fermented bagasse with G. lucidum.

The digestibility value of organic matter has a pattern similar to the digestibility value of dry matter, because indeed most of the dry matter itself is organic material. The results of the addition of urea in the cementation of bagasse with *G. lucidum* showed that the digestibility of organic matter increased with the addition of urea in the bagasse fermentation process. The highest organic matter digestion occurs at 40 days of fermentation, after which there is a decrease in digestibility of organic matter.

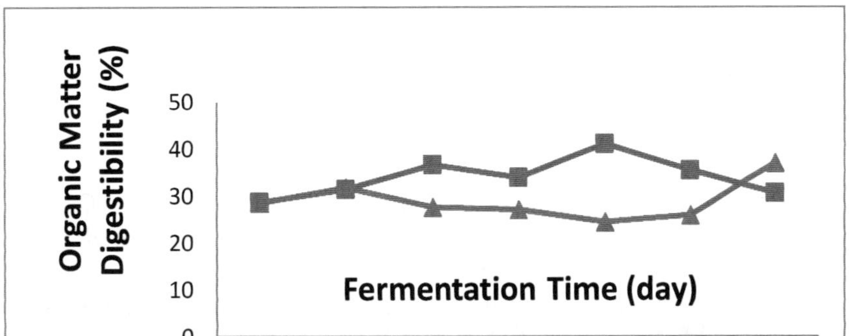

	0	10	20	30	40	50	60
0% Urea	28.53	31.75	27.54	27.06	24.44	25.83	36.94
0.3% Urea	28.63	31.35	36.55	33.79	40.97	35.4	30.58

Figure 10. In vitro digestibility of organic matter from fermented bagasse using *G. lucidum* with the addition of urea at 0, 10, 20, 30, 40, 50 and 60 days.

4. Conclusion

a. The percentage of growth of *Ganoderma lucidum* on the basic substrate of sugarcane with the addition of different energy sources is 100% because in all treatments, *G. lucidum* can grow well only with different growth speeds, ranging from 10 days to 60 days.
b. Urea can be used 0.3% as nitrogen source in fermentation process using *G. lucidum*,
c. The highest VFA content of fermented bagasse was 98.25 mM in the length of fermentation up to 60 days with 0.3% urea with the NH3 value was 15.99 mg%.

d. The highest dry matter and organic digestibility occurs in bagasse which is fermented with the addition of using *Ganoderma lucidum* which is at 40 days fermentation time with dry matter digestibility value 43.39% and organic matter digestibility value 40.97%.

5. References

[1] F. Agustin, T. Toharmat, D, Evvyernie, D. Taniwiryono, S. Tarigan. 2012. Effect of Supplemental Organic Chromium and Fungi *Ganoderma lucidum* on Milk Production and Immune Response in Lactating Cows, in *Proc. The 2nd International Seminar.* Feed Safety for Healthy Food. paper 22. p 156-163.

[2] F. Agustin, T. Toharmat, D. Evvyernie, D. Taniwiryono, S. Tarigan. 2012. Blood Profile of Lactating Cows Supplemented with Organic Chromium and *Ganoderma lucidum.* Pakistan J. of Nutrition. Vol. 11 (10). Pp 991-997.

[3] Xu Z., X. Chen, Z. Zhong, L. Chen, Y. Wang. 2011. *Ganoderma lucidum* Polysaccharides: Immunomodulation and Potential Anti-Tumor Activities. Am. J. Chin. Med., 39 (1): 15-27.

[4] G. C. Yen, J Y. Wu. 1999. Antioxidant and radical scavenging proepecties of extract from *Ganoderma lucidum.* Food chem. 65: 375-379.

[5] Y. H. You, Z B. Lin3. 2003. Antioxidant effect of *Ganoderma* polysaccharide. *Acta Pharm. Sinica*. 38: 85-88.

[6] J. Sun., H. He, B. J. Xie. 2004. Novel antioxcidant peptides from fermented mushroom *Ganoderma lucidum.* J Agric Food Chem. 52. 6645 – 6652.

[7] Z. X. Yang., Y. Y. Sou, & W. An. 2006. Studies on the capability of *Ganoderma lucidum* rich in Chromium. J. CEPS.

[8] F. Agustin, T. Toharmat, D, Evvyernie, D. Taniwiryono, S. Tarigan. 2013. The Incorporation of Chromium in Rice Straw Fermented with *Ganoderma lucidum.* Journal of Animal Science and Technology. Vol. 36. No.1. pp 64-69.

[9] S. T. Chang, & P. G. Miles. 2004. *Mushrooms: Cultivation, Nutritional Value, Medical Effect, and Environmental Impact.* 2nd Ed. London: CRC Press. Boca Raton.

[10] A. K. Misra, AK, A. S. Mishra, M. K. Tripathi, R. Prasad, S. Vaithiyanathaan, & R. C. Yalchmola. 2007. Optimization of solid state fermentation of mustard (*Brassica campestriss*) straw for production of animal feed by white rot fungi (*Ganoderma lucidum*). Asian-Aust J Anim Sci 20 (2): 208-213.

[11] F. Agustin, T. Toharmat, D. Evvyernie, D. Taniwiryono, S. Tarigan. 2010. The Incorporation of Chromium by *Ganoderma lucidum* in Oil Palm by-Product as Substrate. Journal of Animal Science and Technology. Vol. 33. No. 1.

[12] De Vries R. P., Visser J. 2001. *Aspergillus* enzymes involved in degradation of plant cell wall polysaccharides Microbiol. Mol. Biol. Rev. 65, 497–522.

[13] Couto SR, Sanromán MA. 2005. Application of solid-state fermentation to ligninolytic enzyme production. Biochem. Eng. J., 22 (3): 211-219.

[14] Couto SR, López E, Sanromán MA. 2006. Utilization of grape seeds for laccase production in solid-state fermentors. J. Food Eng., 74 (2): 263-267.

[15] V. Madhavi and S. S. Lele. 2009. Laccase: Properties and Aplications. BioResource 4 (4) pp 1694-1717.

[16] C. F. Turston. 1994. The Structure and Function of Fungal Laccases. Review Article. Microbiology, 140 pp 19-25.

[17] Placido, J and Capareda, S. 2015. Ligninolytic Enzymes: A Biotechnological Alternative for Bioethanol Production. Bioresources and Bioprocessing. 2: 23.

[18] Baldrian P. 2006. Fungal laccases-occurrence and properties. *FEMS Microbiology Reviews.* 30 (2): 215–242.

[19] Sanches C. Lignocellulosic Residue; biodegradation and Bioconversion by Fungi. Biotechnol. Adv. 27 (2): 185-194.

[20] Hildén K., Hakala T. K., Maijala P., Lundell T. K., Hatakka A. 2007. Novel thermotolerant laccases produced by the white-rot fungus *Physisporinus rivulosus*. *Applied Microbiology and Biotechnology*. 77 (2): 301–309.
[21] Sharma P., Goel R., Capalash N. 2007. Bacterial laccases. *World Journal of Microbiology and Biotechnology*. 23(6):823–832.
[22] Diamantidis G., Effosse A., Potier P., Bally R. 2000. Purification and characterization of the first bacterial laccase in the rhizospheric bacterium *Azospirillum lipoferum*. *Soil Biology and Biochemistry*. 32(7):919–927.
[23] Alberts J. F., Gelderblom W. C. A., Botha A., van Zyl W. H. 2009. Degradation of aflatoxin B1 by fungal laccase enzymes. *International Journal of Food Microbiology*. 135(1):47–52.
[24] Isroi, Millati R, Syamsiah S, Niklasson C, Cahyanto MN, Lundquist K. 2011. Biological Pretreatment of Lignocelluloses with White-Rot Fungi and Its Applications: A Review. Bioresources. 6 (4): 5224-59.
[25] P. J. Van Soest PJ. 2006. Rice straw, the role of treatmens to improve quality. *Anim Feed Sci Tech* 130 (3-4): 137-171.
[26] F. Agustin, T. Toharmat, D, Evvyernie, D. Taniwiryono, S. Tarigan. 2013. The Incorporation of Chromium in Rice Straw Fermented with *Ganoderma lucidum*. Journal of Animal Science and Technology. Vol. 36. no. 1. pp. 64-69.
[27] F. Agustin, Y. Marlida, Erpomen. 2015. The Fermentation of Bagasse with Fungi *Ganoderma lucidum* and The Evaluation of the In vitro Digestibility. Proc. Nasional Seminar, Paper 18, p124-132.
[28] K. A. Gomez, & A. A. Gomez. 1995. *Statistical Procedure for Agriculture Research.* Indonesia University Press.
[29] J. Buswell, J. A. Y. Cai and S. Chang. 1995. Effect of nutrient and manganese on manganese peroxidase and laccase production by Lentinula (Lentinus edodes). FEMS microbial Letr. 128: 81-88.
[30] Chang S. T, Buswell J. A. *Ganoderma lucidum* (Curt.: Fr.) P. Karst. 1999. (Aphyllophoromycetideae): A mushrooming medicinal mushroom. Int J Med Mushrooms. 1: 139–46.
[31] Wasser S. P, Coates P, Blackman M, Cragg G, Levine M, Moss J, White J. 2005. Encyclopedia of Dietary Supplements. Reishi or Lingzhi (*Ganoderma lucidum*). New York: Marcel Dekker. pp. 680–90.
[32] Boh B, Berovic M, Zhang J, Zhi-Bin L. 2007. Ganoderma lucidum and its pharmaceutically active compounds. Biotechnol Annu Rev. 13:265–301.
[33] De Souza W. R. 2003. Submerged Cultivation of *Ganoderma lucidum*, *Food Technol. Biotechnol. 41* (4): 371–382.
[34] Habijanic J, Berovic M. 2000. The relevance of solid-state substrate moisturing on Ganoderma lucidum biomass cultivation. Food Technol Biotechnol. 38: 225–8.
[35] Kim H. M, Park M. K, Yun J. W. 2006. Culture pH affects exopolysaccharide production in submerged mycelial culture of Ganoderma lucidum. Appl Biochem Biotechnol. 134: 249–62.
[36] P. Shrestha, B. Joshi, J.Joshi, R. Malla, and L. Sreerama. 2016. Isolation and Physicochemical Characterization of Laccase from *Ganoderma lucidum*-CDBT1 Isolated from Its Native Habitat in Nepal. Biomed Res Int. 3238909.
[37] Yang F. C, Liau C. B. 1998. The influence of environmental conditions on polysaccharide formation by Ganoderma lucidum in submerged cultures. Process Biochem. 33: 547–53.
[38] Aro N, Pakula T, Penttila M. 2005. Transcriptional regulation of plant cell wall degradation by filamentous fungi. FEMS microbiology reviews. 29 (4): 719-39.
[39] Lee S, Ha, JK, Kang, HS, McAllister T, Cheng K-J. 1997. Overview of energy metabolism, substrate utilization and fermentation characteristics of ruminal anaerobic fungi. Korean Journal of Animal and Nutritional Feedstuffs. 21: 295–314.

[40] Lobarzewski J. 1990. The Characteristics and Functions of the Peroxidases from Trametes-Versicolor in Lignin Biotransformation. J Biotechnol. 13 (2-3): 111-7.
[41] Galliano H, G. Gas, J. L. Sevis and A. M. 1991. Boudet: Lignin degradation by Rigidoporus lignosus involves synergistic action of two oxidizing enzymes: Mn peroxidase and laccase. Enzyme Microb Technol. 13: 478-82.
[42] Coll PM, Fernandez-Abalos JM, Villanueva JR, Santamaria R, Perez P. 1993. Purification and characterization of a phenoloxidase (laccase) from the lignindegrading basidiomycete PM1 (CECT 2971). Appl Environ Microbiol. 59: 2607-2613.
[43] Bourbonnais R, Paice MG. 1990. Oxidation of non-phenolic substrates: An expanded role for laccase in lignin biodegradation. FEBS Lett 1267: 99-102.
[44] Wong, D. W. S. 2009. Structure and Action Mechanism of Ligninnolytic Enzyme. Appl. Biochem. Biotechnol. 157 (2): 174-209.
[45] Lundel T. K, Makela M. R, Hilden K. 2010. Lignin-modifing enzyme in filamentous basidiomycetes-ecological, functional and philogenetic review. J. Basic Microbiol. 50 (1): 5-20.
[46] Martínez AT, Speranza M, Ruiz-Dueñas FJ, Ferreira P, Camarero S, Guillén F, Martínez MJ, Gutiérrez A, del Río JC. 2005. Biodegradation of lignocellulosics: microbial, chemical, and enzymatic aspects of the fungal attack of lignin. Int Microbiol. 8 (3): 195-204.
[47] D'Souza TM, Boominathan K, Reddy CA. 1996. Isolation of laccase gene specific sequences from white rot and brown rot fungi by PCR. Appl Environ Microbiol. 62: 3739-3744.

The utilization of different binders for pelleted native chicken ration based on coconut meat waste supplemented with mannanolytic thermophilic bacteria and thermostable mannanase: a physical characteristic of pelleted native chicken ration

Harnentis[1], Yetti Marlida[1], and Robi Amizar[1]

[1] Department of Animal Nutrition and Feed Technology, Faculty of Animal Science, Andalas University, Padang, West Sumatera, Indonesia.

Abstract. This research aim was to know the influence of addition of binders (bentonite, tapioca flour, gambier liquid waste, and onggok) on physical characteristics of pelleted native chicken ration based on coconut meat waste supplemented with mannanolytic thermophilic bacteria and thermostable mannanase. A completely randomized design was applied in this experiment with five treatments and five replicates. Moisture content, angle of response, bulk and compacted bulk densities, specific gravity, compaction rate, and pellet durability were determined. Data were analyzed by analysis of variance according to completely randomized design and Duncan's Multiple Range Test. Data indicated that moisture content, bulk density, compact bulk density and pellet durability affected significantly by binders, while angle of response, specific gravity and compaction rate had no affected by the presence of binders. Moisture content, bulk density, compacted bulk density, pellet durability of ration with binders were significantly different with ration without binder. Ration with gambier liquid waste binder had the highest value for bulk density and compacted bulk density, i.e. 1.36 g/cm3±0.011 and 1.46 g/cm3±001 respectively, while the value of pellet durability in ration with bentonite and tapioca flour binders were higher than others (98.63±0.11 and 98.49±0.05). It is concluded that the addition of binders in native chicken ration based on coconut meat waste supplemented with mannanolytic thermophilic bacteria and thermostable mannanase pelleted were able to improve the physical characteristic of native breed chicken ration.
Keywords – pelleted native chicken, coconut meat waste, mannanolytic thermophilic bacteria, thermostable mannanase.

1. Introduction

Indonesian local poultry include native chickens whose existence is well known to the public have a diverse appearance. Consumer demand for native chicken needs already very high. This can be seen from the increase in population growth and demand for native chickens every year [1]. The problem in developing domestic chicken is the high cost of feed. Feed is the biggest component and the success factor of a farm business. The quality and quantity of animal feed is often an obstacle that must be faced in an effort to meet basic life needs, growth, production and reproduction of livestock. For this reason, we need to find alternative sources of feed ingredients that have sufficient nutritional value, low prices, easy to obtain and safe for consumption by livestock.

Coconut meat waste is a by-product of coconut milk processing industry that has the potential to be used as animal feed, especially native chicken, because the production is abundant and has sufficient nutritional value. The use of coconut meat waste as animal feed, especially poultry, is still limited, because it is not palatable, voluminous and high in hemicellulose fiber content namely mannan and galactomannan. Many studies show that the administration of probiotics and enzymes can improve feed quality, so that it can reduce the limited use of feed, especially native chicken feed. A decrease in the hemicellulose content of coconut meat waste from 21.54% to 17.72% BK, using 800 U / kg of thermostable mannanase enzymes [2]. Supplementation of thermophilic mannanolytic bacteria (Bacillus sp. SM-1.4) 1010 CFU/kg in broiler rations based on coconut meat waste-based on can improve intestinal histomorphology, nitrogen retention, and metabolizable energy [3].

Currently poultry feed is switched from mash to pellet form, because it considered very effective. It is generally accepted that compared to mash, the feeding of pellets improves feed consumption [4], reduce the amount of wasted feed [5], make feed more homogeneous, reduce the time and use of energy to eat [6], besides giving it relatively easy and not dusty, increasing the bulkdensity of feed, improving feed flow ability, prolonging storage. and facilitate transportation [7].

Thermostable enzymes and thermophilic bacteria are needed in a feed product if the process of making these feeds uses high temperatures such as in making pellets. The physical form of pellets is strongly influenced by the type of material used, the size of the printer, the amount of water, pressure and the method after processing and the use of binders to produce pellets with a strong, compact and sturdy structure so that the pellets are not easily broken. The adhesive serves as an adhesive between all raw materials so that the feed that is made becomes more compact and stable. The adhesive commonly used by animal feed factories is synthetic adhesives such as bentonite, and lignosulfonate [8]. Synthetic adhesives are relatively expensive so we need to look for alternative adhesive materials from cheap local feed ingredients such as tapioca flour, onggok, and gambir liquid waste.

Besides being cheap, easy to obtain, it is also a natural material that tends to be safer to add to feed. The aim of this research is to produce pelleted native chicken ration based on coconut meat waste supplemented with Bacillus sp. SM-1.4 and thermostable mannanase using low cost starch based natural binders and synthetic binder (bentonite clay) leads to analyze and choose best pellets keeping in view the physical properties of pellets. Selection of binder can play the important role to achieve the same desired properties of pellets without steam conditioning and preheating of biomass material.

2. Materials and methods

Bacteria Preparation. a Pure culture of Bacillus sp. SM-1.4 isolated from hot springs South Solok [9], grown in media nutrient broth (NB) and then incubated at 60°C, for 24 hours. Once it is done counting the number of bacteria by a method using the Standard Plate Count Agar, to obtain the number of bacteria 1010 CFU/ml. The bacteria are then stored in a bottle.

Production of Enzymes. Mannanase thermostable enzyme produced by thermophilic mannanolityc bacteria (Bacillus sp. SM-1.4) [9]. Bacillus sp. SM-1.4 were grown in a medium containing (g/L). 40.9 copra meal, NH4NO3 0.5, Na2HPO4 7.54, NaH2PO4 2.32, MgSO4.7H2O 0.4, FeSO4.7H2O 0.02, CaCl2.2H2O 0.06, yeast extract (pH 7.0). Inoculum used for the production of the enzyme is Bacillus sp. SM-1.4 were cultivated in NB medium at a temperature of 600°C on a shaker water bath (120 rpm) for 12 hours. Enzymes produced using Minifors tool Benchtop Bioreactor with a capacity of 5 L. Medium inoculated with 10% inoculum (v/v) and incubated at 600°C for 18 hours, then centrifuged at a speed of 5,000 rpm for 5 minutes. The supernatant obtained is used as a crude enzyme in this study after the first test of enzyme activity.

Experimental design. This research was carried out using experimental method using Complete Design with 5 treatments and 5 replications. The treatment consists of: D0: diet without binder; D1: diet with bentonite binder 1.5%; D2: diet with tapioca flour binder 1.5%; D3: diet with gambier 1.5% liquid waste binder. D4: diet with 1.5% onggok binder. Diets were formulated in iso-nitrogenous. The nutritional content of the ration adjusted to the needs of protein for the growth of native chicken, which is 19% and the metabolic energy is 2850 kcal/kg [10] as presented in Table 1. The five diets imposed are described in Table 1.

Table 1. Composition of experimental diets.

Ingredients (%)	Experimental Diets				
	D0	D1	D2	D3	D4
Yellow corn	23.0	23.0	23.0	23.0	23.0
Coconut meat	30.0	30.0	30.0	30.0	30.0
waste	9.8	9.8	9.8	9.8	9.8
Fish meal	19.7	19.7	19.7	19.7	19.7
Soybean meal	11.5	11.5	11.5	11.5	11.5
Japfa BRI	5.5	5.5	5.5	5.5	5.5
Rice bran	0.5	0.5	0.5	0.5	0.5
Top mix					
Total	100	100	100	100	100
Binder	Without binder	Bentonite (1.5%)	Tapioca flour (1.5%)	Gambir liquid waste (1.5%)	Onggok (1.5%)
Calculated					
Crude protein (%)	19.01	19.01	19.01	19.01	19.01
ME (kcal/kg)	2857.00	2857.00	2857.00	2857.00	2857.00
Crude fiber (%)	9.39	9.39	9.39	9.39	9.39
Lipid (%)	9.49	9.49	9.49	9.49	9.49
Ca (%)	0.89	0.89	0.89	0.89	0.89
$P_{available}$ (%)	0.42	0.42	0.42	0.42	0.42
Lysine (%)	1.29	1.29	1.29	1.29	1.29
Methionine (%)	0.49	0.49	0.49	0.49	0.49
Tryptophan (%)	0.24	0.24	0.24	0.24	0.24

2.1. Binder preparation

Binder used in this research is bentonite was purchased from market Bogor. Tapioca flour bought in the traditional market Padang, West Sumatera. Gambier liquid waste is obtained from traditional gambir production in Payakumbuh, West Sumatera. Onggok is obtained by separating starch from cassava. Cassava after being grated then added water is homogenizedthen squeezed using gauze until the starch runs out. The remaining starch (onggok) is dried in the sun for 2 days, then milled with a blender For this purpose, each binder of 150 grams was boiled at 80°C in 5.0 liters of water. Binders are ready to be mixed with feed ingredients.

2.2. Preparation of pellets production

Coconut meat waste is obtained in the traditional market, dried for 12 hours by help of the sunlight, then finely ground. Other ration composition materials such as corn, soybean meal, and fish are also finely ground using a hammer mill to flour. Coconut meat waste before being mixed constituents of diet others added a mixture of crude enzyme mannanase with a dose of 800 U/kg of coconut meat waste and thermophilic bacteria (Bacillus sp. SM-1.4) 1010 CFU/kg diet [11]. The preparation of diets using materials and compositions as shown in Table 1, and formed pellet using 3 mm die with a length of 0.7-1.3 cm and then dried for 8 hours, then tested the physical quality of pellets in the Laboratory of Technology and Feed Industry, Faculty of Animal Science.

2.3. Stack angle [12]

Stack angle measurement is done by dropping 500 grams of material at a height of 32.5 cm through a funnel on a flat plane. As a flat plane, white paperboard is used. Stack diameter is twice as high as the fall of material, while to measure height is done with calipers. The weight of the material used is 500 grams.

2.4. Bulk density [13]

Bulk density is calculated by inserting a material with a certain weight into a 500 ml measuring cup with the help of a funnel. In addition, avoidance of shocks in the measuring cup during measurement. Bulk density expressed in units of g / cm^3. Bulk density was calculated as the ratio of the weight of the material (g) to the volume of space (cm^3).

2.5. Compacted bulk density [13]

Compacted bulk density determined in the same way as bulk density determination, but the volume of material is read after the compaction process by shaking the measuring cup by hand until the volume does not change again. Stack compaction density is expressed in units of g / cm3. Compacted bulk density was calculated as the ratio of the weight of the material (g) to the volume of space after compacting.

2.6. Specific gravity [13]

The Specific gravity is measured using the Archimedes Law principle, that is by measuring the change in volume in the measuring cup (500 ml) after added the determined amount of distilled water (200 ml) and material known to its mass (100 grams) into the measuring cup. Stirring is done in a measuring cup to accelerate the loss of air between the particles of the ration during the measurement. Changes in the volume of distilled water are the actual volume of material. Specific gravity was calculated as the ratio of the weight of the material (g) to change in the volume of distilled water (cm^3).

2.7. Pellet durability

Pellet durability was determined in a Holmen Pellet Tester. Clean pellet samples (200 g), with no fines, were rapidly circulated in an air stream around a perforated test chamber for 30 s. Fines were removed continuously through the perforations (2 mm in diameter) during the test cycle at 60 milbar. After the test cycle, the subject pellets were ejected and weighed manually. The pellet durability index (PDI) was calculated as the ratio of retained pellet weight after agitation to initial pellet sample weight.

2.8. Pellet hardness

Pellet hardness was tested in a Stable Micro Systems Texture Analyzer using the method described [14]. Individual pellets a pressure piston and a bar, and by increasing pressure applied by means of the pressure piston, the force (Newton) needed to break the pellets was determined.

2.9. Statistical analysis

The data obtained were analyzed statistically using diversity analysis according to Completely Randomized Design [15]. The differences between treatments were tested using Duncan's Multiple Ranget Test (DMRT).

3. Results and discussion

3.1. The effect of treatment on the moisture content of pelleted native chicken ration

The effect of binders on the moisture content of pelleted native chicken ration presented in Table 2. The moisture content of pelleted native chicken ration significantly ($P < 0.05$) affected by binderts. The moisture content of pellet with gambir liquid waste significantly higher than other binders. The moisture content of pellet with bentonite had no significant with the without binder, but significantly ($P < 0.05$) lower than other binders. The highest moisture content is obtained on pellets with gambier liquid waste binder, while the lowest water content is seen on pellets with benotnite and without binder. The high level of moisture content of pellet with gambier liquid waste binder is caused due to gambier liquid waste containing tannin which has a stronger binding strength than other binders so that it inhibits the release / release of water that is between the particles of the material when it is dried in the same time.

Low water content in pellet with bentonite binder compared to other binders caused by heating on bentonite before being mixed with other feed ingredients, causes bentonite to be activated so that it can remove water molecules contained in the pellets. Bentonite before used must be activated first because in the initial state it only has the lowest adsorption ability. Activation of bentonite with the addition of hot water aims so that water is bound between molecule in the pellet ration can evaporate, so the porosity increases. Bentonite which binds to hot water can increase the ability of its absorption so can reduce the moisture content in the pellets. While low moisture content in pellet without binder caused the pores between the particles of the material are not tightly bound so that the water will be more volatile when drying.

Table 2. The mean of physical characteristic of pelleted native chicken ration in each treatment.

Variables	Treatments				
	D0	D1	D2	D3	D4
Moisture content (%)	8.34 ± 0.06^c	8.48 ± 1.02^c	9.40 ± 0.91^b	10.75 ± 3.60^a	9.05 ± 1.25^b
Angle of response (°)	32.05 ± 3.28	33.06 ± 1.69	32.82 ± 1.92	34.40 ± 2.12	32.88 ± 1.95
Bulk Density (g/cm^3)	1.28 ± 0.004^d	1.31 ± 0.004^c	1.34 ± 0.005^b	1.36 ± 0.011^a	1.31 ± 0.004^c
Compacted Bulk Density (g/cm^3)	1.38 ± 0.04^c	1.42 ± 0.04^b	1.45 ± 0.03^a	1.46 ± 0.01^a	1.41 ± 0.04^b
Specific gravity (g/cm^3)	1.26 ± 0.03	1.32 ± 0.07	1.35 ± 0.08	1.38 ± 0.09	1.40 ± 0.11
Hardness (N/cm^2)	173.84 ± 1.59^b	237.43 ± 2.7^a	213.43 ± 2.13^a	217.33 ± 1.41^a	226.81 ± 1.30^a
Durability (%)	95.82 ± 0.23^c	96.61 ± 0.11^b	98.49 ± 0.05^a	98.63 ± 0.11^a	96.86 ± 0.26^b

: ^{abc} The mean with different superscripts at the same row differs statistically ($P < 0.05$).

3.2. Effect of treatment on stack angle of pelleted native chicken ration

The effect of binders on the of stack angle of pelleted native chicken ration presented in Table 2. The use of different binders in pelleted native chicken ration had no significantl ($P > 0.05$) to the stack angle of pelleted native chicken ration. The value of the bulk angle in the research ration ranged 32.05°-34.40°.Bulk angle value on pelleted native chicken based on coconut meat waste supplemented with thermophilic bacteria and thermostable mannanase in this study included in the category of easy rations that are easy to flow on a sloping plane or a certain height that is, at the bulk angle range of 30-38° [16]. The using binders (bentonite, tapioca, gambier liquid waste and onggok) and witout binder does not show an effect on the bulk angle of pelleted native chicken ration. The size of the bulk angle is affected by particle size, shape and characteristics, moisture content, specific gravity and bulk density [12]. P article size affects bulk angle, ie the smaller the particle size, the higher the bulk angle [17]. Pellets that have a high bulk angle will reduce flow rate of pellet and will cohere so that the pellets are not free to move [18]. Pelleted ration can be categorized in solid form ration, where solid form rations have bulk angles ranging between 20° and 50° [12]. The value of the bulk angle in pelleted ration of native chicken in this study not much different that reported [19] with bulk angles values of commercial ration and pelleted basal ration of starter broiler with tapioca, bentonite and onggok binders which ranges from 33.43°-35.14°.

3.3. Effect of treatment on bulk density of pelleted native chicken ration

The effect of binders on the bulk density of pelleted native chicken ration presented in Table 2. The bulk density of pelleted native chicken ration significantly ($P < 0.05$) affected by binderts. Addition of different binder in native chicken ration significantly ($P < 0.05$) affected bulk density pelleted native chicken ration based on coconut meat waste supplemented with thermophilic bacteria and thermostable mannanase. Bulk density of pelleted native chicken ration with liquid gambier waste binder significantly ($P < 0.05$) higher compared to others and without binder, then followed by pellet with tapioca flour binder. Addition of bentonite and onggok binders had no significant ($P > 0.05$) to bulk density and the lowest obtained without a binder. The high density of pelleted ration with gambier liquid waste is caused by moisture content of pelleted ration is higher than that with other binders. The value of bulk density in this study getting lower with increasing moisture content, because the constituent ingredients of pellets will swell with higher moisture content causing the volume of space needed to be greater [20]. Materials that have low bulk density (< 0.45 g/cm^3) takes time to flow in a vertical direction for longer, on the contrary with materials that have a larger bulk density (> 0.5 g/cm^3) [13]. The results of bulk density in this study showed that the use of bentonite, tapioca flour, gambier liquid waste, onggok and without binder had a bulk density above 0.5 g/cm^3. The higher value of bulk density causes the volume of space loaded by pellets to be smaller. Materials with high bulk density can save costs for packaging and storing materials [21]. The bulk density of pelleted native chicken ration was higher than that reported [22], where the value of bulk density on synthetic binders pellets obtained was 0.64-0.66 g/cm^3.

3.4. Effect of treatment on compacted bulk density of pelleted native chicken ration

The effect of binders on the compacted bulk density of pelleted native chicken ration presented in Table 2. Compacted bulk density of pelleted native chicken ration significantly ($P < 0.05$) affected by binderts. Compacted bulk density on pellets with gambier liquid waste binder was not significantly different ($P > 0.05$) with tapioca flour binder, but significantly ($P < 0.05$) higher compared to other binders. Pellets with bentonite adhesives give no significant effect ($p > 0.05$) on stack compaction density and the lowest is obtained on pellets without binders. the difference between the value of compacted bulk density is caused by pellet water content [23], specific gravity, also influenced by the intensity and method of compaction which is conducted. Compaction on feeds with high specific gravity will increase compaction density [24]. The higher the value of compacted bulk density, the smaller the volume of space occupied by the pellet. The value of compacted bulk density on pelleted native chicken ration based on coconut meat waste supplemented with thermophilic bacteria and thermostable mannanase higher than that reported [22], which is equal to 0.69 - 0.70 g/cm^3 in broiler ration pellets with synthetic binder.

3.5. Effect of treatment on specific gravity of pelleted native chicken ration

The effect of binders on the of specific gravity of pelleted native chicken ration presented in Table 2. The use of different binders in pelleted native chicken ration had no significantl ($P > 0.05$) to the specific gravity of pelleted native chicken ration. The average specific weight of local chicken pellets with a binder of 1.26 ± 0.03 - 1.40 ± 0.11 g/cm^3. There is no influence of different binders on specific gravity of pelleted native chicken based on coconut meat waste supplemented with thermophilic bacteria and thermostable mannanase caused by compaction in the same pellet machine so that the space between particles is not different [25]. Rations consisting of particles of different specific gravity are large enough that this mixture is unstable and tends to be easily separated again [13]. Specific gravity values in the pellet ration in this study indicate level equal level of convenience in transportation and storage space capacity. The specific gravity of pellets in this study was not different from that reported [19], which ranged from 1.27 to 1.40 g/cm^3 on starter broiler ration pellets with bentonite, tapioca and onggok binders, while [26] reported that the average specific density of pellets with lignosulfonate synthetic and bentonite binder were 1.37 g/cm^3 and 1.40 g/cm^3.

3.6. Effect of treatment on hardness of pelleted native chicken ration

The effect of binders on the hardnessof pelleted native chicken ration presented in Table 2. The hardness of pelleted native chicken ration significantly ($P < 0.05$) affected by binderts. The average pellet hadrness of pelleted native chicken ration with different binders ranged from 173.84 ± 1.59 - 237.43 ± 2.7 N/cm^2. Using the different binders was significantly ($P < 0.05$) improved pellet hardness.

Pellet hardness in different binders, resistance to pellet breakdown due to pressure in bulk bins will also increase. Hardness of pellets with bentonite, tapioca flour, gambier liquid waste, and onggok binder significantly (P <0.05) higher than pellets without binders. This is probably influenced by the starch content of tapioca flour and onggok binder, while gambier liquid waste contained tannin which function as binder. That pellet hardness in pelleted broiler ration based on wheat and soybean meal range between 31.4 – 42.8 N [27].

3.7. Effect of treatment on of durability index of pelleted native chicken ration

The effect of binders on the durability index of pelleted native chicken ration presented in Table 2. Durability index of pelleted native chicken ration significantly (P <0.05) affected by binderts. Durability index of pelleted native chicken ration using different binders ranged from 98.50% - 99.12%, the highest obtained in ration with gambier liquid waste and tapioca binders, then followed by pelleted ration with bentonite, onggok and without binder. Durability index on pellet with gambier liquid waste binder had no significantly different (P> 0.05) with tapioca binders, but significantly (P <0.05) higher than the durability of pelleted ration with bentonite, onggok and without binders. This is due to gambir liquid waste containing tannin which has a higher adhesion than pellets with bentonite, onggok and without binder. while tapioca flour has starch with a higher amilopectin content than amylose. Pelleted native chicken ration based on coconutmeat waste supplemented with mannanolytic thermophylic bacteria and thermostable mannanase in this study had a high durability index, this may also caused by the work of the thermostable nannanase enzyme which breaks down mannan in coconut meat waste to produce mannosa which in turn increase gelatinization when mixing with hot binder (50-60oC) before pelleting Pellet durability is considered high when the calculated value is above 80%, while if the value is between 70% and 80%, and low when the value is below 70% [28]. The results of durability index pelleted native chicken ration in this study were higher than those reported [29], the shrimp feed pellet ration used 5% CMC binder, 5 and 10% tapioca binder, and 5% seaweed ranged between 92.26 - 94.44%.

4. Conlusion

It was concluded that using binders (bentonite, tapioca flour, gambier liquid waste, and onggok) could improved the physical quality of pelleted native chicken ration based on coconut meat waste supplemented with thermophilic bacteria and thermostable mannanase,while tapioca flour and gambier liquid waste showed best physical quality for pellet binders.

5. References

[1] Bakrie, B., D. Andayani, M. Yanis, dan Zainuddin. 2003. Pengaruh Penambahan Jamu ke Dalam Air Minum terhadap Preferensi Konsumen dan Mutu Karkas Ayam Buras. Prosiding. Seminar Nasional Teknologi Peternakan dan Veteriner 'IPTEK untuk meningkatkan Kesejahteraan Petani melalui Agribisnis Peternakan yang Berdaya Saing'. Pusat Penelitian dan Pengembangan Peternakan. Bogor.

[2] Harnentis. 2015. Isolation and Production of Mannanase Enzyme by Thermophilic Bacteria from Hot Spring and Its Application to Increase The Quality of Poultry Feed. PhD. Thesis, Postgraduate. Andalas University. Padang.

[3] Harnentis, and E. Syahruddin, 2016. The effect of steam temperature and supplementation of thermophilic mannanolytic bacteria on gut histomorphology, nitrogen retention, and metabolizable energy of broiler fed coconut residu-based diets. Indonesian Journal of Anim Sci. Vol. 14 (02).

[4] Shavice Sarrestani, T., N. Nabiri., M, J, Agah., and H, Norollahi. 2006. Effect of pellet and mash diets associated with byozyme enzyme on broiler performance. International Journal. of Poult. Sci. 5(5): 485-490.

[5] Behnke, K. C., and R, Scott Beyer. 2004. Effect of feed processing on broiler performance. PhD. Disertation, Kansas State University, Manhattan.

[6] Jensen, L.S. 2000. Influence of pelleting on the nutritional needs of poultry. Asian-Aust. Journal. Anim. Sci, 13: 35-46.
[7] Greenwood, M.W., and R.S. Beyer. 2003. Effect of manufacturing practices on nutrient availability and feed quality. Proceeding of 30th Annual Carolina Poultry Nutrition Conference, Raleigh, NC. pp: 7-16.
[8] Retnani, Y., N. Hasanah., Rahmayeni., and N. Herawati. 2010. Physical properties test of pellet broiler chicken rations added onggok binder through water spraying process. Journal Agripet. Vol 11 (1): 11-18.
[9] Harnentis., Yetti Marlida, Yose Rizal and Maria Rndo Mahata. 2013. Isolation, characterization and production of mannanase from thermophilic bacteria to increase the feed quality. Pakistan Journal of Nutr. 12 (4). 360-364, 2013 ISSN 1680-5194.
[10] Nawawi. 2011. Pakan Ayam Kampung. Penebar Swadaya. Jakarta.
[11] Harnentis and Syahruddin. 2016. Pengaruh Temperatur Steam dan Suplementasi Bakteri Mannanolitik Termofilik terhadap Histomorfologi Usus, Retensi Nitrogen dan Energi Metabolisme Ransum (Pellet) Broiler Berbasis Ampas Kelapa. Jurnal Peternakan Indonesia. Vol. 18 (1).
[12] Khalil. 1999b. Effect of moisture content and particle size on changes in physical behavior of local feed ingredients: stack angle, threshold and hygroscopic factors. Media Peternakan. Vol 22 (1): 33-42.
[13] Khalil. 1999a. Effect of moisture content and particle size on changes in physical behavior of local feed ingredients: bulk density, compacted bulk density and spesific gravity. Media Peternakan. Vol 22 (1): 1-11.
[14] Svihus, B., K. H. Kløvstad, V. Perez, O. Zimonja, S. Sahlstrom, R. B. Schuller, W. K. Jeksrud, and E. Prestløkken. 2004. Physical and nutritional effects of pelleting of broiler chicken diets made from wheat ground to different coarsenesses by the use of roller mill and hammer mill. Anim. Feed Sci. Technol. 117:281–293.
[15] Steel, R.G.D. and J.H. Torrie, 1995. Principles and Procedures of Statistics a Biometric Approach McGraw Hill, New York.
[16] Fasina, O. O., and S. Sokhansanj. 1993. Effect of moisture content on bullhandling properties of alfalfa pellets. Canadian Agric. Engine. 35(4):269-279 (Abstr.).
[17] Mujnisa. A. 2007. Physical properties test of milled corn on various particle sizes. Animal Nutrition and Feed Bulletin. Vol 6 (1): 1-9.
[18] Geldart, D., M.F. Mallet and N. Rolfe. 1990. Assessing the flowability of powders using angle of respon powder. Handling and Processing 2. (4):59-74.
[19] Retnani, Y., L. Herawati and S. Khusniati. 2011. Physical characteristics on crumble ration of broiler starter using tapioca, bentonite and onggok binders. Journal ITP. Vol 1 (2): 88-97.
[20] Suadnyana, I. W. 1998. Pengaruh kandungan air dan ukuran partikel terhadap perubahan sifat fisik protein local. Skripsi Fakultas Peternakan . Institut Pertanian Bogor. Bogor.
[21] A/S Niro. 2005. Bulk Density. http://www.niro.com/niro/cmsdoc.nsf/webdoc/ndkw6u9atz.
[22] Retnani, T., Y. Harmiyanti., D. A.P. Fibrianti., and L. Herawati. 2009. The effect of using synthetic binder on physical quality of chicken ration. Journal Agripet. Vol 9 (1): 1-9.
[23] Sayekti, W. B. R. 1999. Karakteristik Sifat Fisik Berbagai Varietas Jagung. Skripsi. Fakultas Peternakan. Institut Pertanian Bogor. Bogor.
[24] Gautama. 1998. Sifat Fisik Pakan Lokal Sumber Energi, Sumber Mineral, serta Sumber Hijauan pada Kadar Air dan Ukuran Partikel yang Berbeda. Skripsi. Fakultas Peternakan. Institut Pertanian Bogor. Bogor.
[25] Rahmayeni. 2002. Uji Sifat Fisik Pakan Ayam Broiler Starter Bentuk Pelet dengan Penambahan Perekat Onggok. Skripsi. Fakultas Peternakan. Institut Pertanian Bogor. Bogor.
[26] Harmiyanti. 2002. Uji Sifat Fisik Ransum Ayam Broiler Bentuk Pellet dengan Penambahan Perekat Lignosulfonat dan Bentonit dengan Beberapa Proses Pengolahan. Skripsi. Fakultas Peternakan. Institut Pertanian Bogor. Bogor.

[27] Abdollahi., M.R., and V. Ravindra. 2013. Influence of pellet length on pellet quality and performance of broiler starters. Poutry Science Association, Inc. p. 516-522.
[28] Colley Z, O. O., Fasina., D. Bransby ., and Y. Y. Lee. 2006. Moisture effect on the physical characteristics of switchgrass pellets. T ASAE. 49: 1845-1851
[29] Wulansari. R., Y. Andriani., and K. Haetami. 2016. The use of binder types on the physical quality of shrimp ration. Jurnal Perikanan kelautan. Vol VII (2): 140-149.

Acknowledgments

This experiment was supported by the Excellent Research Colleges Funding Program provided by Directorate General of Higher Education, Ministry of National Education of Republic of Indonasia. I am very grateful to the Dean of the Faculty of Animal Science and the Rector of the Andalas University who have given me chance to compete with others in obtaining this funding. My special thanks is also directed to the Director General of Higher Education and the Ministry of National Education who have provided the funding.

The effects of leguminous supplementation on ammoniated rice straw based completed feed on nutrient digestibility on *in vitro* microbial protein synthesis

Mardiati Zain[1], Rusmana W S Ningrat[1], Erpomen[1], Ezi Masdia Putri[2], Malik Makmur[2]

[1]Department of Animal Nutrition and Feed Technology, Faculty of Animal Science Andalas University, Padang, West Sumatera, Indonesia.
[2]Post Graduated Student, Faculty of Animal Science Andalas University, Padang, West Sumatera, Indonesia.
Corresponding Author: mardiati@ansci.unand.ac.id, live DNA:62.19662.

Abstract. Legume is the best source of protein to ruminant. Tannin content in legumes can provide protein by pass for ruminant. This study was conducted to determine the effect supplementation of legumes (*Leucaena leucocephala*) as substitute of concentrate on nutrient digestibility and microbial protein synthesis of the completed feed based on ammoniated rice straw. The study was conducted in an *in vitro* used a randomized block design where there were 3 treatments dan 5 replications. The treatments were A. 40% ammoniated rice straw + 60% concentrate, B. 40% ammoniated rice straw + 50% concentrate + 10% *Leucaena leucocephala*, C. 40% ammoniated rice straw + 40% concentrate +10, + 20% *Leucaena leucocephala*. The results showed that the addition of of *Leucaena leucocephala* had increased the digestibility of DM, OM, CP and microbial protein synthesis ($p < 0.05$). Increased doses of *Leucaena leucocephala* until 20% decreased digestion of nutrition compared with dose 10% but was still higher than control. In conclusion, the supplementation of *Leucaena leucocephala* on completed feed based on ammoniated rice straw *in vitro* can increase nutrient degradability. Supplementation of 10% and 20% *Leucaena leucochepala* were suitable to be used for further studies, therefore *in vivo* experiment is required to study the effects on animal production.
Keywords – digestibility, *Leucaena leucocephala,* completed feed, ammoniated rice straw.

1. Introduction

Feed was one of the determinants of livestock productivity, so the availability of good quality feed was a requirement for livestock development in an area [1]. Cattle breeding systems that were still traditional, where feeding is only dependent on rice straw and native grass, caused the need for nutrients was not fulfilled. Addition of leguminous which had high quality in low quality rations such as rice straw will increase livestock productivity. Legumes such as *Leucaena leucochepala* was a high-protein content with a good amino acid profile, had mineral content, and low crude fiber [2]. It also had tannin content act as added value because it can function to protect the excess of protein in the rumen (by pass protein). Therefore, the amount of protein that can absorb by the small intestine becomes higher and also had the potential to reduce methane production in the rumen [3]. The high use of *Leucaena leucocephala* in the ration can also have a negative effects on livestock due to excess nitrogen in the rumen [4] and the presence of mimosine, antinutrients, that can cause poisoning in livestock [5] [6]. Based on the description above this study was conducted with the aim to determine the effect of *Leuchaena leucochepala* as substitute for concentrate in ration of ammoniated straw on nutrient digestibility and microbial protein synthesis.

2. Research methods

This research was carried out in the Laboratory of Ruminant Nutrition, Faculty of Animal Science, Andalas University. This experiment carried out Randomized Block Design with three treatments and five groups as replications. The treatments were A. 40% ammoniated rice straw + 60% concentrate, B. 40% ammoniated rice straw + 50% concentrate + 10% *Leucaena leucocephala*, C. 40% ammoniated rice straw + 40% concentrate + 20% Leucaena leucocephala. Concentrate consist of rice brain, palm kernel cake, cassava and mineral. Chemical composition of each treatment can be seen in Table 1. Ruminal fluid was obtained from a cannulated steer. The *in vitro* digestibility measurement was conducted according to [7]. Fermentation tubes contained of 50 ml of rumen fluid and 200 ml McDougall buffer solution. Three fermentation tubes that did not contain substrate were also incubated and used as blanks.Tubes were incubated for 48 h at a temperature of 39°C in a shaker water bath. After incubation for 48 h, fermentation activity was stopped by immersion in iced water to stop the activities of microbial activity. Rumen fluid pH was measured with pH meter. Tubes were then centrifuged at 1500 rpm for 30 min and the supernatant was removed. Residu samples were oven dried at 60°C for 24h and stored for later chemical analysis according to [8] for dry matter (DM), organic matter (OM), Crude protein (CP), and microbial protein shyntesis based on [9].

Statistical Analysis: Data were analyzed by ANOVA using a completely randomized design with subsampling. Differences among means were tested using LSD [10].

Tabel 1. The Ingredient composition and the chemical composition (% dry matter).

Ingredient composition (%)	Diet		
	A	B	C
Ammoniated rice straw	40	40	40
Concentrate	60	50	40
Leucaena leaf meal	-	10	20
Total	100	100	100
Chemical composition			
CP	10.93	12.21	13.48
TDN	67,79	68,20	68.60
CF	5,07	4,91	4.74
NDF	46.73	44.49	42.26
ADF	49,31	51,53	52,01
Lignin	36,77	40.04	43,26

Notes: TDN = total digestible nutrient, CP = Crude protein, NDF = Neutral Detergent Fiber, ADF = Acid Detergent Fiber, CF = Crude Fat.

3. Result and discussion

3.1. Digestibility nutrien and microbial protein synthesis

Digestibility of nutrient was an indicator of the use of feed by livestock. Table 2 showed that the treatments had a significant effect on the digestibility of dry matter, organic matter, crude protein and microbial protein synthesis (P <0.05).

Table 2. Nutrien digestibility and microbial protein synthesis.

No.	Treatments	Digestibility			Microbial protein synthesis (mg/100 ml)
		DM	OM	CP	
1	A	58,83^{c}	59.50^{b}	57,83^{b}	74.33^{c}
2	B	69.00^{a}	69.70^{a}	71.23^{a}	137.17^{a}
3	C	64.33^{b}	66.50^{a}	66,51^{b}	88.04^{b}
	SE	0.57	1.20	1,3	1.7

Notes: Different superscripts (^{a,b,c}) in the same row indicate statistically significantly different effects (P<0.05).

Supplementation of 10 and 20% *Leucaena leucocephala* in ammoniated rice straw diet signifinantly increased the digestibility of dry matter, organic matter, crude protein and rumen microbial protein synthesis compared without *Leucaena leucocephala* supplementation. This happens because the content of nutrents from treatment B and C also increases compared to treatment A. So that the availability of nutrients for rumen microbes is also more. Besides that, *Leucaena leucocephala* also contains sulfur and calcium which are needed by rumen microbes to develop, especially cellulolytic bacteria and fungi [11]. [12] [13] said that Leucaena as a protein supplement can improve the quality of feed. Supplementation 10% of *Leucaena leucocephala* provides better digestibility than 20% supplementation (treatment C). The decrease in nutrient digestibility in treatment C is caused more by the high lignin content of the ration. Lignin binds to cellulose and hemicellulose and cannot be digested by enzymes produced by microbes. The higher the lignin content in the diet, the nutrient digestibility of diet will decreased as explained by [14]. Besides lignin, the tannin content which is slightly higher in this treatment could reduced digestibility because tannin will bind protein and carbohydrate feed so it cannot be degraded in the rumen. Tannin bond with protein can be released in an acidic condition according to the opinion of [15] also states that tannins in the rumen are able to bind proteins and carbohydrates so that the digestibility in the rumen can be reduced.

Table 2 also showed that the addition of *Leucaena leucocephala* significantly affects the digestibility of the crude protein. Protein degradation described a number of proteins that are degraded in the rumen which is a supply of feed protein as a source of N for rumen microbes as well as an indication of a number of diet protein that are not degraded in the rumen entering the small intestine. Protein digestibility increased with the addition of *Leucaena luecocephala* in the ration. This increament is caused by the increase in diet protein content because *Leucaena leucocephala* is a feed ingredient with good protein content [16]. Protein digestibility in the rumen decreases with the addition of 20% *Leucaena luecocephala*. This decrease was caused by protein ration in treatment C mostly came from *Leucaena luecocephala* which was bound with tannin so that it was not degraded in the rumen as stated by [2] that Leucaena has tannin content which can provide added value because it can protect the over-protein overhaul in rumen (by pass protein) so that the amount of protein that can be absorbed by the small intestine becomes higher.

Microbial protein synthesis was a description of microbial growth in the rumen. Increased microbial protein synthesis means that the microbial population in the rumen also increases. Table 2 showed that the addition of *Leucaena luecocephala* increases significantly microbial protein synthesis in the rumen. This was due to the availability of nutrients for optimal

IOP Conf. Series: Earth and Environmental Science **287** (2019) 012018 doi:10.1088/1755-1315/287/1/012018

microbial growth and the presence of the most important microbial growth factors such as sulfur and calcium minerals in *Leucaena leucochepala* especially for cellulolytic microbial growth and fungi [11]. Increased microbial protein synthesis was also associated with a decrease in the population of protozoa in the rumen. The presence of saponins and tannins in legumes including *Leucaena luecocephala* could reduced the number of protozoa in the rumen [17]. The reduced number of protozoa in the rumen will increase the bacterial population so that microbial protein synthesis will also increase. This is in line with research conducted by [18] [19] and [20].

4. Conclusion

Supplementation of *Leucaena luecocephala* as concentrate substitution on completed feed based on ammoniated rice straw *in vitro* can increase degradability and microbial protein synthesis. Supplementation of 10% and 20% *Leucaena luecocephala* were suitable to be used for further studies, therefore *in vivo* experiment is required to study the effects on animal production.

5. References

[1] Retnani Y, Kamesworo S, Khotidjah L, Saenab A. 2010. Pemanfaatan Wafer Limbah Sayuran Pasar Untuk Ternak Domba. Seminar Nasional Teknologi Peternakan dan Veteriner, 2010 Agustus 2-3; Bogor, Indonesia. Bogor (ID): Pusat Penelitian dan Pengembangan Peternakan. 503-510.

[2] Panjaitan, T. S., 2002. Mengenal Potensi Lamtoro Hibrida F2 sebagai Sumber Pakan Hijauan Ternak. BPTP Nusa Tenggara Barat.

[3] Jayanegara, A., Wina, E., Soliva, C. R., Marquardt, S., Kreuzer, M., Leiber, F. 2011. Dependence of forage quality and methanogenic potential of tropical plants on their phenolic fractions as determined by principal component analysis. Animal Feed Science and Technology. 163: 231-243.

[4] Calsamiglia, S., Ferret, A., Reynolds, C., Kristensen, N., & Van Vuuren, A. 2010. Strategies for optimizing nitrogen use by ruminants. *Animal. 4*(7), 1184-1196.

[5] Ghosh, M. K., Samiran, B. 2007. Mimosine toxicity a problem of Leucaena feeding in ruminants. Asian Journal of Animal and Veterinary Advances. 2: 63-73.

[6] Dalzell, S. A., Burnett, D. J., Dowsett, J. E., Forbes, V. E., Shelton, H. M. 2012. Prevalence of mimosine and DHP toxicity in cattle grazing *Leucaena leucocephala* pastures in Queensland, Australia. Animal Production Science. 52: 365-372.

[7] Tilley J M A & Terry R A. 1963. A two-stage technique for the in vitro digestion of forage crops. J. Brit. Grassland Soc. 18:104-11.

[8] AOAC, 2007.Official Method of Analysis.Association of Official Analytical Chemist, Washington D.C.

[9] Makkar, H. P. S., O. P. Sharma, R.K. Dawra and S. S. Negi. 1981. Simple determinination of microbial protein in rumen liquor. J. Dairy Sci 65: 2170-2173.

[10] Steel, R. G. D. & J. H. Torrie.1993. Principles and Procedures of Statistics: A Biometrical Approach. McGraw-Hill, New York.

[11] Aregheore, E. M. 1999. Nutritive and antinutritive value of some tree legumes used in ruminant livestock nutrition in Pacific island countries. Journal of South Pacific Agriculture. 6: 50- 61.

[12] Jabbar, M. A., L. Reynolds, A. Larbi and J. Smith, 1997. Nutritional and economic benefits of Leucaena and Gliricidia as feed supplements for small ruminants in humid West Africa. Trop. Anim. Health Prod., 29: 35-47. PMID: 9090017.

[13] Aregheore, E. M. and D. Perera. 2004. Effect of supplementation of a basal diet of maize stover with *Erythrina variegata*, *Gliricidia sepium* or *Leucaena leucocephala* on feed intake and digestibility by goats. Trop. Anim. Health Prod., 36: 175-189. PMID: 14998316.

[14] Hariadi BT, Santoso B. 2010. Evaluation of tropical plants containing tannin on in-vitro methanogenesis and fermentation parameters using rumen fluid. J Sci Food Agri. 90:456–461.
[15] Patra, A. K & J. Saxena. 2010. Review A new perspective on the use of plant secondary metabolites to inhibit methanogenesis in the rumen. J. Phytochem. 71: 1198–1222.
[16] Galindo, J., Marrero, Y., Ruiz, T., González, N., Díaz, A., Aldama, A., Moreira, O., Hernández, J., Torres V., Sarduy, L. 2009. Efecto de una mezcla múltiple de leguminosas herbáceas y Leucaenaleucocephala en la población microbiana y productos fermentativos del rumen de añojos mestizos de Cebú. Revista Cubana de Ciencia Agrícola. 43: 256-264
[17] Ningrat, R. W. S., Mardiati Zain, Erpomen and Heny Suryani. 2016. Effects of doses and different sources of tannins on *in vitro* ruminal methane, volatile fatty acids production and on bacteria and protozoa populations. Asian J. AnimSci 11: 47-53.
[18] Carulla, J. E., M. Kreuzer, A. Machmüller & H. D. Hess.2005. Supplementation of Acacia mearnsii tannins decreases methanogenesis and urinary nitrogen in forage-fed sheep. Aust. J. Agric. Res. 56: 961-970.
[19] Herawaty, R., N. Jamarun, M. Zain, Arnim and R. W. S. Ningrat. 2013. Effect of Supplementation *Sacharonyces cerevisiae* and *Leucaena leucocephala* on Low Quality Roughage Feed in Beef Cattle Diet. Pakistan. J. Nut. 12 (2): 182-184.
[20] Zain M, J. Rahman, Khasrad and Erpomen, 2016. Supplementation of Saccharomyces cerevisiae and Sapindus rarak in diet based of oil palm frond (OPF) on nutrient digestibility and daily weight gain of goat. Asian J. Anim. Vet. Adv., 11: 314-318.

Acknowledgments
This study was supported by Professor Research Grant Cluster by BOPTN Andalas University contract No. 33/UN.16.17/PP.KP.RGB/LPPM/2018 The research would not have been possible without the cooperation of my post graduated students and technical assistance of Laboratory of Ruminant Nutrition, Faculty of Animal Science of Andalas University.

Comparison of *in vitro* digestibility and rumen fluid characteristics between the tithonia (*Tithonia diversifolia*) with elephant grass (*Pennisetum purpureum*)

Novirman Jamarun, Roni Pazla, Mardiati Zain and Arief

Faculty of Animal Science, Andalas University, Padang, West Sumatera, Indonesia.
Email: novirman55@gmail.com.

Abstract. The objective of this study was to evaluate the effects of the tithonia (*Tithonia diversifolia*) and elephant grass (*Pennisetum purpureum*) on *in vitro* digestibility and rumen fluid characteristics. This research was carried out using a randomized block design with 3 treatments (the level of tithonia and elephant grass) and 5 replications. The following treatments were performed: T1 = 100% tithonia; T2 = 100% elephant grass; T3 = 50% tithonia + 50 % elephant grass. The data were subjected to an analysis of variance (ANOVA), and differences between the treatment means were tested using Duncan's Multiple Range Test (DMRT). The parameters measured were as follows: dry matter digestibility (DMD) (%), organic matter digestibility (OMD) (%), pH, volatile fatty acid (VFA) (mM) and NH_3. The results revealed that DMD, OMD and VFA were significantly ($P<0.01$) increased by the 100 % tithonia (T1), pH were lowest ($P<0.01$) in T1. However, T1 and T3 non-significantly affected ($P>0.05$) NH_3 Concentration. 100% tithonia resulted in the best *in vitro* digestibility and rumen fluid characteristics.
Keywords – *In vitro, Tithonia diversifolia, Pennisetum purpureum,* digestibility, rumen fluid characteristics.

1. Introduction

Exploration of sources of ruminant animal feed is usually needed for the main work of the world of nutritional nutrition lately. Production limitations are needed by scientists to find alternatives that can be used by farmers in their application. One of the feed commodities that can be used as alternative feed is tithonia plants. There are still few of these plants that are used for ruminant feed because these plants have been mostly used in agriculture, namely composting with sufficient content such as Nitrogen (N) and Phosphorus (P) which are quite high [1]. High N and P content is very good for ruminants. High N content includes high crude protein content, while very high P content is needed by rumen microbes for growth [2] [3]. Both of these things are needed by the rumen microbes to grow and develop. Rumen microbes are the main engine of ruminants.

There is a problem in the use of P on the tithonia because P is not available due to being bound to phytic acid. Phytic acid is the most antinutrient in tithonia [4]. In monogastric livestock the presence of phytic acid in feed ingredients is a problem because there is no enzyme that can break down phytate in the digestive tract, so that it will be a threat to livestock health. Ruminants have the advantage of monogastric in the process of phytic degradation. Rumen microbes are able to produce pytase enzymes to break P bonds in phytic acid so that p can be available to livestock and become the main mineral supporting microbial growth [5]. This study aims to evaluate the digestibility and rumen fluids characteristics of tithonia and elephant grass and evaluate the effect of the combination level of the two feed ingredients.

IOP Conf. Series: Earth and Environmental Science **287** (2019) 012019 doi:10.1088/1755-1315/287/1/012019

2. Materials and methods

For this study, Analysis of in vitro digestibility and rumen fluid fermentability was conducted in the Laboratory of Dairy Nutrition of the Bogor Agricultural Institute. Samples of T and EG were collected around the town of Padang. All samples were finely ground and mixed according to the treatment combinations. The parameters measured in this study were DM digestibility (DMD), OM digestibility (OMD), NH_3 concentration, total VFA content and pH. The chemical composition of *Tithonia diversifolia* and elephant grass is presented in Table 1.

Tabel 1. The chemical composition of *Tithonia diversifolia* and elephant grass.

Composition (%)	Feed Material	
	Elephant Grass	*Tithonia diversifolia*
Dry Matter	21.23	25.57
Crude Protein	10.88	22.98
Crude Fiber	32.77	18.17
Fat	2.48	4.71
Ash	10.54	15.99
NDF	66.57	61.12
ADF	41.71	40.15
Cellulose	34.18	34.59
Lignin	6.29	4.57

Source: Nutritional Analysis of Ruminant Nutrition Laboratory Faculty of Animal Husbandry Andalas University (2017).

2.1. *In vitro procedure*

The *in vitro* procedure in this study followed that of [6] Rumen fluid was taken from a cow fistula at Bogor, LIPI. The fermenter tube was filled with 0,5 g of sample and 40 mL of McDougall solution was added. The tube was placed in a shaker bath at 39^0C, filled with 10 mL of rumen fluid and shaken with CO_2 for 30 second, the pH was checked (6,5-6,9) and the sample was covered with a ventilated rubber cap and fermented for 48 h. After 48 h, the rubber cap of the fermenter tube was removed and 2-3 drops of $HgCl_2$ were added to kill the microbes. The fermenter tube was centrifuged at 5.000 rpm for 15 min. The substrate was separated into a precipitate layer at the bottom and a clear supernatant at the top. The supernatant was removed, the resulting sediment was centrifuged at 5.000 rpm for 15 min and 50 mL of 0,2% pepsin-HCl solution was added. This mixture was then reincubated for 48 h without a rubber cap. The remaining digested residue was filtered using Whatman filter paper No. 41 (identified by weight) with the help of a vacuum pump. The precipitate on the filter paper was placed in a porcelain dish that was placed in a 105EC oven for 24 h. After 24 h, the porcelain cup + filter paper + residue was removed, inserted into a desiccator and weighed to determine the DM content. Furthermore, the ingredients in the cup were placed in a kiln or in an electric furnace for 6 h at 450-600^0 C and weighed to determine the amount of organic material. Residues originating from fermentation without feed ingredients were used as a control. The proximate analysis of materials and residues followed the [7] and [8] procedure.

2.2. *Rumen fluid characteristics*

The measured rumen fluid characteristics were as follows: Ruminal pH, NH_3 –N concentration and VFA content. Fermenter tubes were each filled with 0.5 g of sample and 40 mL of buffer solution and 10 mL of fresh rumen fluid (ratio of 4:1) were added. After the tube received CO_2, the tubes were closed with a rubber ventilator. Tubes were inserted into the fermenter shaker water bath at a temperature of 39^0 C and incubated for 4 h, followed by analyses of pH, NH_3 and VFA. The ruminal pH was measured using a pH meter. The NH_3 -N concentration was measured using the micro diffusion (Conway) method and the total VFA content was measured using the steam distillation method.

3. Result and discussion

Data on digestibility values and characteristics of rumen fluid treatment are presented in Table 2.

Table 2. Data on digestibility values and characteristics of rumen fluid treatments.

Parameters	Treatments		
	T1	T2	T3
Digestibility (%)			
Dry matter (DMD)	58,56^{a}	55,99^{b}	57,12^{c}
Organic Matter (OMD)	55,46^{a}	53,99^{b}	54,72^{c}
Rumen Fluid Characteristics			
pH	6,78^{a}	6,79^{b}	6,80^{c}
VFA (mM)	125,88^{a}	87,53^{b}	112,15^{c}
NH_3 (mM)	22,48^{ab}	20,41^{a}	22,69^{b}

Means in the same row with different letter (a,b and c) are significant (P<0,01).

Table 2 shows that the use of 100% tithonia plants (T1) was very significant (P <0.01) increasing DMD and OMD. The lowest DMD and OMD were obtained on 100% elephant grass (T2). The high level of DMD and OMD in tithonia plants is due to the tithonia nutrient content better than elephant grass. This can be seen from the level of protein, the amount of amino acids and mineral content. [9] reported that tithonia contains a lot of amino acids. [1] added that the tithonia leaves contain high levels of N, P, K, Ca and Mg minerals. The level of lignin titone is lower when compared to elephant grass. Amino acids found in tithonia are needed by rumen microbes in their growth, while mineral content such as P and Mg greatly stimulates the rumen microbes to grow and develop [10]. The development of rumen bacteria will be positively correlated with an increase in DMD and OMD [11]. Lower lignin in plants allows more bacteria to degrade feed so that DMD and OMD are higher in this treatment. [11] reported a positive correlation between decreasing lignin and increasing DMD and OMD. The combination between tithonia and elephant grass in treatment T3 is better than treatment using elephant grass (T2). This is due to the contribution of the nutrient content of the tithonia plant to be able to supply the rumen microbial needs better so that it increases DMD and OMD in this treatment.

The pH value in this study is the pH value that is still normal for the growth and development of rumen microbes. According to [12] the optimal pH range for cellulose digestion is 6,4-7. If the pH of the rumen fluid drops to below 6 it will inhibit microbial activity. Furthermore, [13] stated that pH greater than 7,1 can reduce the microbial population drastically so that the energy produced is low. Conditions for the rumen microbes to be able to perform activities optimally when the rumen pH is in normal conditions is 6-6,9 [14]. [15] reported that a mixture of several levels of elephant grass combination with different tithonia produced a normal pH value of 6,67-6,68. pH at T1 treatment showed the lowest pH value. Low pH in this treatment is due to the amount of nutrient content of the VFA produced. High VFA content will reduce pH. In accordance with the opinion of [16] that the higher the VFA value, the more other organic acids will be produced so that the rumen fluid pH will be low.

The highest VFA value (P <0.01) is found in T1. VFA describes the amount of carbohydrates degraded by rumen microbes. The growth and development of rumen microbes is thought to be very good in T1 due to a better supply of nutrients in the form of amino acids and minerals from tithonia plants. As reported by [9] that Tithonia contains a lot of essential amino acids for the growth of microbes such as methionine, leucine, isoleucine and valine. Rumen liquid VFA production can be used as a benchmark for feed fermentability, where the higher the level of feed fermentability, the greater the VFA produced [17].

The highest NH3 concentration was produced in T3 treatment but not significantly different (P> 0.05) with T1. The high content of crude protein in the tithonia and the number of amino acids contained in the tithonia will positively correlate to NH_3 concentrations. NH_3 concentration is one indicator to determine feed fermentability which is related to protein digestibility, rumen microbial activity and population [18]. NH_3 in all three treatments is still within normal limits. [19] states that ammonia concentration to support rumen microbial growth is 6 mM – 21 mM. NH_3 and VFA which increase linearly will form more microbial proteins so that they have a positive effect on livestock digestibility and productivity.

4. Conclusion

100% tithonia resulted in the best *in vitro* digestibility and rumen fluid characteristics.

5. References

[1] Jama, B. A., C. A. Palm., R. J. Buresh., A. I. Niang., C. Gachengo., G. Nziguheba., and B. Amadalo. 2000. *Tithonia diversifolia* as a Green Manure for Soil Fertility Improvement in Western Kenya: aReview. Agroforestry Systems. 49: 201-221.

[2] Pazla, R., M. Zain., I, Ryanti and A. Dona. 2018. Supplementation of Minerals (Phosphorus and Sulfur) and Saccharomyces cerevisiae in a Sheep Diet Based on a Cocoa By-product. *Pak. J. Nutr.,* 17: 329-335.

[3] Zain, M., J. Rahman and Khasrad. 2014. Effect of Palm Oil by Products on *In Vitro* Fermentation and Nutrient Digestibility. *Anim. Nutr. Feed Technol.,* 14: 175-181.

[4] Fasuyi, A. O., F. A. S. Dairo and F. J. Ibitayo. 2010. Ensiling wild sunflower (*Tithonia diversifolia*) leaves with sugar cane molasses. *Livest. Res. Rural Dev.,* Vol. 22, No. 3.

[5] Lamid, M. 2012. Karakterisasi Enzim Fitase Asal Bakteri Rumen *(Actinobacillus sp dan Bacillus pumilus)* dan Analisis SEM terhadap Perubahan Struktur Permukaan Dedak Padi untuk Ransum Ayam Broiler. Universitas Airlangga. (Unpublished).

[6] Tilley, J. M and R. A. Terry. 1969. A Two Stage Technique for *In Vitro* Digestion of Forage Crops. J Br. Grassland Society 18 (2): 104–111.

[7] AOAC. 1995. Official Methods of Analysis. 15th Edition. Association of Official Analytical Chemists. Association of Official Analytical Chemists, Washington, D.C.

[8] Van Soest, P. J., J. B. Robertson and B. A. Lewis. 1991. Methods for Dietary Fiber, Neutral Detergent Fiber and Non-Starch Polysaccharides in Relation to Animal Nutrition. *J. Dairy Sci.* 74: 3583-3597.

[9] Oluwasola, T. A. and F. A. S. Dairo. 2016. Proximate composition, amino acid profile and some anti-nutrients of *Tithonia diversifolia* cut at two different times. *African Journal of Agricultural Research.* Vol. 11(38), pp. 3659-3663.

[10] Febrina, D., N. Jamarun., M. Zain and Khasrad. 2016. The Effects of P, S and Mg Supplementation of Oil Palm Fronds Fermented by *Phanerochayte chrysosporium* on Rumen Fluid Characteristics and Microbial Protein Synthesis. *Pakistan Journal of Nutrition* 15(3): 299-304.

[11] Jamarun, N., M. Zein, Arief and R. Pazla, 2018. Populations of rumen microbes and the *in vitro* digestibility of fermented oil palm fronds in combination with Tithonia (*Tithonia diversifolia*) and elephant grass (*Pennisetum purpureum*). Pak. J. Nutr., 17: 39-45.

[12] Erdman, R. A. 1988. Dietary Buffering Requirements of Lactating Dairy Cows. A Review. *J. Dairy Sci.* 71:3246-3246.

[13] Van Soest, R. J. 1982. Nutritional Ecology of the Ruminant Metabolism Chemistry and Forage and Plant Fiber. Cornell University, Oregon, USA.

[14] Kamra, D. N. 2005. Rumen Microbial Ecosystem. Special Section: Microbial Diversity. *Current Science.* 89(1):124-135.

[15] Jamarun, N., Elihasridas., R. Pazla and Fitriyani. 2017. *In Vitro* nutrients digestibility and rumen fluid characteristics of the combination Tithonia (*Tithonia difersivolia*) and napier grass (*Pennisetum purpureum*). Proceedings of the 3th Nasional Seminar on Cows and Buffalo, Oktober 4-5, 2017, Padang, Indonesia.

[16] Pazla, R., N. Jamarun., M. Zain and Arief. 2018. Microbial protein synthesis and *in vitro* fermentability of fermented oil palm fronds by *Phanerochaete chrysosporium* in combination

with tithonia (*Tithonia diversifolia*) and elephant grass (*Pennisetum purpureum*). Pak. J. Nutr.,17: 462-470.

[17] Jamarun, N and M. Zain. 2013. Dasar Nutrisi Ruminansia. Penerbit Jasa Surya. Padang.

[18] Muhktarudin dan Liman. 2006. Penentuan Tingkat Penggunaan Mineral Organik untuk Memperbaiki Bioproses Rumen pada Kambing secara *In Vitro. Jurnal Ilmu-Ilmu Peternakan Indlrfesia.* 8(2): 132-140.

[19] Mc. Donald, P., R. A. Edwards, J. F. D. Greenhalgh and C. A. Morgan. 2010. Animal Nutrition. 7th Edition. Longman. Scientific and Technical John Willey and Sons. Inc. New York.

Acknowlecgments

This research was funded by the Andalas University Profesor Grant 2018 and was part of the Ph.D. dissertation of Roni Pazla in Animal Nutrition, Faculty of Animal Science, Andalas University.

IOP Conf. Series: Earth and Environmental Science **287** (2019) 012020 doi:10.1088/1755-1315/287/1/012020

The effect of cocoa pod fermentation with mol of rumen content on fiber fraction component and in vitro digestibility

Nurhaita[1], Neli Definiati[1], Syahro Ali Akbar[2], and Urip Santoso[3]

[1] Department of Animal Science, Faculty of Agriculture, University of Muhammadiyah, Bengkulu, 38119, Indonesia.
[2] Department of Animal Science, Faculty of Agriculture, University of Muhammad Yamin, Solok, West Sumatera, 27321, Indonesia.
[3] Department of Animal Science, Faculty of Agriculture, University of Bengkulu, 38119, Indonesia.
Corresponding author: nurhaita@gmail.com

Abstract. This study aimed to evaluate the fiber fraction component and in-vitro digestibility of cocoa pod husk fermented by local microorganisms (MOL), namely the rumen contents. The research used completely randomized design with 5 treatments of 4 replications each. The dose of MOL to ferment pod cocoa was as follows A: 0 ml/kg of substrate, B: 3 ml/kg of substrate C: 6 ml/kg substrate, D: 9 ml/kg of substrate, and E: 12 ml/kg of substrate. The parameters measured were the fiber fraction component of fermented cocoa pod and in vitro digestibilities of dry matters and NDF. The results showed that the dose of MOL significantly ($P<0.05$) affected The NDF, ADF and hemicellulose contents, but it had no effect on cellulose and lignin contents ($P>0.05$). In vitro dry matters and NDF digestibilities were significantly ($P<0.05$) affected by the treatments. In conclusion, fermentation by MOL at a dosage of 12 ml/kg substrate reduced fiber fraction component and increased in–vitro digestibilities of dry matter and NDF.
Keywords – cocoa pod fermentation, rumen content, fiber fraction, in vitro, digestibility.

1. Introduction

Cocoa pod has enough potential as alternative feedstuffs to substitute grass, because the production of cocoa is numerous and concentrated in specific areas. An increased cocoa plantation will annually produce cocoa pod waste more. In 2015, an extensive cocoa plantation in Bengkulu Province has reached 13,125 hectares with a production of 4,039 tons/year [1], and produce a cocoa pod waste as much as 3.029,25 tons.

Cocoa pod waste has not been used, and cause environmental pollution. In addition, pod cocoa is usually allowed to pile up around the cocoa plant, which is a very good place for the development of *Phytoptora palmifora* causing black pod disease. The usage of cocoa pod as animal feedstuffs will provide a solution to address the problem of environmental pollution and conserve the environment. It also can break the chain of disease transmission for cocoa plants.

Cocoa pod has a good nutritional value for animal feedstuffs particularly ruminants. However, a high lignin content (31.51%) and *theobromine*, which is toxic become a limiting factor in its use as animal feedstuffs. To optimize the usage of cocoa pod as feedstuffs, they should be processed. There are some processing techniques which has beed proven to improve the value of the by product [2] [3]

[4] [5] [6] [7] [8] [9] [10] in order to improve the digestibility and eliminate toxins contained in the cocoa pod, the cocoa pod could be fermented.

Fermentation is the process of processing the material with the help of microbes capable of breaking down complex components into simpler forms, such as cellulose and hemicellulose to glucose [11]. Fermentation waste fiber materials ranging frequent lately, because in addition to more easily and cost are also safer and more environmentally friendly than the use of chemicals. The feedstuffs that undergo fermentation usually have better nutritional value of its origin, caused by microorganisms that break down the components of the complex into simple substances that are so easy to digest. In addition, fermentation can also increase the crude protein feed ingredients, improve palatability because it produces a fragrant smell and eliminate toxins, microorganisms can also synthesize certain vitamins such as riboflavin, vitamin B 12, pro-vitamin A and other growth factors. Thus, the fermentation treatment is expected to improve the quality of waste palm fronds for the better.

Fermentation with local microorganisms is one alternative and lately often used. Local microorganisms in the form of a solution are fermented from a variety of waste materials. The local microorganism solution contains bacteria and fungi which could degrade organic materials. The most benefit of local microorganism is no cost for local microorganism which can be made from the waste of fruits and vegetables, livestock waste, slaughterhouse waste or household waste. In addition, the manufacturing process is easy and applicable. In this study, local microorganism is derived from rumen contents. [12] showed that there are 8 thermophilic bacteria isolated from MOL rumen contents including gram-positive bacteria. Fermented cocoa pod with rumen contents is expected to improve the quality of the cocoa pod and eliminate toxic compounds. [13] showed that fermented cocoa pod with local microorganisms of rumen contents proved to increase the nutrient value of cocoa pod.

2. Materials and methods

2.1. Local microorganism manufacture of rumen contents

Local microorganism was made from rumen contents as a source of microorganisms and enriched with coconut water and palm sugar as a source of energy. The materials used were 10 liters of coconut water was filtered, 2 kg of palm sugar that has been dissolved and 2 kg of rumen contents. All the ingredients put into a container and then closed. The container was perforated to insert the tube. Another container was filled with water. Connect the first container with a small tube into the second container to to drain the gas formed, incubated for 10 days.

2.2. Fermentation of cocoa pod

Cocoa pod was chopped into 3-5 cm size, dried under the sun until its moisture content of 60%. Cocoa pod was then added rice bran as much as 10% and 1% sugar and local microorganisms according to each treatment and mixed. the cocoa pod was then put into a plastic bag, tied and incubated for 7 days. After 7 days of incubation, they were then opened, and measured pH, fungi, texture and smell. After that, they were then dried under the sun, and ground into flour. The results of this fermentation were then analyzed their nutrient content.

2.3. The experimental treatments

The present study used completely randomized design with 5 treatments of 4 replications each. The dose of MOL to ferment pod cocoa was as follows A: 0 ml/kg of substrate, B: 3 ml/kg of substrate C: 6 ml/kg substrate, D: 9 ml/kg of substrate, and E: 12 ml/kg of substrate. The content of fiber fractions was measured by the method of [14], whereas in vitro digestibility was measured using the Daisy^{II} method as described by ANKOM Technology Corporation.

3. Results and discussion

3.1. The content of fiber fractions

The content of fiber fractions of fermented cocoa pod is presented in Table 1.

Table 1. The content of fiber fraction of fermented cocoa pod.

Treatment	Parameters (%)				
	NDF	ADF	Hemicellulose	Cellulose	Lignin
A	70.66^a	60.96^a	9.71^a	26.65	34.31
B	69.18^a	60.23^a	8.95^a	26.14	34.09
C	68.44^a	59.79^a	8.65^a	25.23	33.82
D	68.16^a	59.39^a	8.77^a	26.66	32.73
E	61.75^b	55.17^b	6.57^b	23.66	31.51
SE	0.34	0.22	0.27	0.44	0,56

The different superscripts in the same column indicate significantly different (P <0.05).

The unfermented cocoa pod had 71.06% NDF, 60.90% ADF, 10.17% hemicellulose, 25.04% cellulose and 35.29% lignin. The content of fermented fiber fraction tends to lower than the unfermented cocoa pod.

Experimental results showed that the dosage of MOL significantly (P <0.05) affected the contents of NDF, ADF and hemicellulose, but not significantly (P> 0.05) affected the content cellulose and lignin. The content of NDF, ADF and hemicellulose in E group treatment was significantly lower than other treatments.

The reduction in the fraction of the content of fiber in fermented cocoa pod occurred due to the microorganisms converted the complex molecules into simpler compounds. In this case, the bacteria contained in MOL converted the NDF, ADF into hemicellulose and cellulose, finally into glucose. This indicates that the MOL of rumen contents contains bacteria that have cellulolytic activity which can break down the components of fiber fraction.

Fraction fibers which are composed of NDF, ADF hemicellulose, and cellulose is a component of plant cell walls that can be digested by ruminants and is a major source of ruminant feed. However, the lignin binds the hemicellulose and cellulose to form lignohemicellulose and lignocellulose. This cause hemicellulose and cellulose could not be degraded by rumen microbes. Cell wall structure of cocoa pod have lignified further and its lignin has crystallized making it difficult to digest.

The reduction in NDF, ADF and hemicellulose contents of fermented cocoa pod reached 13.10% (71.06% vs 61.75), 9:41% (60.9% vs 55.15%) and 35.40% (10.17% vs 55.17%), respectively when compared with the unfermented cocoa pod. The largest decrease occurred in the treatment E, whereas in other treatment groups did not significantly differ from the unfermented cocoa pod. This suggests that a dose of MOL for fermentation should be higher, which means it contains more bacteria to degraded fiber components more.

3.2. The digestibility of dry matter and NDF

The digestibility of dry matter and NDF of fermented cacao pod are presented in Table 2.

Table 2. The digestibility of dry matter and NDF of fermented cacao pod (%).

Treatment	Parameters	
	Dry matter	NDF
A	42.97^a	16.14^a
B	42.65^a	17.09^a
C	42.65^a	18.55^a
D	42.44^a	19.55^a
E	55.89^b	35.67^b
SE	0.74	1.34

The different superscripts in the same column indicate significantly different (P <0.05).

Experimental results showed that the treatments significantly affected (P<0.05) the dry matter and NDF digestibilities. It was shown that E had higher dry matter and NDF digestibilities than A, B, C, and D (P<0.05), whereas A, B and C and D did not differ significantly (P>0.05) to each other. Dry matter and NDF digestibilities in E was increased at a level of 31.69% and 121.0%, respectively. This phenomenon may closely relate to a decrease in the content of fiber fractions (NDF, ADF and hemicellulose) as a result of the activity of microorganisms, which convert the complex molecules into simple molecules during the fermentation process.

Digestibility is a reflection of the quality of the feedstuffs, which describes what percentage of substances that are digested and what percent is excreted through feces. Substances contained in the feedstuffs are not entirely available to the body, part of them will be excreted through the feces. The higher digestibility of feedstuffs will result the higher value of feedstuffs for livestock.

The values of dry matter and NDF digestibilities of fermented cacao pod were low, ranging from 42% -55% and 16.14% -35.67%, respectively.

4. Conclusion

In conclusion, fermentation by MOL at a dosage of 12 ml/kg substrate reduced fiber fraction component and increased digestibilities of dry matter and NDF.

5. References

[1] Dinas Perkebunan, 2017. Direktorat Statistik Perkebunan Indonesia, Komoditas Kakao 2015-2017. http://ditjenbun.pertanian.go.id/tinymcpuk/gambar/file/statistik/2017/Buku-Kakao-2015-2017.pdf. Accessed 18 Sept 2018.

[2] Akbar, S. A., N. Jamarun, R. Saladin dan Mardiati, 2005. Pengaruh fermentasi dan defaunasi tandan kosong sawit terhadap kandungan gizi, kecernaan dan karakteristik cairan rumen in-vitro. Jurnal Ilmiah Ilmu-ilmu Peternakan Vol VIII no 2 : 132-144.

[3] Astuti, T., Y. S. Amir., F. Yelni and Isyaturriyadhah. 2014. The Result of Biotechnology by Local Microorganisms to Banana Peel on Rumen Fluid Characteristics as Ruminant Feed. Journal of Advance Agricuture Technology. Vol 1 no 1: 28-31.

[4] Nurhaita, N. Jamarun, R. Saladin, L Warly dan Mardiati Z. 2007. Efek beberapa metoda pengolahan limbah daun kelapa sawit terhadap kandungan gizi dan kecernaan secara *in-vitro.* J. Ilmu-Ilmu Pertanian Indonesia Ed. Khusus Dies natalis ke 26 Unib. No 2. P: 139-144.

[5] Nurhaita, 2008. Evaluasi dan Pemanfaatan Daun Kelapa Sawit dalam Ransum Ternak Ruminansia. *Disertasi* Program Pascasarjana Universitas Andalas, Padang.

[6] Nurhaita, N. Jamarun, L. Warly dan M. Zain. Nurhaita, 2010. Kecernaan ransum domba berbasis daun sawit teramoniasi yang disuplementasi sulfur, fosfor, dan daun ubi kayu. J. Media Peternakan Vol 33 No. 3. P: 144 -149.

[7] Nurhaita, N. Definiati, R. Zurina dan Edi, E. 2011. Nilai gizi dan kecernaan pelepah sawit fermentasi (evaluasi secara in vitro). Prosiding Seminar Nasional Peternakan "Prospek dan Potensi Sumber Daya Ternak Lokal dalam Menunjang Ketahanan Pangan Hewani". Fakultas Peternakan UNSOED Purwokerto, 15 Oktober 2011.

[8] Nurhaita, W. Rita, N. Definiati dan R. Zurina, 2012. Fermentasi Bagase Tebu dengan *Neurospora sitophila* dan Pengaruhnya Terhadap Nilai Gizi dan Kecernaan Secara in Vitro. Jur. Embrio Vol 5 No 1 P:1-7.

[9] Nurhaita, W. Rita, Ruswendi dan Robiyanto. 2014. Optimizing Rumen Bioprocess through Supplementation of Microbe Precursor Nutrient in Ammoniation of Palm Oil Frond-Base Cattle Ration. Journal of Advanced Agricultural Technologies Vol. 1, No. 1, P: 10-13.

[10] Zain. M., Jamarun, Suryahadi dan Nurhaita. 2006. Fermentabilitas dan kecernaan *in-vitro* serbuk sabut kelapa yang difermentasi dengan mikrob rumen. Jurnal Ilmiah Ilmu-ilmu Peternakan. Vol. IX No 1: 39 – 49.

[11] Winarno, F. G., S. Fardiaz and D. Fardiaz. 1980. Pengantar Teknologi Pangan. PT. Gramedia. Jakarta.

[12] Astuti, T., Y. Amir, Irdawati and U. Santoso, 2016. Nutritional improvement of palm oil fronds for ruminant feedstuffs using a local biotechnological approach. Pak. J. Nutr., 15: 450–454.
[13] Nurhaita, and D., Neli and Suliasih, Suliasih. 2016. Fermentation with local microorganism to improve pod cacao quality as ruminants feed. In: Proceeding ISEPROLOCAL. Badan Penerbitan Fakultas Pertanian Universitas Bengkulu, Bengkulu, Indonesia, pp. 218-221.
[14] Goering, H. K. and P.J. Van Soest. 1970. Forage Fiber Analysis. Apparatus, Reagent, Procedures and Some Applications. Handbook. ARS-USDA, Washington, D.C.

Acknowledgement

The study was supported by Ditlitabmas, Direktorat Jenderal Pendidikan Tinggi through Hibah Bersaing Program year 2015-2016. The study would not have been possible without the cooperation of LABTIAP – BPPT, Serpong.

Effect of addition garlic flour as feed additive in digesta viscosity, microflora, and intestinal characteristic of native chicken crossbred

Sjofjan O[1], Irfan H D[1], Halim M H[1], Hanitawati[2], and Teguh H[2]

[1] Lecturer of Animal Nutrition Department, Animal Science Faculty, Brawijaya University
[2] Student of Animal Science Faculty, Brawijaya University
E-mail: osofjan@yahoo.com

Abstract. The research purpose was to determine the garlic flour effect as the feed additive on digesta viscosity, microflora, and intestinal characteristic of broiler. The research method was used completely randomized design with 5 treatments and 5 replicates. The materials used for this research were 300 unsex day old chicks with native chicken crossbreed strain with average body weight 40.11±9.13 g/head. The treatments used for research were dietary with T0 (basal feed), T1 (basal feed + 0.25% garlic flour), T2 (basal feed + 0.30% garlic flour), T3 (basal feed + 0.35% garlic flour), T4 (basal feed + 0.40% garlic flour). The parameters observed were digesta viscosity, intestinal characteristic bacteria (lactic acid bacteria, Escherichia coli, and Salmonella sp.), and (villus length, crypt depth, and villus surface). The data analysis was the analysis of variance (Anova) and continued by Duncan Multiple Range Test. The results showed that using the garlic flour effect as feed additive has significant difference (P<0.05) on viscosity digesta, intestinal characteristic (crypt depth and villus surface) (Salmonella sp.) and significantly different (P<0.01) (Lactic acid bacteria and Escherichia coli). The addition of 0.45% garlic flour gave the best effect on digesta viscosity, microflora, and intestinal characteristic of native chicken.
Keywords – garlic, microflora, intestinal characteristic, villus, native chicken crossbred.

1. Introduction

Sources of protein can be obtained from livestock products e.g. meat. There are several commodities that produce meat product e.g. chicken, cattle, and goat. However, chickens are given biggest sources of protein resources from white meat products. The quality of native chicken carcass mostly from several factors. The feed is the effect to the quality of native chicken. The using feed additive is one method to improve the quality of feed. The native chicken has lower fat content than broiler. The productivity of native chicken crossbred is low due to the extensive method rearing. The antibiotics are given as growth promoter but antibiotics cause bacterial resistance and residue in the carcass [1].

Nowadays, feed additive use as substitute from antibiotics due it's safer. The used of feed additive can increase the absorption nutrient in the native chicken that will affect the internal organs. The indicator performance (feed intake, body weight, feed conversion ratio, mortality, and production index) and enzyme activities (protease and amylase), intestinal characteristic (villus total and villus height) [2]. Garlic can be used as feed additive to optimizing the absorption. The garlic flour as feed additive producing the allicin and diallyl sulfide that can inhibit microbial pathogen growth.

Probiotics, prebiotics, and medicinal plants as natural feed additives are currently used in poultry diets to enhance the intestinal characteristics and immune response of chickens [3]. According to the [4] stated the garlic (*Allium sativum*) has bioactive components like sulfur containing compounds (Allin, Diallysulfides, and Allicin) that act as antibacterial, antifungal, anti-parasite, antiviral, antioxidant, antithrombotic, ant cancerous, and vasodilator characteristics. Garlic has several beneficial effects on chickens having antimicrobial, antioxidant as well as antihypertensive properties. These functions were attributed to bioactive components [4].

2. Materials and methods

2.1. Materials

The principal equipment for the research are litter with 100x100x70 cm, brooder, hygrometer, thermometer, analytic scale, the postal cage with closed house, feed bunk, and drink bunk. The principal materials are 300 unsex day old chicks with native chicken crossbred strain with average body weight 40.11±9.13 g/head produced by local breeder in Blitar, East Java Province.

2.2. Methods

The research method was used completely randomized design with 5 treatments and 5 replicates. The replicates consist 12 heads of native chicken crossbred with total (300heads). The level of dietary and garlic flour were:

T_0= (basal feed)
T_1= (basal feed + 0.25% garlic flour)
T_2= (basal feed + 0.30% garlic flour)
T_3= (basal feed + 0.35% garlic flour)
T_4= (basal feed + 0.40% garlic flour)

2.3. Variables

The variables observed were digesta viscosity, intestinal characteristic bacteria (lactic acid bacteria, Escherichia coli, and Salmonella sp.), and (villus length, crypt depth, and villus surface).

2.4. Data Analysis

The data analysis using analysis of variance (Anova) and continued by Duncan's Multiple Range Test [5].

3. Results and discussion

3.1. The garlic flour in the digesta viscosity and intestinal characteristic bacteria

Table 1. Digesta viscosity and intestinal characteristic bacteria.

CODE	Variables			
	Digesta viscosity*	Villus length*	Crypt depth* (μm)	Villus surface* (μm)
T_0	$10.78{\pm}0.35^{a}$	$485.71{\pm}92.46^{b}$	$140.87{\pm}17.42^{b}$	$759.11{\pm}95.75^{b}$
T_1	$11.23{\pm}0.4^{a}$	$381.60{\pm}49.59^{a}$	$148.46{\pm}8.57^{b}$	$630.52{\pm}119.44^{b}$
T_2	$10.48{\pm}1.64^{a}$	$453.77{\pm}27.40^{ab}$	$142.50{\pm}17.6^{b}$	$787.05{\pm}53.23^{b}$
T_3	$17.64{\pm}2.21^{c}$	$406.96{\pm}35.19^{ab}$	$121.96{\pm}11.63^{a}$	$794.50{\pm}32.14^{b}$
T_4	$14.45{\pm}4.48^{b}$	$532.02{\pm}93.24^{ab}$	$148.02{\pm}9.23^{b}$	$1073.49{\pm}99.92^{c}$

** Superscript showed significantly different (0.01).
*Superscript showed significant different (0.05).

3.1.1. The garlic flour effect in digesta viscosity

Based on table 1 result from the research on the digesta viscosity activities showing the T_3 higher are (17.64±2.21) with (basal feed + 0.35% garlic flour) and lower T_0 are (10.78±0.35) (basal feed). The treatment from lower to higher is T_0, T_2, T_1, and T3. The average viscosity digesta are significant different (P<0.05). The used garlic flour is impacted to the viscosity digesta.

The viscosity digesta are fluidal system in the intestinal that compared between the gradient of the digestive tract. The higher of the viscosity digesta has negative effect to the efficiency digestive tract due to slowly the endogenous enzyme diffusion that react with the substrate and nutrient in intestinal wall [6]. There are several factors that affect to the digesta viscosity e.g. pH [7].

According to [8] stated if the viscosity of digesta higher the native chicken is one product which has a high nutritional value that can be accepted while public awareness of the native chicken which is a conventional food and medicine due to its bioactive components form of functional food which able to prevent the disease. That depressed absorption of nutrients, including amino acids are more common such as this might cause by an increase in digesta viscosity and a reduction in the movement of nutrients through a dense of unstirred water layer rather than changes in the transport capacity of the mucosal membranes [9].

3.1.2. The garlic flour effect in villus length

Based on table 1 result from the research on the villus length showing the T_4 higher are (532.02±93.24) with (basal feed + 0.40% garlic flour) and lower T_1 are (406.96±35.19) (basal feed + 0.25% garlic flour). The treatment from lower to higher is T_1, T3, T_2, and T4. The average villus length are significant different (P<0.05). The used garlic flour is impacted to the villus length due to the level of garlic given.

The viscosity digesta are fluidal system in the intestinal that compared between the gradient of the digestive tract. Stated the higher of the viscosity digesta has negative effect to the efficiency digestive tract due to slowly the endogenous enzyme diffusion that react with the substrate and nutrient in intestinal wall [6]. There are several factors that affect to the digesta viscosity e.g. pH. The garlic flour is affect to the variable due to bioactive compound that inhibit and decreasing the microbial pathogen that villus growth are optimum. The bioactive compound is allicin that decreasing the Aspergillus flavus and Aspergillus parasiticus.

3.1.3. The garlic flour effect in crypt depth

Based on table 1 result from the research on the villus length showing the T_1 higher are (148.46±8.57) with (basal feed + 0.25% garlic flour) and lower T_3 are (121.96±11.63) (basal feed + 0.35% garlic flour). The treatment from lower to higher is T_3, T_2, T_4, and T1. The average crypt depth are significant different (P<0.05). The used garlic flour is impacted to the crypt depth due to the level of garlic flour given.

The garlic flour affected increased due to the bioactive compounds that consists and synergetic with basal feed. That the garlic flour stimulate the growth of the native chicken, increasing the optimizing of internal organ, that affecting to the nutrient absorbing in other wise, has function to balancing the antibody and immune system of chicken [4]. The garlic flour affected to the consistency of the feed increase in otherwise the garlic has hygroscopic ability to absorb the water content.

3.1.4. The garlic flour effect in villus surface

Based on table 1 result from the research on the villus length showing the T_4 higher are (1073.49±99.92) with (basal feed + 0.40% garlic flour) and lower T_1 are (630.52±119.44) (basal feed + 0.25% garlic flour). The treatment from lower to higher is T_1, T_0, T_2, T_3, and T4. The average villus surface are significant different (P<0.05). The used garlic flour is impacted to the villus surface due to the level of garlic flour given. That villus surface are depend on the several factors e.g. environment, feed, bioactive compounds [10]. The increasing of villus surface are indirect with the flour but by bio active compounds from garlic.

3.2. The garlic flour in the digesta viscosity and intestinal characteristic bacteria

Table 2. The garlic flour effect in Lactic acid bacteria, E. coli, and Salmonella sp.

CODE	Variables		
	Lactic acid bacteria**	Escherichia coli** (log CFU/g)	Salmonella sp.*
T_0	$8.33{\pm}0.50^{ab}$	$4.44{\pm}0.61^{a}$	$5.29{\pm}0.36^{ab}$
T_1	$9.05{\pm}0.96^{b}$	$4.07{\pm}0.54^{a}$	$4.52.{\pm}1.29^{ab}$
T_2	$8.95{\pm}0.39^{b}$	$5.46{\pm}0.45^{b}$	$6.51{\pm}0.95^{b}$
T_3	$7.68{\pm}0.17^{a}$	$4.42{\pm}0.32^{a}$	$4.29{\pm}1.08^{a}$
T_4	$8.92{\pm}0.67^{b}$	$4.12{\pm}0.33^{a}$	$5.17{\pm}0.32^{ab}$

** Superscript showed significantly different (0.01).
*Superscript showed significant different (0.05).

3.2.1. The garlic flour effect in Lactic acid bacteria, E. coli, and Salmonella sp.

Based on table 2 result from the research on the villus length showing the T_1 higher are ($9.05{\pm}0.96^b$) with (basal feed + 0.25% garlic flour) and lower T_3 are (7.68±0.17) (basal feed + 0.35% garlic flour). The treatment from lower to higher is T_1, T_2, T_4, T_0, and T3. The average villus surface are significantly different (P<0.01). The used garlic flour is impacted to the lactic acid bacteria due to the level of garlic flour given. That intestinal villus have a correlation with the capacity to absorb the chicken diet, the longer and wider the intestinal villus the more efficient nutrient absorption process [10].

The active substances in garlic increase the Lactobacillus sp bacteria to decrease the number of the Escherichia coli with the lowest are T_3 are (4.12±0.33) and Salmonella sp T_3 are (4.29±1.08) in the intestinal native chicken. However, the garlic flour is relatively useful in maintaining the balancing between non-pathogenic and pathogenic bacteria. The garlic powder has antibacterial properties due to the greater presence of an active substance. The key of succeeded chicken performance are depend on the number bacteria pathogenic and non-pathogenic in gut [11].

4. Conclusions

The addition garlic flour gave the best effect on digesta viscosity, microflora, and intestinal characteristic of native chicken. There was no negative effect on the characteristics of small intestine and the number of lactic acid bacteria, Escherichia coli, and Salmonella sp.

5. References

[1] Sjofjan, O. 2003. Kajian probiotik (*Aspergillus niger* dan *Bacillus spp*) sebagai imbuhan Pakan dan implikasi efeknya terhadap mikroflora usus serta penampilan produksi ayam petelur. Disertasi. Universitas Padjajaran. Bandung.
[2] Sjofjan, O., M. H. Natsir, dan T. Ardianti. 2015. Efek penggunaan probiotik kultur campuran dalam air minum terhadap karakteristik dan mikroflora usus ayam petelur. J. Ilmiah Bio. 1: 52-58.
[3] Sjofjan, O., E. Widodo, and V.A. Soffa. 2012. Pengaruh penggunaan fitobiotik dalam bentuk tanpa dan terenkapsulasi sebagai aditif pakan terhadap aktivitas enzim usus ayam pedaging. JIIP.
[4] Dharmawati, S., N. Firahmi, dan Parwanto. 2013. Penambahan tepung bawang putih (Allium sativum L) sebagai feed additive dalam ransum terhadap penampilan ayam pedaging. J. Ziraah. 38 (3): 5-12.
[5] Steel, R.G.D. dan J.H. Torrie. 1992. Prinsip dan prosedur statistika. PT. Gramedia Jakarta.
[6] Natsir, M. H., E. Widodo dan Muharlien. 2016. Penggunaan Kombinasi Tepung Kunyit (*Curcuma domestica*) dan Jahe (*Zingiber officinale*) Bentuk Enkapsulasi dan Tanpa

Enkapsulasi Terhadap Karakteristik dan Mikroflora Usus Ayam Pedaging. Buletin Peternakan. 40 (1): 1-10.

[7] Sjofjan, O., M. H. Natsir, dan D. Y. Primacitra. 2014 Pengaruh Penambahan Probiotik (*Lactobacillus sp*) dalam pakan terhadap energi metabolis, kecernaan protein, dan aktivitas enzim burung puyuh. J. Ternak Tropika. 15 (1): 74-79.

[8] Jein, R. L., Achmanu, O. Sjofjan, and M. Najoan. 2013. Egg Internal Quality and n-3 fatty acids of native chicken fed on skipjack fish (*Katsuwonus pelamis l*) industrial waste containing feed. Int. Poult. Sci. 12 (8): 484-488.

[9] Jet, S. M., H. Soetanto, O. Sjofjan, and B. Tulung. 2013. The effect of native gedi leaves (*Abelmoschus manihot L. Medik)* of northern Sulawesi-Indonesia as a source of feedstuff on the performance of broilers. Int. J. Biosci. 3 (1): 82-91.

[10] Mozin, S., D. Rosyidi, O. Sjofjan, and E. Widodo. 2015. The effect of shallot (*Allium* ascalonicum *L.*) by-product as an antibacterial and alternative phytobotic on characteristics of small intestine of broiler. Livestock. Rsch. for rural dev. 27 (4): 1-8.

[11] Septiani, A., O. Sjofjan, and Irfan H. Djunaidi. 2016. Effect of some kinds of commercial feed on quantitative and qualitative production performance of broiler chicken. Buletin peternakan. 40 (3): 187-196.

Quality and fermentation characteristic of corn stover - rubber cassava (*Manihot glaziovii* M.A) combination silage

W Kurniawan, A Bain, Syamsuddin, M Abadi and Y N Sandy

Faculty of Animal Science, Universitas Halu Oleo, Kendari, Indonesia.
Corresponding Author: kurniawan.widhi@uho.ac.id.

Abstract. Utilization of agricultural by-products as animal feed has enormous potential. The abundant availability and low prices make it a potential source of feeding commodities. However, seasonal availability and low quality of feed ingredients from agricultural waste becomes an obstacle to making it a quality feed with continuous availability. This research was conducted using Corn Stover (CS) combined with Rubber Cassava Leaves (RC) with different proportions (90: 0, 80:10, 70:20, and 60:30) as Silage and yellow corn (CY) as additives of the remaining 10%. Silage fermented by using a 1-liter silo tube model for 21 days of fermentation. The parameters observed were silage pH, dry matter content (DM), Fleigh score and crude protein content (CP). The results showed that the use of RC affects the pH and CP silage produced. The pH of silage becomes more acidic when there is an RC combination compared to the control due to WSC availability, as the BAL energy source substrate enough contained in the RC. It also occurs in CP content that linearly increases (8.47-16.04) when the proportion of RC increases because RC has a high CP content. Meanwhile, the use of RC up to 30% level did not affect the content of DM silage which resulted in the perfect silage Fleigh score (94.7-98.8). Thus, the use of RC as a CS silage combination material will be able to improve the quality of silage so as to increase the quality feed from agricultural by-products.
Keywords – silage, characteristic, quality, fermentation, fleigh.

1. Introduction

Utilization of agricultural by-products as animal feed has enormous potential. The abundant availability and low prices make it a potential source of feeding commodities. However, seasonal availability and low quality of feed ingredients from agricultural waste become an obstacle to making a quality feed with continuous availability. Therefore, it is necessary to process the agricultural waste material so it can improve the quality of its nutrients. Improving the quality of agricultural waste nutrients can be done through feed processing technology that has been widely used. Making silage as a feed preservation technology is not enough, this is because most of the agricultural waste has low quality and requires additional other ingredients that can improve its quality. Forage plant species suitable as silage feedstock should have high dry matter production in the field, high digestibility, low buffering capacity, and higher water-soluble carbohydrate (WSC) content [1].

The combination of Corn Stover with Cassava leaves has high protein content for silage materials and will be able to improve the quality and continuity of feed from corn farming. However, the use of forage feed that has a high protein (N) content, as a silage material, has the potential to constrain buffering capacity, where the pH of the silage will be difficult to decrease due to the alkali character of N. Buffering capacity refer to the number of moles of H^+ that must be added to 1 L of solution to decrease the pH by 1 unit [2]. Forages buffering capacity can be defined as the degree to which forage material resists changes in pH. All forages have different buffering capacities. Fresh forage with a high buffering capacity will require more acid to reduce its pH than forage with a low buffering capacity. Buffering capacity in forage and protein concentrate are 5.6 and 4.1 times higher than BC in grains [3].

This study was conducted to evaluate the Quality and Fermentation characteristic of Silage from Corn Stover and Cassava Leaves (*Manihot glaziovii* M.A.) combination.

2. Materials and Methods

Corn stover and rubber cassava leaves were chopped about ± 2 cm and wilted for ± 24 hours to reduce moisture content up to 60% [4], which is the optimal moisture content for fermentation. The DM content which is less than 20% will cause the risk of higher decay and loss of DM during ensilage process [5].

Table 1. Nutrient information of silage materials*.

Materials	Nutrient	
	Dry Matter (DM) %	Crude Protein (CP) %DM
Corn Stover (CS)	31.90	8.30
Rubber Cassava Leaf (RC)	22.37	26.44
Yellow Corn Meal (CY)	87.36	7.58

*Nutrition and Feed Science Laboratory Proximate Analysis of Animal Science Faculty, UHO 2018.

The material that has been mixed is put into a jar (silo tube) and compacted with to reduce oxygen levels in silo, so that the aerobic respiration phase gets shorter. The silos were closed and tightened using duct tape to be air tight (anaerobic). Silages were stored in a storage room that is protected from direct sunlight for 21 days [6], in general the lowest pH of the silage has been reached (phase V) [7].

This research was conducted using 80 days old Corn Stover (CS) combined with rubber Cassava Leaves (RC) (*Manihot glaziovii* M.A.) with different proportions (90: 0, 80:10, 70:20, and 60:30) as silage and yellow corn (CY) as additives of the remaining 10%. The research was design based on Complete Randomized Design, using 4 replications for each treatment. The data obtained were tabulated and analyzed using variance analysis. If the treatment has a significant effect, it will be continued with different test treatments using the Duncan Multiple Range Test (DMRT) posttest [8]. The observed parameters were:

1. Silage pH, obtained by juicing 5 grams of silage sample on 50 mldistilled water
2. Dry matter content (DM), obtained through proximate procedures,
3. *Fleigh* score (FS), calculated using the formula:
   $FS = 220 + [(2 \times DM (\%)) - 15] - (40 \times pH)$ [9],
4. Crude protein content (CP), obtained through proximate procedures.

3. Results and Discussions

3.1. Silage pH

Silage is one of the techniques of forage preservation at a certain moisture content levels through a microbial fermentation process by lactic acid bacteria called ensilage in a place called silo [10]. The degree of acidity or also called pH is one indicator to determine the quality of silage. The results showed that the inclusion of RC leaves significantly ($p < 0.05$) affected the pH of silage (Table 2.). CS silage combined with RC leaves has a lower pH compared to CS silage (4.45; 4.36; 4.40 vs. 4.75). A good silage pH is between 3.5 and 4.5 [11]. Addition of rubber cassava leaves to CS silage simply provides a substrate for lactic acid bacteria (LAB) to produce lactic acid, so therefore a decrease in pH occurs. A lower pH value is said to be due in the addition of RC with different percentage levels.

Rubber cassava leaves have high WSC (8.93% of DM) [12], which can be used by LAB as an energy source during the ensilage process. Low pH levels will inhibit the growth of undesirable bacteria (*Clostridium* and *Enterobacterium*), yeast and fungi that can cause damage [13].

3.2. Silage Dry Matter

The combination of corn stover and rubber cassava leaf levels did not have a significant effect ($p > 0.05$) on the silage DM content. The average of silage DM content of combination corn stover and rubber cassava leaves respectively CS 35.33%; CSRC1 36.65%; CSRC2 35.78% and CSRC3 35.80%. The result of this study proves that the fermentation process occurs in good conditions. This is indicated by the same value of silage DM content or relatively the same value between control and silage treatment.

The stable DM content in the range of 35-36% is considered to be influenced by the pH level of silage which is classified as an ideal pH so that the ensilage process will occur normally. Faster of pH decline, more silage dry matter will be maintained during the fermentation process [14].

3.3. Silage Fleigh Score

Fleigh score is a characteristic index of silage fermentation based on silage DM content and pH. *Fleigh* score between 85 -100 indicated silage in very good; 60 - 80 is good; 55 - 60 is moderate, 25-40 is fair, and less than 20 is low quality [9]. The combination of CS and RC in different levels did not have a significant effect ($p > 0.05$) on the *Fleigh* score. This is suspected as the implication of the DM content in all treatments that are relatively the same range (35.33 - 36.65%) and fairly stable pH of about 4, so that made the silages *Fleigh* score relatively the same. This phenomenon explained that the *Fleigh* score will be high if the silage has high of DM content and low of pH [15]. Silage DM content which reaches 36.65% reflects that the ensilage process is able to maintain and preserve the material, while the low pH value will give an idea that the ensilage process is in normal condition. The result of *Fleigh* score's calculation is in the range of 94.7 - 98.8, it is explained that silages are classified as very good quality.

Table 2. Silage Quality and Fermentation Characteristic

Parameters	Treatments			
	CS	CSRC1	CSRC2	CSRC3
pH	4.75 ± 0.14^{b}	4.45 ± 0.14^{a}	4.36 ± 0.14^{a}	4.40 ± 0.15^{a}
DM (%)	35.33±2.25	36.65±2.93	35.78±2.44	35.80±1.22
Fleigh	94.7±7.70	98.8±13.90	97.6±11.90	97.6±4.70
CP (%)	8.47 ± 0.50^{c}	12.93 ± 1.20^{b}	12.2 ± 1.60^{b}	16.04 ± 1.40^{a}

Different superscript in the same line means significantly different ($p < 0.05$).
CS: 90% Corn Stover + 10% Corn Meal, CSRC1: 80% Corn Stover + 10% Corn Meal + 10% Rubber Cassava Leaves, CSRC2: 70% Corn Stover + 10% Corn Meal + 20% Rubber Cassava Leaves, and CSRC3: 60% Corn Stover + 10% Corn Meal + 30% Rubber Cassava Leaves.

3.4. Silage Crude Protein

Important events in the aerobic phase are proteolysis of forage protein which reaches 50% become amino acids, ammonia and amines. The enzyme's activity on proteolysis would stop when the acidity level increases. This phase should be passed as quickly as possible to preserve the silage nutritional content [16]. A significant increase in the CP content of silage proves that the higher the percentage of RC leaf inclusion linearly will increase the silage CP content. The increase in CP has thought due to the addition of non-legume protein sources forage such as rubber cassava leaves into corn stover silage. The low pH of silage (pH no more than 4) is successful in preventing CP from proteolysis in ensilage, which could reduce the activity of protein degradation by protein-degrading enzymes. The rate of proteolysis depends on the rate of decrease of pH in ensilage [16]. The acidic environment in silage will reduce the activity of protein-degrading enzymes. The pH value that goes down at the start of ensilage is very useful to prevent the protein degradation. Optimal protease activity is at pH 4-7 depends on the used material [17]. The proteolysis process occurs during the ensilage if the level of acidity has not been achieved [18]. Crude protein content decreased from 0.8% to 0.6% at the beginning of the ensilage process [19].

4. Conclusion

Based on the research results, it was found that the combination of corn stover silage (60%) and rubber cassava leaves (30%) did not only increase the value of crude protein, *Fleigh* score and dry matter silage but also it has a lower pH than other treatments.

5. References

[1] Demirel R, Akdemir F, Saruhan V, Demirel D S, Akinci C and Aydin F. 2011. *African Journal of Agricultural Research.* Vol. 6(24), pp. 5469-5474.
[2] Segel, I. H. 1976. *Biochemical calculation 2nd Edn* (John Wiley and Sons New York-US).
[3] Moharrery, A. 2007. *American Journal Animal Veteriner Science* 2: 72-78.
[4] Umiyasih, U. and Wina, E. 2008. *WARTAZOA* 18(3): 127-136.
[5] Despal, Permana, I. G, Safarina S. N. and Tatra A. J. 2011. *Media Peternakan* 43: 69-76.
[6] Komalasari, Liman and Syahrio, T. Y. S. 2015. *Peternakan Terpadu* 3(2): 31-35.
[7] Schroeder, J. W. 2004. *Quality Forage* (NDSU Extension Service North Dacota-US).
[8] Steel, R. G. D. and Torrie J. H. 1980. *Principles and procedure of statistics* (McGraw-Hill Book Co. New York-US).
[9] Ozturk, D., Kizilsimsek, M., Kamalak, A., Canbolat, O., and Ozkan, C. O. 2006. *Asian-Australasian Journal of Animal Science* 19(4): 526-532.
[10] McDonald, P., Edwards, R., and Greenhalgh, J. 2002. *Animal Nutrition 6th* (Scientific and Tech John Willey & Sons. Inc.New York-US).
[11] Subekti, E. 2009. *MEDIAGRO* 5(2): 63–71.
[12] Rijali, D. H. 2010. Kualitas silase daun singkong, daun ubi jalar dan daun lamtoro yang dipanen pada waktu berbeda (Bachelor Thesis) Bogor Departemen Ilmu Nutrisi dan Teknologi Pakan Fakultas Peternakan Institut Pertanian Bogor.
[13] Heinritz, S. 2011 Ensiling suitability of high protein tropical forages and their nutritional value for feeding pigs (Diploma Thesis) University of Hohenheim. Stutgart.
[14] Ward, R. T. 2008. *Fermentation analysis of silage: use and interpretation* (Cumberland Valley Analytical Services Inc Hagerstown-US).
[15] Yosef, E., Carmi, A., Nikbachat, M., Zenou, A., Umiel, N., and Miron, J. 2009. *Animal Feed Science and Technology* 152: 1–11.
[16] Schroeder, J. W. 2013. *Silage fermentation and preservation Quality Forage* (NDSU Extension Service North Dacota-US).
[17] Sloner, D. and Bertilsson, J. 2006. *Animal Feed Science and Technology.* 127: 101-111.
[18] Sun, Z. H., Liu, S. M., Tayo, G. O., Tang, S. X., Tan, Z. L., Lin, B., He, Z. X., Hang, X. F., Zhou, Z. S. and Wang, M. 2009. *Animal Feed Science and Technol*ogy. 152: 219-231.
[19] Givens, D. I., and Rulguin, H. 2004. *Animal Feed Science and Technol*ogy. 114: 1-18.

Influence of addition of chicken manure and tithonia on quality of liquid fertilliser made from organic waste

Zulkarnaini and Syafrizal

Animal Science Study Program, Agriculture Faculty, Tamansiswa University, Padang, West Sumatera, Indonesia.

Abstract. Indonesian's waste production continues to grow as the population increases. Organic waste such as leaves and food waste generally decomposes and can be composted simply with the use of barrels and buckets. Effective Microorganism 4 bioactivator can accelerate this process. This research was conducted to discover if addition of Tithonia diversifolia and chicken manure could increase the nutrients from liquid fertilizer and solid compost. Complete randomized design with 4 treatments and 4 replications was used as follows: A = 100% organic waste (control), B = 60% waste + 28% tithonia + 12% manure, C = 60% waste + 32% tithonia + 8% manure and D = 60% waste + 36% tithonia + 4% manure. The results showed that the addition of manure and tithonia increased N, P and K in liquid fertilizer. The highest increase in N and P occurred with treatment D, and K with treatment C.
Keywords – composter, liquid fertilizer, Manure, Tithonia diversifolia

1. Introduction

Indonesian's waste production continues to grow as the population increases. Organic waste such as leaves and food waste generally decomposes and can be composted simply with the use of barrels and buckets. The manufacture of solid or liquid organic fertilizers is a decomposition process using microbial activity and so the speed of decomposition and quality of compost depends on the condition and types of microbes that are active during the composting process [11]. These microbes may be aerobic bacteria or fungi [9]. The optimum conditions for microbial activity include aeration and provision of adequate growing media and food sources for microbes. One technology to accelerate the fermentation process is the use of Effective Microorganism 4 (EM4) bio-activator which consists of a culture of decomposing bacteria. EM4 is applied as an inoculant to increase the diversity and population of microorganisms in soil and plants and also increase growth and improve the quality and quantity of crop production. It can also be added to composting organic waste to hasten the composting process [8].

Liquid organic fertilizer (LOF) is a decomposition product from organic waste. It can be used to quickly overcome nutrient deficiencies in the soil and is not subject to nutrient leaching. Compared to liquid fertilizers from inorganic materials, LOF are generally less likely to damage soil and plants even though they are used frequently. In addition, this fertilizer contains a binding material, so that the fertilizer solution applied on the soil surface can be directly absorbed by plants. LOF types include liquid manure, biogas-making liquids, and liquid fertilizers from organic waste [1].

Tithonia is a common weed found throughout the tropics that has a high nutrient content including 3.50% N, 0.37% P, and 4.10% K and is good for increasing crop production (Hartatik, 2007). These three chemicals along with calcium (Ca), magnesium (Mg) and sulfur (S) are important plant nutrients that are abundant in chicken manure [2].

In this study the effectiveness of addition of Tithonia diversifolia and chicken manure to LOF is investigated.

2. Materials and methods

Organic waste from the market was used along with EM4, chicken manure and Tithonia. Completely randomized design with 4 treatments (A = 100% organic waste (control), B = 60% waste + 28% tithonia + 12% manure. C = 60% waste + 32% tithonia + 8% manure. D = 60% waste + 36% tithonia + 4% manure) was used with 4 replications, one in each of 16 composters. The variables measured were N content (determined by the Kjedahl method), P (extraction method) and K (Flame photometer).

Table 1. Composition of materials used in treatments A, B, C and D.

Material	Treatment (%)			
	A	B	C	D
Organic waste	100	60	60	60
Tithonia	0	28	32	36
Manure	0	12	8	4

3. Results

3.1. Nitrogen (N) content

Nitrogen in a fertilizer is important in the production of plant proteins [5]. The content of N in LOF after 28 days of fermentation is seen in Table 2.

Table 2. Average LOF nitrogen content from each treatment.

Treatment	Average Nitrogen Content (%)
A	3.04^{b}
B	3.29^{b}
C	3.01^{b}
D	4.76^{a}

Description: Different superscripts show significant differences (P <0.01).

Analysis of variance showed that only treatment D showed significantly higher levels of nitrogen than the control treatment. A post hoc test with DMRT confirmed that there was no significant difference between treatments A, B, C. The activity of microorganisms during the fermentation of organic materials reduces the nitrogen content of LOF so a more efficient fermentation process may lead to lower nitrogen content (Notohadiprawiro, 1999). Micro-organisms contained in the chicken manure will aid in this fermentation process. However, Wulandari et al. (2015) states that an increase in nitrogen as a protein decomposition product can result from the decomposition process. These N values are significantly higher than those the study by [3] who found 0.10% at day 10 and 0.07% on day 42 from pure organic waste without additions.

The high levels of N in treatment D showed that this lower level of chicken manure was more effective in producing high nitrogen levels. These N levels are within the Indonesian Government standards for organic fertilizers, biological fertilizers and soil enhancers, namely N macro nutrient quality standards of 3-6% [6].

3.2. Phosphorus (P) content

The highest average phosphorus was found in the LOF after treatment D (2.37%) and the lowest after treatment A (1.32%) as shown in table 3.

Table 3. Average phosphorus level in LOF after each treatment.

Treatment	Average Phosphorus (%)
A	1.32 [b]
B	1.69 [ab]
C	1.98 [a]
D	2.37 [a]

Description: Different superscripts show significant differences (P <0.01).

Analysis of variance showed that treatment had a very significant effect (P <0.01) on phosphorus levels. A post hoc test with DMRT showed significant differences between A and D, and, B and D but not between A and B, B and C, or C and D. Higher levels of Tithonia and lower levels of chicken manure resulted in higher P content in the resulting LOF. The higher levels of P are associated with organic acids results from the decomposition of Tithonia. This improvement is due to the way the organic materials are broken down by microorganisms so releasing phosphorus. The carboxyl or hydroxyl (COOH / OH) functional groups in the organic acids in this plant act as competing anions to phosphorus anions in mineral exchange complexes so that available phosphorus increases (Supriadi,2003). High phosphorus content can be influenced by high nitrogen content, increasingly nitrogen will result in multiplication of microorganisms which leads to greater breakdown of phosphorus increasing available phosphorus content [10]. None of these phosphorus levels reach the government standard for organic fertilizers, biological fertilizers, and soil enhancers, namely 3-6% of P2O5 [6].

The highest phosphorus content, which came from treatment D, was 2.37% which is higher than that from [4] study (0.39% phosphorus) of a compost made from fresh leaves, possibly due to the materials being used. The amount of nutrients in organic fertilizers varies greatly depending on the raw material used [7].

3.3. Potassium content

The average content of potassium depended on the treatment (Table 4).

Table 4. Average potassium levels in LOF after each treatment.

Treatment	Average Potassium (%)
A	2.94 [a]
B	3.02 [a]
C	3.11 [b]
D	3.03 [a]

Description: Different superscripts show significant differences (P <0.01).

Analysis of variance showed that treatment significantly affected potassium levels of LOF (P <0.01). a post hoc test with DMRT showed that C and B treatments, C and D treatments, C and A treatments, were significantly different, but no other differences between treatments were significant. Treatment C was clearly superior to the other treatments for K production. This difference is caused by different raw material content (manure, Tithonia diversifolia and organic waste) resulting in different potassium. All these LOF using manure and Tithonia with market waste fulfilled Indonesian government standards for organic fertilizers, biological fertilizers, and soil enhancers, namely K2O macro nutrient content of 3-6% [6].

4. Conclusion

The results showed that the addition of manure and tithonia increased the Nitrogen, Phosphorous and Potassium content of LOF. The highest increase in N and P elements occurred with treatment D (lowest manure content), while the highest increase in potassium with Treatment C (second lowest manure content). This suggests that while manure can increase the availability of these macronutrients in LOF, supplementation with sufficient tithonia may be even more important.

5. References

[1] Hadisuwito, S. 2007. Membuat Pupuk Kompos Cair. Agromedia Pustaka. Jakarta.

[2] Musnamar. 2003. Pupuk Organik Cair dan Padat, Pembentukan dan Aplikasi. Penebar Swadaya. Jakarta.

[3] Nurjazuli., A. Awiyatul., C. Juliana., K. D. Pertiwi., K. Samosir., P. Prasetyawati., S. Pertiwi. 2016. Teknologi Pengolahan Sampah Organik Menjadi Kompos Cair (Organic Waste Treatment Technology Toward Liquid Compost). Seminar Nasional Sains dan Teknologi Lingkungan II e-ISSN 2541-3880 Padang, 19 Oktober 2016.

[4] Ovianti, F. 2015. Pengaruh Pemberian Pupuk Organik Cair Daun Gamal Terhadap Pertumbuhan Tanaman Sawi. Skripsi Fakultas Tarbiyah dan Keguruan UIN Raden Patah. Palembang.

[5] Parman, S. 2007. Pengaruh Pemberian Pupuk Organik Cair terhadap Pertumbuhan dan Produksi Kentang (Solanum tuberosum L.) Bulletin Anatomi dan Fisiologi 15 (2): 11-18.

[6] Permentan No. 70/permentan/sr.140/10/2011. Tentang Pupuk organik, pupuk hayati dan pembenah tanah.

[7] Simamora, S dan Salundik. 2006. Meningkatkan Kualitas Kompos. Agromedia Pustaka. Jakarta.

[8] Suparman, M. 1994, EM4 Mikroorganisma Yang Efektif, KTNA, Sukabumi.

[9] Unus, S. (2002). Pupuk Organik Kompos dari Sampah, Bioteknologi Agroindustri. Bandung: Humaniora Utama Press.

[10] Yuli, A., Hidayati, T.B, Benito., A. Kurnani., E. T, Marlina., E. Harlia. 2011. Kualitas Pupuk Cair Hasil Feses Sapi Potong Menggunakan Saccaromyces cereviceae (Liquid Fertilizer Quality Produced by Beef Cattle Feses Fermentation Using Saccaromyces cereviceae). Jurnal Ilmu Ternak. Vol. 11 No. 2., 104-107.

[11] Yuwono, T. 2006, Kecepatan Dekomposisi dan kualitas Kompos Sampah Organik, Jurnal Inovasi Pertanian. Vol. 4, No. 2.

The chemical characteristics of yoghurt (Lactobacillus fermentum MGA40-6 and Streptococcus thermophilus) with additional puree from *Senduduk* fruit (Melastoma malabathricum, L.)

Afriani Sandra, Yulianti Fitri Kurnia, Ade Sukma, and Endang Purwati

Department of Animal Science, Andalas University, Padang, West Sumatera, Indonesia.
Email: sandraafriani@gmail.com; yulianti_fk@yahoo.com.

Abstract. The purpose of this study was to know the effect of adding the puree of *senduduk* fruit on the chemical characteristics of yoghurt. The experiment is conducted by using a Completely Randomized Design (CRD) by adding different concentrations of senduduk fruit's puree into yoghurt; A (0%), B (1.5%), C (3%), D (4.5%), dan E (6%). The results of this study with adding the puree of senduduk fruit into yoghurt can significantly increased the total titratable acidity, protein concentration and antioxidant activity. The experiments also show the decrease on the pH value, but showing not significant differences at the water content and ash concentration of the yoghurt. In conclusion, adding 4.5% (D) of senduduk fruit puree into yoghurt is the best to produce a consumtive yoghurt with (1.54%) total titratable acidity, (5.18%) protein concentration, (56.72%) antioxidant activity, (4.4) pH, (77.59 %) water and (1.22%) ash.
Keywords – Lactobacillus fermentum MGA40-6, yoghurt, senduduk fruit puree, and chemical.

1. Introduction

Milk and other dairy products are familiar found in the today's market as food product. The variations of dairy products are available as a way to extend the storage life of the milk, because milk contains complete nutritional components which can easily be damaged, especially by microbes, if not handled properly.

Yoghurt is made by using Lactic Acid Bacteria (LAB) as the starter. The LAB is able to convert glucose to become lactic acid. The processed milk products using LAB are one of the applications of biotechnology which exploits bacteria. They have therapeutic or better known as *functional* terms.

LAB isolation and identification have been carried out by many researches to improve the quality of yoghurt processing and other processed food products. LAB can be isolated from *dadih* of Lima Puluh Kota District, which obtain the LAB type of Lactobacillus fermentum MGA40-6 [1]. LAB application on other fermented products like yoghurt has not been done, therefore this study explores the making of yoghurt by using the Lactobacillus fermentum MGA40-6 bacteria and Streptococcus thermophilus.

The nutritional value and function of consumptive yoghurt can be improved by adding useful components such as the puree of *senduduk* fruit into it. The fruit is a local fruit which is not optimally utilized yet. This fruit can easily found in many regions. Beside its wide availability, there are more functional benefits can be obtained from this fruit. The fruit contains anthocyanins which help the compound products produce more antioxidants.

The results of the existing studies prove that the use of 5% *senduduk* fruit's puree produces the best hard candy characteristics [2], and adding 6% of it produces the best Nagka straw jam [3]. The experiment of making yoghurt by using LAB Lactobacillus fermentum MGA40-6 and Streptococcus thermophilus with a starter of 5% and adding of the fruit's puree with varied concentrations between 0%, 1.5%, 3%, 4.5%, and 6 %, in terms of color and taste, are acceptable for the panelists, but the effect on the physicochemical properties of the produced yoghurt remains unknown.

This study is aimed at determining the effect of adding senduduk's puree into yoghurt (Lactobacillus fermentum MGA40-6 and Streptococcus thermophilus) on its physicochemical properties such as pH level, total acid titration, water content, ash concentratrion, and the antioxidant activity.

2. Methods

This study was carried out by using a Completely Randomized Design method. The puree of *senduduk* fruit is added in several different concentrations (0%, 1.5%, 3%, 4.5%, and 6%) with 3 replications for each. The data were analyzed by using the Analysis of Variant or ANOVA which was followed by further DMRT tests with a 5% confidence interval [4]. The model is designed as follows: $Yij = \mu + Pi + \varepsilon ij$.

This research was carried out in several stages which include the making of the starter by using Lactobacillus fermentum MGA40-6 and S.thermophilus and the evaluation of physicochemical properties of the yoghurt [5].

3. Results and discussion

Treatment	pH	Total Titratable Acidity (%)	Water Content (%)	Protein (%)	Ash (%)	Antioksidan Activity (%)
A	5.30^{a}	0.86^{a}	79.43^{a}	4.21^{a}	1.09^{a}	13.35^{a}
B	4.80^{b}	1.52^{b}	77.12^{a}	5.25^{b}	1.23^{a}	19.71^{a}
C	4.40^{c}	1.53^{b}	77.76^{a}	5.31^{b}	1.21^{a}	48.00^{b}
D	4.40^{c}	1.54^{b}	77.59^{a}	5.18^{b}	1.22^{a}	56.72^{c}
E	4.70^{d}	1.66^{b}	79.19^{a}	4.41^{ab}	1.21^{a}	68.28^{d}

3.1. pH analysis

The analysis of variant (ANNOVA) shows that adding the senduduk's puree created a significant effect with the value of $P < 0.05$ at the pH value of the yoghurt. Adding the senduduk's puree up to 4.5% (D) can reduce the pH value of the produced yoghurt, but at a concentration of 6% added puree (E), the pH value increases again which is caused by the lower the pH, the lesser microbes will live there. Therefore, in general, lactic acid content produced by the fermentation and the effect of the acid from senduduk's puree on the treatment of pH value E (6%) became relatively smaller compared to the pH value of yoghurt with the added puree in treatment B, C, and the lower pH value after the fermentation will cause less microbes to survive [6].

The results shows that the increasing of the puree concentration decreases the pH of the yoghurt. This is caused by the effect of sugar contained in *senduduk*'s fruit toward LAB activity in producing lactic acid. The LAB (L. fermentum MGA40-6 and S. thermophillus) exploits the milk carbohydrate and *senduduk*'s puree in the fermentation into lactic acid. That during the fermentation process, the LAB will utilizes the existing carbohydrates until it forms lactic acid, until the pH level is decreased and the acidity is increased [7]. The *senduduk* fruit contains 7.75% carbohydrates [8].

The pH of *sedudук*'s puree (4.1) also contribute to the pH of the yoghurt, so the pH value will decrease along with the addition of the puree concentration. The pH value from the study has matched the common pH of yoghurt. The minimum pH of yoghurt is 4.3, so it is conluded that the results of this study have met the standard of yoghurt fermented drinks [9].

3.2. Total titratable acidity

The experiments' results showed that the mean value of Titrated Total Acid (TTA) obtained is ranging between 0.86 - 1.66. TTA is stated by the percentage of lactic acid. Lactic acid is the largest component formed from the fermentation of milk into yoghurt. The increase of lactic acid in this study is consistent with its decrease of pH value, thus the TTA of the results showed an inversion to the pH value. The TTA value is inversely proportional to the pH value [10]. The results show that TTA has met the standard of yoghurt fermented drinks which is 0.5-2% (% lactic acid)[11].

3.3. Water content

The results of statistic analysis showed that the yoghurt without the additional senduduk's puree implies an insignificant results with the value of $P > 0.05$. However, in general, the results of the water content analysis in the table above shows that the yoghurt water content without additional senduduk's puree and the one with additional puree showed that the water content is decreased in almost all treatments. The decrease of water content on the yoghurt with additional senduduk's puree is caused by adding of puree which reduce the pH value of the yoghurt, so that the protein will experience coagulation. The condition is indicated by the increasing formations of coagulants or clots which lead to the increase of total solidity. The acidic pH, the yoghurt protein undergoes coagulation [12].

3.4. Protein

The increase of yoghurt protein concentration along with the addition of senduduk's puree to yoghurt causes condition of yoghurt to become acidic. Thus, lowering the pH value, so this condition will affect the hydrolysis of the yoghurt protein, which means lowering the pH value, so this condition will affect the hydrolysis of the yoghurt protein. Acidity formed by the activity of yoghurt bacteria so that the total solidity increases and the texture of the yoghurt produced becomes thicker. The acidic pH condition the yoghurt protein will experience coagulation which form more clots or coagulants [12].

Yoghurt protein have increased along with the addition of senduduk's puree. It is also caused by the phenol content in the senduduk fruit's puree which is bonded with the peptide from the hydrolyzed protein, so that the protein of the produced yoghurt is higher. The interaction between polyphenols and milk protein or polysaccharides will cause an increase in protein levels.

The high protein content of yoghurt with addeed senduduk's puree is resulted by the protein contribution from the senduduk fruit it self. The senduduk fruit contains protein as much as 1.46% / 100 grams. Yoghurt produced from this study has met the quality standards. Requires a minimum protein content of yoghurt [13].

3.5. Ash concentration

The analysis on the ash concentrations show that the addition of senduduk's fruit puree made no significant effect on the yoghurt. In Table 1, It can be seen that the highest average ash content is 1.23 (Treatment B) and the lowest is 1.09 (Treatment A). The results table also implies that adding different senduduk's fruit puree concentrations as in treatments A, B, C, D, and E lead to insignifinact difference. Thus, it can be used for the real measurements, but the general results shown in the table above implies that the ash content of yoghurt with different additional senduduk's purees concentration are all in average gray level, or it is not showing signifincant improvement at each treatments. This is due to the more senduduk's puree is added also increase the amount of chemical components available in the yoghurt. One of the many chemical components which can be used as indicators for determining ash levels is minerals. The minerals contained in fresh iron senduduk fruit are 2.60 mg / 100 g, manganese 12 mg / 100 g, calcium 1.92 mg / 100 g and copper 2.60 mg / 100 g [14].

3.6. Antioxidant activity

The analysis of the antioxidant activity shows that the addition of the senduduk puree into the yoghurt makes a significant effect on the value of its antioxidant activity. The highest antioxidant activity was shown by treatment E (68.2) and the lowest antioxidant activity was found in treatment A (13.35%). This implies that the more senduduk's puree is added into the yoghurt, the higher will the antioxidant activity be. Senduduk fruit contains anthocyanin which triggers the increase of activity in the produced yoghurt. The fruit contains anthocyanins which have reddish purple pigments which can act as antioxidants [15].

4. Conclusion

The addition of the senduduk's fruit puree into the yoghurt increases total titrated acid, protein content and antioxidant activity, yet decreases the pH value. However, the water content and ash content show no significant differences. Adding senduduk puree into yoghurt as much as 4.5% (D) is best method for producing yoghurt.

5. References

[1] Kurnia, Y.F., H. Purwanto., E. Purwati. 2016. The potensial of dadiah from 50 kota district, west Sumatera as a probiotic food based on Total lactic acid bacteria. Prosiding. International conferace.
[2] Rahayu, R.M. 2015. Penambahan sari buah senduduk (Melastoma malabatricum, L, auct non L) terhadap karakteristik mutu hard candy. Skripsi. Universitas Andalas. Padang.
[3] Marisa. 2015. Pengaruh penambahan ekstrak buah senduduk (Melastoma malabatricum, L) terhadap karakteristik mutu selai jerami nangka (Arthocarpus heterophilyllus, L). Skripsi Fakultas Teknologi Pertanian. Universitas Andalas Padang. 67 hal.
[4] Steel, C.J dan J.H.Torrie 1995. Prinsip dan Prosedur Statistik. PT Gramedia, Jakarta.
[5] [AOAC] Association of Official of Analitycal Chemist. 2006. Official Methods of Analysis of The Association of Official Agriculture Chemist. Inc. Washington D.C.
[6] Winarno F.G dan Fernandez., I.E. 2007. Susu dan produk fermentasinya.Mbrio Press, Bogor.
[7] Rahayu. 2009. Perkembangan terkini penggunaan probiotik dalam industri susu . Foodreview Indonesia. 4 (6): 30-33.
[8] Nayak, J. and U.C Uday. 2015. Analysis of some nutritional properties in eigh wild edible fruith of odisha, India. Int. J. Curr Sci 14: 55-62.
[9] Nummer. 2002. Fermenting Yoghurt At Home. http://yoghurt.html.
[10] Usmiati, S dan N.Richana. 2011. Potensi bakteriosin dari Lactobacillus sp. Galur SCG 1223 sebagai biopreservatif pada daging segar. Jurnal Teknologi Pasca Panen Pertanian Vol. 7 (2).
[11] SNI.1992.SNI Yoghurt (SNI 01-2981-1992.1992). Badan Standarisasi Nasional.
[12] Widodo. 2003. Bioteknologi Industri Susu. Depok : Lacticia Press.
[13] Badan Standardisasi Nasional. 2009. SNI 01- 2981-2009. Syarat Mutu Yogurt. Standar Nasional Indonesia, Jakarta.
[14] Nayak, J. and U.C Uday. 2015. Analysis of some nutritional properties in eigh wild edible fruith of odisha, India. Int. J. Curr Sci 14: 55-62.
[15] IPTEK. 2009. Yoghurt. http: // iptek.apjii. or. Id/artikel/pangan/IPB/Yoghurt.pdf.

Acknowledgment

Author express gratitude to the Institute of Research and Community Service, Andalas University for the Cluster Young Lecturer No. 10/UN.16.17/PP.RDP/LPPM/2018 and the Animal Product Technology Laboratory in the Faculty Animal Science, Andalas University, who provided facilities for research.

The characteristics of organic fertilizer made of cow feces using the Indigenous Micro-Organisms (IMO) from raw manures

D Novia, A Rakhmadi, E Purwati, I Juliyarsi, R Hairani, and F Syalsafilah

Department of Animal Science, Andalas University, Padang, West Sumatra, Indonesia
Email: dnovia@ansci.unand.ac.id

Abstract. Raw manures are very potential if it can be processed into Indigenous Micro-Organisms (IMO) and organic fertilizers. It has sustainable availability in large number, yet if it is not processed into the useful matter, it causes water and air pollution. IMO is a kind of biological products which can be used as decomposers in the production of organic fertilizers, yet it is almost unknown to the public. The IMO of raw manures is cheaper, easier to produce and more economical than IMO of commercial products such as EM4 or others. The making of the IMO used the experimental method called Complete Randomized Design (CRD) in which 5 treatments were given by using raw manures from the different animal such as cow, goat, rabbit, quail, and chicken with 4 replications for each. The result IMO was further examined for the moisture content, pH value, total plate count, lactic acid bacteria count, fungal count, and N. Then the IMO was applied in the production of organic fertilizer from cow feces with 21 days fermentation time, then it is further examined for the value of moisture content, pH, N, P, K, and C/N ratio. If the results of variance analysis (ANOVA) showed a significant effect then it is proceeded by Duncan's test. Based on the research results, it is obtained that IMO of quail feces produce organic fertilizer with the highest N, P, K content.
Keywords - fermentation, cow feces, raw manures, IMO, organic fertilizer.

1. Introduction

Raw manures is the farm waste which often becomes a problem because it is usually left piled up and ignored so it causes contamination in the water and air, and affecting global warming. The number of livestock feces is very abundant while it is potential enough to be utilized as organic fertilizer in sustaining an organic farm which products have greater export opportunities. According to Statistics Centre (BPS, 2016), the population of dairy cows in West Sumatera province was 320,044 heads. A cow is able to produce 23.6 kg/day solid feces, whereas every farmer usually has 6 - 7 cows. Therefore, each farmer produces large enough cow feces for about 141.6 kg/day or 4,248 kg/month. Unfortunately, our farmers are using inorganic fertilizers such as urea which is relatively more expensive.

Raw manures, if left out naturally can be utilized as organic fertilizer but this natural process takes longer time. Manures were made by putting it in a black plastic bag that is loosely folded and it was composted for 16 weeks under room temperature in summer [1]. Making organic fertilizer can be accelerated for 2 - 3 weeks by adding activators or decomposers [2]. Decomposers available in the market such as EM4 and other brands require extra cost to get, whereas natural decomposer is cheap and easy to cultivate from the cattle's feces. Some livestock feces are potential decomposers or also known as IMO (Indigenous Micro-Organisms) such as the feces of cow, goat, rabbit, quail, and chicken. Raw manures contain nutrients such as nitrogen, phosphorus, potassium, and water which have different compositions. Rabbit feces contain higher nitrogen and phosphorus, whereas the goat feces contain higher potassium, and the cow feces contain more water.

Feces naturally contain foods that microbes can use for their lives. Naturally, in the stool there is Nitrosococcus sp, Pseudomonas striata, Nitrosomonas sp, Mycorrhiza, Pseudomonas fluorescens, Streptomyces, Trichoderma sp. The types of bacteria that have been identified in the IMO include Rhizobium sp, Azospirillium sp, Azotobacter sp, Pseudomonas sp, Bacillus sp, and phosphate solvent bacteria. The results of pre-research IMO contain lactic acid bacteria, bacteria, and mold. Composting of cow feces using IMO of manures for 21 days produces organic fertilizer with increased N content. The use of organic fertilizers in plants will give a higher nutritional effect than inorganic [3].

1.1. The purpose of the research

The purpose of this research is to produce cheap, easy and quality from the raw manures, and utilizing it to produce organic fertilizers that meet the standards. It is expected that this research result can be applied by farmers, so it can increase their income by replacing expensive chemical fertilizers and improve the soil structure.

2. Materials and methods

2.1. The material of the research

This study used the feces of farm animals such as cows, goats, rabbits, chickens, and quails which have been produced within the maximum time of 24 hours which were collected from the Universitas Andalas, Animal Farm Unit. Other important ingredients are coconut water and sugar which are gathered from the local market in Padang, Indonesia. The types of equipment used in the experiment and analysis were; analytical scales, stainless steel spoon, electric oven, desiccator, porcelain cup, mortar and pestle, cup clasp, funnel, distillation flask, distiller, glass cup, 30 mesh sieve, AAS, including the tools used for IMO fermentation and black plastic bag for the fertilizer fermentation process.

2.2. Methods

IMO were made by using an experimental method called Complete Randomized Design (CRD) of 5x4 treatments scheme on the feces of cows, goats, rabbits, quails, and chickens. The generated IMO was further examined for the value of moisture content, pH, total plate count, total lactic acid bacteria count, and total mold count. The IMO was then applied in the making of organic fertilizer made of cow feces with 21 days fermentation time. The result was then further examined to check the value of N, P, K, C/N ratio and its water content. If the results from the Analysis of Variance (ANOVA) shows significant impact then it will be proceeded into Duncan's test.

2.3. IMO processing [4]

IMO making procedure were 1) the raw manures were weighed as much as 100 g/sample, 2) then the raw manures were put into a water bucket, 3) then mix in sugar as much as 50 g/sample, 4) it was stirred it has no more grains, 4) coconut water was pour into a bucket as much as 500 ml/sample, 5) after it was stirred thoroughly, put it into a large Aqua bottle which are connected through a small pipe to a smaller Aqua bottle filled by mineral water, so it is not exposed into air and kept in anaerobic condition for 14 days and kept away from the sunlight, 6) after 14 days, IMO was produced from manures. It was then analyzed the pH value, moisture content, total plate count, lactic acid bacteria count, fungal count and N.

2.4. Processing organic fertilizer [4]

The procedure of making organic fertilizer that was: 1) cow feces were dried under the sun for 2 days, 2) once it was dry enough, it was filtered with a size of 30 mesh. Then the IMO was weighted and distributed into 20 units of the black plastic bag which weigh 500 g each and placed on black plastic, 3) the generated IMO was diluted with water, in a ratio of 1:10, and then it was homogenized, 4) the IMO was sprayed onto cow feces and stirred until it was evenly distributed. It was then fermented for 21 days and stirred every 3 days, 5) the produced organic fertilizer was further examine to check it was the resulting organic fertilizer was tested moisture content, pH, the levels of N, P, K, and C/N ratio.

3. Result and discussion

3.1. IMO

Based on the results of research on IMO production from different raw manures, the results were shown in Table 1.

Table 1. pH value, moisture content, total plate count (TPC), total lacted acid bacteria (TLAB) count, total fungal count (TFC) and total N of the generated IMO.

Treatments	pH Value	Moisture Content (%)	TPC ($x10^4$ CFU/ml)	TLAB Count ($x10^8$ CFU/ml)	TFC ($x10^{11}$ CFU/ml)	Total N (%)
A	$3.83{\pm}0.03^{d}$	$91.46{\pm}0.07^{c}$	$49.75{\pm}3.69^{a}$	$48.18{\pm}10.70^{b}$	$38.16{\pm}5.96^{b}$	$0.14{\pm}0.04^{b}$
B	$3.88{\pm}0.13^{cd}$	$88.51{\pm}0.41^{d}$	$55.25{\pm}10.21^{a}$	$81.63{\pm}4.85^{a}$	$48.75{\pm}5.74^{a}$	$0.16{\pm}0.02^{b}$
C	$4.28{\pm}0.04^{a}$	$91.27{\pm}0.44^{c}$	$35.93{\pm}7.36^{b}$	$28.74{\pm}4.75^{d}$	$27.93{\pm}4.14^{c}$	$0.08{\pm}0.03^{c}$
D	$3.92{\pm}0.01^{bc}$	$92.21{\pm}1.61^{b}$	$28.66{\pm}10.10^{c}$	$36.48{\pm}5.17^{c}$	$29.41{\pm}4.17^{c}$	$0.25{\pm}0.03^{a}$
E	$3.96{\pm}0.05^{b}$	$95.20{\pm}0.14^{a}$	$39.46{\pm}6.67^{b}$	$42.00{\pm}8.00^{b}$	$28.96{\pm}4.71^{c}$	$0.13{\pm}0.05^{b}$

Note: The mean with different lowercase letters shows a significantly different effect ($P < 0.05$).

Table 1 shows the treatments on several types of manures which were used in the IMO production towards the pH value, moisture content, TPC, total LAB count, total mold count and N. The results from the analysis of variance showed that different treatments on the feces significantly affected the pH value, moisture content, TPC, total LAB count, total fungal count and total N of the generated IMO.

Based on Duncan's further test at treatment B which used goat feces the highest of real and significant differences from other treatments on the number of total LAB count, total fungal count and TPC (not significantly different from treatment A). Treatment B also shows the lowest pH value and moisture content so that the generated IMO was thicker and more distinctive. The treatment D, on the quail feces, shows the highest total N and the lowest TPC so that the available total N allows microbial food reserves during the process of making organic fertilizer.

The IMO is a liquid containing the microorganisms (bacteria) which are useful for plants and help to fertilize the soil such as Rhizobium sp, Azospirillum sp, Azotobacter sp, Pseudomonas sp, Bacillus sp and phosphate solvent bacteria which were produced from the local materials. The IMO solution is potential as the agent to transform organic matter, thus the IMO can be used as a good decomposer [5].

3.1.1. pH value

During the fermentation process in generating the IMO, there was a decrease in feces' pH due to the microorganisms activities in it. Measurement of feces' pH before fermentation was ranged from 7.75 ± 0.13 (cow feces) to 9.73 ± 0.04 (rabbit feces), while the IMO pH was ranged from 3.83 ± 0.03 (IMO of cow feces) to 4.28 ± 0.04 (IMO of rabbit feces) which indicate the decrease of pH value. The decrease of pH value of the feces from neutral-base into acid with the ration between 3.92-5.45 points in the IMO. It was down by almost half, which indicated that microorganisms have been working well under the anaerobic conditions with the pH 3 to 4 [6].

The low pH of cow feces IMO is in line with the pH of cow feces which is also the lowest among other treatments. The low pH of cow feces IMO was caused during fermentation due to the activity of microorganisms which release CO2 and turn into H2CO3 which is easily broken down into H+ + HCO3 ions. The presence of H+ ions will cause acidity or decrease of the pH. Besides the decrease of pH is also caused by lactic acid bacteria which produce lactic acid (acidic) [6]. The pH value of the generated cow feces IMO was lower than the IMO of banana humps 4.2 [7] and was higher than the pH of papaya IMO, which was researched that is 3.68 and the pH value of rabbit feces IMO is lower than the pH of cabbage IMO which is 4.52 [8], and golden snail IMO 6.55[7], tofu MOL 5.00-5.75 [9].

3.1.2. Moisture content

The moisture content of the generated IMO was very high with the value of 88.51±0.41 (IMO of goat feces) up to 95.20±0.14% (IMO of chicken feces). The high moisture content of generated IMO was due to the high ratio of feces and water (1: 5) and the moisture content of each stool is also high. The high moisture content in chicken feces IMO is caused by the high moisture content of chicken feces as the result of a mix of the feces and urine which is difficult to be separated because they are produced from the same place called cloaca [10]. It is different from the ruminant animals such as cows and goats of which the feces and urine are produced from the different place.

The low moisture content of goat feces IMO is caused by the low moisture content of the goat feces in comparison to the cow feces. The moisture content of goat feces is 60%, while the moisture content of cow feces is 85% [11].

3.1.3. Total Plate Count (TPC)

Based on Table 1 the TPC of the generated IMO from raw manures were ranging between $28.66 \pm 10.10 \times 10^4$ (IMO of quail feces) to $55.25 \pm 10.21 \times 10^4$ CFU/ml (IMO of goat feces). The TPC of generated IMO were lower than the IMO generated from coconut fiber which value was ranging between 8.97×10^7 SPK/ml [12] and 29.80×10^8 SPK/ml IMO on the tofu pulp [9].

The process of making the IMO is carried out under anaerobic conditions so that the number of aerobic bacteria is much lower than the total LAB and fungi which growth is not expected because they are pathogenic. The growth of aerobic bacteria desperately needs oxygen in order to develop properly, but fermentation is done by closing the container tightly and flowing the air through the hose into a water-filled bottle.

3.1.4. Total Lactic Acid Bacteria (LAB) count

The highest total LAB count was found in treatment B, which value was reached $81.63 \pm 4.85 \times 10^8$ CFU/ml and the lowest was in treatment C with the value $28.74 \pm 4.75 \times 10^8$ CFU/ml, but it is lower than BAL in honey is 10.43×10^5-31.30×10^5CFU/ml [13].

The high total BAL count generated IMO from goat feces in treatment B was happened because the goats were fed from legume plants which are rich in N and through digestion of ruminants (four stomachs) and the feces were solid with lower moisture content in comparison to other treatments. Thus, the total concentration of the nutrients was also higher which produces optimal growth of the LAB and fungi. Goat feces contain a C/N ratio of 32.65 and an N content of 1.45% [14] which is needed by microbes to grow.

The low total LAB count of rabbit feces IMO in treatment C is influenced by the pH value of the feces were used as an IMO source. The highest pH of rabbit feces used by other feces is 9.76. The analysis results of total LAB count were not found in organic fertilizer with the addition of rabbit IMO, while in fertilizer with addition goats feces IMO were found so that the total colonies on IMO were also high. Indigenous bacteria in cow feces are Lactobacillus sp, Actinomycetes sp [15] while the type of bacteria in poultry feces is the Streptococcus or Lactobacillus genus [16].

In addition, the rabbits' digestion which is simple or monogastric [17], consume vegetables and concentrates to produce low nutrients for LAB growth results in low N content of IMO as a microbial food source so that bacteria in the phase of death and has stopped producing acid lactate. Lactic acid bacteria will produce lactic acid so that there is a decrease in pH [18], but the pH on rabbit IMO is the highest compared to other treatments.

3.1.5. Total Fungal Count

Total fungal count of IMO was highest in treatment B (48.75±5.74x10^{11} CFU/ml) and the lowest was in treatment C (27.93±4.14x10^{11} CFU/ml). The total fungal count of generated IMO from this study were higher than the generated IMO from rumen contents with soybean medium at the 4th day of breeding stage is 39x10^{9} CFU/ml [19], the IMO tofu 5.08x10^{5} SPK/ml [9] and [12] the IMO generated from coconut fiber with a total mold count 11.07x10^{4} SPK/ml. This condition is influenced by the nutrient content of the fungus which is more available within the feces than the coconut fiber which contains only carbohydrates in the form of cellulose, hemicellulose, and lignin which are difficult to be remodeled by microorganisms.

Total fungal count of goat feces IMO in treatment B and the total fungal of rabbits feces IMO were lower. It is in line with the total LAB colonies which were influenced by microbial growth phase, moisture content, and availability of nutrients for growth. The mold in its growth requires lower moisture content than bacteria. Indigenous mold in cow feces is Aspergillus sp. [15]. In line with the moisture content, the treatment of goat feces IMO has the lowest moisture content. This is caused by the digestive system of the goat which absorbs water in the rumen so that the feces is denser.

3.1.6. Total N

Total N of feces IMO were ranged from 0.08±0.03 to 0.25±0.03% (rabbit and quail). The total N of manures IMO is higher than that of tofu pulp which is 0.06% [9] and is lower than research papaya IMO is 0.45% [8] and coconut fiber IMO is 0.53% [12]. This is influenced by the development of microbes which need N for their lives. Rabbits IMO with the lowest total N cause microbes to be lacking in nutrients for their life so that the microbial content in the IMO is the lowest, while in other feces IMO, total N have been widely used by microbes to grow but it is still available in sufficient quantities so that the microbial population is higher and much higher than coconut fiber IMO.

Quail feces IMO have the highest total N of other treatments. This is due to the type of ration from quail are corn and soybean meal which is high in total N [20]. Different feces in the fermentation process will occur the mineralized process of N by microbes which affect the N content of IMO produced.

3.2. Organic fertilizer

Table 2. pH value, moisture content, N, P, K content and C/N ratio of organic fertilizer.

Treatments	pH	Moisture Content (%)	Total N (%)	P Content (%)	K Content (%)	C/N Ratio
A	7.66±0.42^{a}	68.58±0.77^{a}	1.4140±0.13^{c}	1.6682±0.05^{b}	.0.8493±0.04^{b}	24.31±1.73^{a}
B	7.44±0.25^{bc}	68.78±1.32^{a}	1.1129±0.15^{d}	1.8042±0.04^{a}	0.8823±0.19^{b}	25.68±4.56^{a}
C	7.32±0.20^{c}	67.42±0.89^{b}	1.5954±0.10^{b}	1.7936±0.14^{a}	0.8708±0.16^{b}	20.43±0.97^{b}
D	6.80±0.11^{d}	66.68±0.24^{c}	1.6502±0.12^{ab}	1.8264±0.04^{a}	1.2618±0.25^{a}	17.62±1.88^{c}
E	7.52±0.08^{ab}	68.30±1.24^{ab}	1.7894±0.34^{a}	1.6278±0.02^{b}	0.7189±0.05^{c}	19.30±3.77^{bc}

Note : The mean with different lowercase letters shows a significantly different effect (P <0.05).

Treatments of different IMO from raw manures in the production of organic fertilizer from cow feces to pH value, moisture content, N, P, K content and C/N ratio generated data as in Table 2. Results of variance analysis (ANOVA) showed that IMO treatments at different raw manures give significant effect (P<0.05) on pH, moisture content, N, P, K, and C/N ratio. Treatment D which used the quail feces IMO produced the highest P, K content and the lowest pH and C/N ratio. Based on Duncan's test treatment D was significantly different from other treatments for pH, moisture, K content and was not significantly different from C treatment to N content, with B, C treatments to P values, with treatment E to C/N ratio.

3.2.1. pH value

The organic fertilizers made by using livestock feces IMO produces pH with a range of 6.80±0.11 to 7.66±7.66 at the treatment using the quail and cow feces IMO. The pH range of the produced organic fertilizer is close to the neutral range. This is influenced by the overhaul of animal feces into organic fertilizer by IMO. The IMO generated of organic household waste has a pH 8.4 [21]. The pH within several types of organic fertilizers are shown as follow; compost 7.7, bokashi 8.0, vermicompost 8.2, from. manure 7.9, liquid humus 8.0 [22].

The pH of organic fertilizers that use quail feces IMO is closer to neutral than the other treatments that will help accelerate changes by the assistance of the enzymes to become organic fertilizers. The enzyme works optimally at neutral pH while the optimum pH of the enzyme depends on the origin of the enzyme [23].

3.2.2. Moisture Content

The moisture content of organic fertilizers was ranging between 66.68±0.24 to 68.78±1.32%. The moisture content of organic fertilizer was lowest in quail feces IMO and highest in goat feces IMO. The moisture content of organic fertilizer which was produced is still below SNI. According to the SNI, the moisture content of organic fertilizer is 50%. The moisture content of organic fertilizer was higher compared to SNI is caused by the high moisture content of cow feces used for making the organic fertilizer. In line with the ratio of C/N organic fertilizers that use goat feces IMO have the highest C/N ratio. The higher the C/N ratio, the more crumb the fertilizer and the lower the moisture content, the more optimal the work from microbes to break down cow feces into fertilizer. In line with the research on making the rabbit feces Bokashi using EM4, the lower the moisture content, the lower the C/N fertilizer ratio produced [24].

3.2.3. Total N

The total N of organic fertilizer were ranging between 1.1129±0.15% to 1.7894±0.34%. The highest N content is found in the chicken feces IMO and not significantly different from the treatment of quail feces IMO. Total N in dry animal feces which will be used as fertilizer is 0.85%. After being kept for 21 days in anaerobic conditions without the addition of IMO, the total N turns into 0.89% which means that there was an increase in total N. But, it is higher with IMO addition 0.26 - 0.94% of feces which have not been fermented. The total N of organic fertilizer research results has met the standards of SNI 19-7030-2004, which is 0.40%.

The lower total N in the exclusion of organic fertilizers which use goat feces IMO in line with the results of the composting research using several types of feces with the lowest total N of feces of goats <camel <cow <poultry <buffalo [1].

The high total N organic fertilizer from chicken feces IMO results are almost the same as the solid organic fertilizer from biogas sludge which has been improved in quality by adding blood flour which is 1.7894±0.34% [21], and organic fertilizer using Promi bio-activator is 1.76% by 10 days fermentation [11].

The high total N organic fertilizer from chicken feces IMO results are almost the same as solid organic fertilizer from biogas sludge which has been improved in quality by adding blood flour which is 1.7894±0.34% [25] and organic fertilizer using Promi bio-activator is 1.76% by fermentation for 10 days [14].

There was no real difference between the organic fertilizer made of chicken feces IMO and that of the quail feces IMO due to the characteristics of quail feces which has more intense odor compared to other feces because of its high protein content. Besides that, fertilizer made of quail feces is classified as heat fertilizer which is rapidly decomposed so that absorption by plants is faster. Quail manure contains N 1.56%, P 0.02%, K 1.55%, S 1.24% and C/N 11.58, besides quail manure can improve the efficiency of S uptake [26].

3.2.4. P content

The P levels of organic fertilizer were produced ranged from 1.6278±0.02% (chicken feces IMO) to 1.2618±0.25% (quail IMO) as shown in Table 2. The P level of cow feces that will be used as fertilizer 2.08%, after 21 days without the addition of IMO decreases to 1.87%, but with the addition of IMO more down to 0.45-0.82% of the initial content. However, this P content meets SNI 19-7030-2004 which is 0.10%.

The low P content was caused by a decrease in P content in the fermentation process which consumed by microbes. In line with the results of the study [27] the P level of chicken feces was 2.55% and after fermentation for 4-6 weeks the P content becomes 1.93%.

3.2.5. K content

The K content of organic fertilizer research results is ranging from 0.7189±0.05 to 1.2618±0.25% using chicken and quail feces IMO. The K value of cow feces that will be fermented is 0.91% and is changed to 1.12% after 21 days without the addition of IMO. In general, there is an increase in K values in the fermented feces without IMO except in the one using quail feces IMO. In line with the results of the research on making rabbit feces bokashi, there was a decrease in K values with longer fermentation due to the treatment (reversal) that occurred during the fermentation process [24].

During the fermentation process of making organic fertilizers using cow feces, there was a reshuffle with the help of cow feces IMO. The K content of research results is much higher than the results of research on solid organic fertilizer from sludge biogas of cow feces which has been improved with the addition of husk ash which is 0.3416% [25].

3.2.6. C/N ratio

Organic fertilizers with the use of feces IMO in their remodeling process resulted in a C/N ratio ranging from 17.62±1.88-25.68±4.56 namely the lowest quail feces IMO and highest goat's feces IMO. The C/N ratio of organic fertilizer produced has met SNI 19-7030-2004, which is a maximum of 20. This is influenced by the physical properties of different feces. The high C/N ratio of organic fertilizer that used cow feces IMO because it is slow to break down. Goat and cow fertilizers with more solid experience slower decomposition [28].

Quail feces IMO was used with the highest N content causes fermented cow feces to decompose quickly so that the lowest C/N ratio in treatment D. In line with research the total N chicken feces is higher than cow and goats feces cause chicken feces to get faster [28].

4. Conclusions

Raw manures were used in making IMO and on the best result was shown in the treatment using the goat feces IMO with pH 3.88±0.13, moisture content 88.51±0.41%, total plate count $55.25 \pm 10.21 \times 10^4$ CFU/ml, total BAL count $81.63 \pm 4.85 \times 10^8$ CFU/ml, total fungal count $48.75 \pm 5.74 \times 10^{11}$ CFU/ml and total N 0.16±0.02%. After being was applied in the production of organic fertilizer made of cow feces, it produced the best organic fertilizer by using quail feces IMO with pH 6.80±0.11, moisture content 66.68±0.24%, total N 1.6502±0.12%, P 1.8264±0.04%, K 1.2618±0.25%, and C/N ratio 17.62±1.88.

5. References

[1] M. Irshad, A. E. Eneji, Z. Hussain, and M. Ashraf, "Chemical characterization of fresh and composted livestock manures," J. Soil Sci. Plant Nutr., no. ahead, pp. 0–0, 2013.

[2] Setyorini, R. Saraswati, and E. K. Anwar, 2. Kompos - Balittanah. Litbang Dinas Pertanian, 2015.

[3] K. Tabitha, T. Wilson, and P. G.-O. Joseph, "Influence of Organic and Inorganic Manures on Macro-Nutrients, Micro-Nutrients, and Anti-Nutrients in two Amaranth spp in Kiambu County, Kenya," Asian J. Res. Crop Sci., vol. 1, no. 1, pp. 1–17, 2018.

[4] Mulyono, Membuat Mikroorganisme Lokal (MOL) & Kompos dari Sampah Rumah Tangga.

AgroMedia, 2016.

[5] E. Husen and Irawan, "Efektivitas dan Efisiensi Mikroba Dekomposer - Balittanah," Pros. Semin. Nas. Dan Dialog Sumberd. Lahan Pertan. Buku II Teknol. Pengelolaan Sumberd. Lahan Bogor 18-20 Novemb. 2008 Balai Besar Penelit. Dan Pengemb. Sumberd. Lahan Pertan. Badan Penelit. Dan Pengemb. Pertan. Dep. Pertan., pp. 75–91, Nov. 2008.

[6] Y. H. Indriani and W. B. Prasetya, Cara Mudah dan Cepat Buat Kompos. Jakarta: Penebar Swadaya, 2017.

[7] A. A. Suhastyo, I. Anas, D. A. Santosa, and Lestari, "Studies of microbiology and chemical properties of the Local Microorganisms (MOL) used in rice farming by SRI method (System of Rice Intensification," Sainteks, vol. X, no. 2, pp. 29–39, 2013.

[8] S. Handayani, A. Yunus, and A. Susilowati, "Uji kualitas pupuk organik cair dari berbagai macam mikroorganisme lokal (MOL)," EL - VIVO, vol. 3, no. 1, pp. 54 – 60, 2015.

[9] N. W. Marsiningsih, A. A. N. G. Suwastika, and S. W. Sutari, "Analisis kualitas larutan Mol (Mikroorganisme Lokal) berbasis ampas tahu (Analysis of Local Microorganisms Solution Quality Based on Tofu dregs.)," E-J. Agroekoteknologi Trop., vol. 4, no. 3, 2015.

[10] Abun, Denny Rusmana, and Nyimas Popy Indriani, "Penentuan kecernaan ransum mengandung ampas umbi garut (Maranta arundinacea Linn.) pada ayam broiler dengan metode pemotongan," J. Bionatura, vol. 5, no. 3, pp. 227–238, 2003.

[11] M. I. Said, By Product Ternak Teknologi dan Aplikasinya, 1st ed. Bogor: IPB Press, 2014.

[12] P. A. W. Dharma, A. A. N. G. Suwastika, and N. W. S. Sutari, "Study of coconut husk waste utilization become local microorganism fermented solution," E-J. Agroekoteknologi Trop., vol. 7, no. 2, pp. 200–210, Apr. 2018.

[13] M. Mathialagan, Y. S. J. Thangaraj Edward, P. M. M. David, M. Senthilkumar, M. R. Srinivasan, and S. Mohankumar, "Isolation, Characterization and Identification of Probiotic Lactic Acid Bacteria (LAB) from Honey Bees," Int. J. Curr. Microbiol. Appl. Sci., vol. 7, no. 04, pp. 894–906, Apr. 2018.

[14] L. Trivana and A. Y. Pradhana, "Optimalisasi Waktu Pengomposan dan Kualitas Pupuk Kandang dari Kotoran Kambing dan Debu Sabut Kelapa dengan Bioaktivator PROMI dan Orgadec (Time Optimization of the Composting and Quality of Organic Fertilizer Based on Goat Manure and Coconut Coir Dust using PROMI and Orgadec Bio-Activator)," Sain Vet., vol. 35, no. 1, pp. 136–144, 2017.

[15] Idham, Sudiarso, N. Aini, and Y. Nuraini, "Isolation and identification on microorganism decomposers of Palu local cow manure of central Sulawesi Indonesia.," J. Degraded Min. Lands Manag., vol. 3, no. 4, pp. 625–629, 2016.

[16] L. Nazef, Y. Belguesmia, A. Tani, H. Prévost, and D. Drider, "Identification of Lactic Acid Bacteria from Poultry Feces: Evidence on Anti-Campylobacter and Anti-Listeria Activities," Poult. Sci., vol. 87, no. 2, pp. 329–334, Feb. 2008.

[17] F. Lebas, . "Feeding behaviour in the rabbit," III Int. Rabbit Prod. Symp. Villareal Port., p. 19 pp, Nov. 2005.

[18] R. Setiarto and N. Widhyastuti, "Effect of Lactic Acid Bacteria Fermentation Lactobacillus plantarum B307 to Proximate Levels and Amylography Modified Tacca Flour (Tacca leontopetaloides)," J. Ilmu Pertan. Indones., vol. 21, pp. 7–12, Apr. 2016.

[19] F. Firdaus, B. P. Purwanto, and Salundik, "Dosis penggunaan MIkroorganisme Lokal (MOL) ragi tempedan isi rumen untuk pengomposan," J. Ilmu Produksi Dan Teknol. Has. Peternak., vol. 02, no. 1, pp. 257–261, Jan. 2014.

[20] N. Nuraini, A. Djulardi, and M. Endo Mahata, Pakan non konvensional fermentasi untuk unggas. Lembaga Pengembangan Teknologi Informasi dan Komunikasi (LPTIK) Universitas Andalas, 2017.

[21] A. Niño, A. Rivera, and A. Ramírez, "Production of organic fertilizer with Macro-Micronutrients from the solid waste generated at home," Eur. J. Exp. Biol., vol. 2, no. 1, pp. 199–205, 2012.

[22] M. I. Saldaña y Hernández, R. Gómez-Álvarez, M. del C. Rivera-Cruz, J. D. Álvarez-Solís, J. M. Pat-Fernández, and C. F. Ortiz-García., "The influence of organic fertilizers on the chemical properties of soil and the production of Alpinia purpurata," Cien Inv Agr, vol. 41, no. 2, pp. 215–224, 2014.
[23] D. Y. Primacitra, O. Sjofjan, and M. H. Natsir, "Pengaruh penambahan probiotik (Lactobacillus sp.) dalam pakan terhadap energi metabolis, kecernaan protein dan aktivitas enzim burung puyuh," J Ternak Trop., vol. 15, no. 1, pp. 74–79, 2014.
[24] D. Kurniawan and S. Kumalaningsih, "The effect of effective microorganism 4 (EM4) volume addition 1% and fermentation time on the quality of bokashi fertilizer made from rabbit feces and jackfruit waste," vol. 2, no. 1, pp. 57–66, 2013.
[25] E. Febriyanti, E. Purwati, and Apriman, "The Effect of Addition of Some Organic Material in the Manufacture of Solid Organic Fertilizer of Biogas Sludge from Dairy Faeces on N, P and K Content," J. Peternak. Indones. Indones. J. Anim. Sci., vol. 14, no. 1, pp. 270–278, 2012.
[26] H. Widijanto, N. Anditasari, and Suntoro, "Efficiency of S uptake and rice yield with quail manure and inorganic fertilizer in paddy soil (season II)).," Sains Tanah – J. Ilmu Tanah Dan Agroklimatologi, vol. 8, no. 1, pp. 61–70, 2011.
[27] H. Yulipriyanto, "Karakteristik pengomposan limbah organik kotoran ayam fase thermofilik pada lingkungan alami menggunakan Indore PIT Methode," Semin. Nas. MIPA 2006 Dengan Tema" Penelit. Pendidik. Dan Penerapan MIPA Serta Peranannya Dalam Peningkatan Keprofesionalan Pendidik Dan Tenaga Kependidikan" Yang Diselenggarakanoleh Fak. Dan Ilmu Pengetah. Alam UNY Yogyak., p. 13, Aug. 2006.
[28] B. Satata and M. E. Kusuma, "The effect of three kind manure (cow, chicken, and goat) to the vegetative growth and production of Brachiaria humidicola," J. Ilmu Hewani Trop., vol. 3, no. 2, pp. 5–9, 2014.

Acknowledgments

The award and acknowledgment were addressed to those who have supported the implementation of this competitive grant research, especially to LPPM as a funder through BOPTN contract No. 22/UN.16.17/PP.RD/LPPM/2018, dated March 12, 2018, then Rector, Chairman of LPPM, Dean of Animal Science Faculty, Andalas University.

Physicochemical properties and total plate count of raw salted eggs with blanching and different concentration of aloe vera solution during the salting process

D Novia and I Juliyarsi

Department of Animal Science, Andalas University, Padang, West Sumatra, Indonesia
Email: dnovia@ansci.unand.ac.id.

Abstract. Salted egg is a typical popular snack in some societies in Indonesia. However, it contains high NaCl and cholesterol. One way to cope with it is by utilizing aloe vera which has the chemical composition of a complete and well-known traditional medicine. The purpose of this study is to determine the physicochemical properties and the bacterial colonies forming raw salted eggs with different temperature and concentration of aloe solution for 5 days salting process. This study used 2×3 factorial design with three replications, factor A (without blanching and blanching) and factor B concentration of aloe vera solution (2, 8 and 14%). According to the research, there was an interaction between the process of the different heating and concentrations on NaCl, yolk and albumen pH, cholesterol and total plate count raw salted eggs were produced. The heating process of aloe vera increases the content of NaCl, albumin pH, cholesterol and bacterial colonies formed. Treatment with blanching at 14% concentration effectively produced lowest moisture content, yolk pH, cholesterol, and highest ash content, NaCl content, Albumen pH, total solids and total plate count the best raw salted eggs.
Keywords – raw salted eggs, NaCl content, cholesterol, total plate count, blanching.

1. Introduction

An area which famous for salted egg in West Sumatra is Sicincin [1], while the most famous area for salted eggs in Java is Brebes. The developments of salted eggs making are also supported by the increase of duck population in West Sumatra. The data in the 2007-2009 shows that its population increased from 1 million in 2007 to 1.19 million in 2009. The duck which is mostly kept by people is laying ducks.

For people who have cholesterol problems, blood clots, and heart disease, the salted egg should not be consumed excessively since it has the high content of cholesterol content. Cholesterol contained by duck egg yolk of Pitalah is 239.81 mg/dl [2]. Salted egg cholesterol level can be lowered by implementing 2 parts of the salting process which can make the salted egg cholesterol level becomes 89.30 mg/dl [3]. In addition to lowering its cholesterol is to make the herbal salted egg.

Salted herb is a salty snack made from eggs. It is a source of vegetable protein which also has the effect of herbal for those who consume them. Research on salted egg with the effects of herbs by adding natural ingredients in the treatment process which has been done is by using onion skin [4], ginger and white turmeric extract [5], garlic oil which has antibacterial activity [6], gambier liquid waste [7],[8], and aloe vera [9].

Aloe vera has a function to increase the thyroid function and autoimmunity[10], extend the shelf life of chicken nuggets stored in the refrigerator [11], antifungal and anti-ochratoxigenic activities

[12], diabetic ulcers [13], antibacterial and ant biofilm potential [14], immunomodulatory, gastroprotective, hypoglycemic anticancer, anti-inflammatory properties, as well as wound and burn healing [15]. This main function of Aloe vera is affected by the compounds of acemannan, aloin, saponins, enzymes, amino acids, carbohydrates, vitamins, minerals which will affect the salting process of salted eggs produced.

The heating process or blanching solution of aloe vera was done in the temperature of 80°C. On the other hand, the salted egg making without the heating process was done through a process of salted herbal aloe vera diffusion through the shell pores. The specific objective of this research is to produce salted egg with herbal quality and preferably through the utilization of aloe vera and concentration different blanching in salted egg salting.

2. Research methods

This study uses duck egg from Tegal (Anas javanica) with bluish green colored egg shells, aged up to 48 hours as many as 72 items weighing 65-70 grams obtained from duck breeders in Piai. Meanwhile, the aloe vera was obtained from Koto Parak yard, Pauh District, Padang, Indonesia. The other materials are husk ash, salt, and water. The adjuvant for the analysis is distilled cholesterol kit. In addition, the equipment used was an analytical scale and spectrophotometers.

The method used was an experimental method using a factorial randomized block design. If the result of the variance is influential then further test of Duncan's is conducted by using pattern of 2x3 for 3 replications factor A (without blanching and with blanching) and factor B concentration of aloe vera (2, 8, and 14%) was observed at day 5 on moisture, ash, salinity, pH and salted egg yolk and white, total solid, total cholesterol content and bacterial colonies.

3. Results and discussion

Table 1. Total mean of bacterial colony physicochemical and raw salted egg salting for 5 days.

Parameter Observed	Treatment					
	Without Blanching (A1)			Blanching (A2)		
	2% (B1)	8% (B2)	14% (B3)	2% (B1)	8% (B2)	14% (B3)
Moisture content (%)	62.11	63.72	57.14	62.60	60.83	55.89
Ash content (%)	1.86	1.85	1.81	2.17	1.84	2.35
NaCl content (%)	1.06^{c}	1.47^{bc}	1.11^{bc}	1.52^{ab}	1.52^{ab}	1.53^{a}
Yolk pH	6.80^{a}	6.61^{c}	6.70^{b}	6.60^{c}	6.80^{a}	6.60^{c}
Albumen pH	6.91^{c}	8.33^{b}	6.81^{c}	8.79^{a}	8.40^{b}	8.41^{b}
Cholesterol content (mg/dl)	80.27^{cd}	84.77^{bcd}	128.00^{abc}	135.37^{ab}	148.47^{a}	56.13^{d}
Total solids	38.70	38.28	48.18	38.98	40.92	46.51
Total plate count ($x10^2$CFU/g)	2.67^{a}	1.33^{b}	2.33^{ab}	2.00^{b}	2.67^{a}	3.33^{a}

Description: The average of the different small superscript indicates that the interaction is significantly different ($P < 0.05$).

The analysis result of variance shows that the interaction between blanching and concentration of aloe vera shows a different solution of the levels of NaCl, pH yellow yolk and egg white, cholesterol levels and total bacterial colonies producing the raw salted eggs. The best treatment is A2B3 which are liquid aloe vera in blanching, the concentration of 14%, lowest moisture content, yolk pH, cholesterol, and highest ash content, NaCl content, Albumen pH, total solids, and total plate count (Table 1).

Along with the high concentration of aloe vera blanching solution, it lowers the water content, cholesterol, egg white pH and increases the ash content, total solids and total bacterial colonies. This is caused by the fiber and enzymes contained in aloe vera. The fiber content of the aloe vera solution in the blanching will be perfectly deposited so that its concentration of 14% is optimized in its function as an adsorbent that absorbs water from the eggs and cholesterol in the yolk. Thus, the water content of the cholesterol is the lowest. In addition, the diffusion process of mineral and NaCl in the salting solution is also not disturbed because the fibers are sediment so that the ash and NaCl are highest for the salted egg which is made through blanching process. Also, according to [16], aloe vera has anti-nutrition compounds; aloin and saponins which in the extraction for 60 minutes need to be heated in the temperature of 80°C. This anti-nutrition inhibits the substances and increases the antioxidant activity [16].

The treatment is effective in lowering cholesterol without blanching with a concentration of 2% ie A1B1 treatment. Along with increasing the concentration of the solution, using aloe vera also increases the blood cholesterol levels of salted eggs produced. It is affected by the content of enzyme in aloe vera solution which is still active (without blanching). The blanching process will inactivate the enzymes to work, avoids the degradation of the protein, improved the molecular weight of the protein and the color [17]. The process of adsorption of cholesterol in the treatment without blanching at the lowest concentration of 2%, with the increase in concentration will increase the enzyme activity that would interfere the absorption of cholesterol so that the cholesterol content increases.

Salted egg making without going through the blanching process produces lower salt levels than the making which uses the blanching process. It is also affected by the activity in the enzyme that inhibits the absorption of NaCl.

4. Conclusion

From the research conducted, it can be concluded that in order to get the herbal salted egg by using aloe vera, the best treatment is through salting process for 5 days using aloe vera solution at blanching and the concentration of 14%.

5. References

[1] D. Novia, I. Juliyarsi, and S. Melia, "Quality improvement and production of business group salted eggs in sicincin, padang pariaman regency," *LOGISTA - J. Ilm. Pengabdi. Kpd. Masy.*, vol. 2, no. 1(Jan-Jun), pp. 1–14, Apr. 2018.

[2] F. Yunenshi, "Pengaruh Pemberian Probiotik (Pediococcus pentosaceus) Asal Fermentasi Kakao Hibrid terhadap Penurunan Kolesterol Telur Itik Pitalah," *Tesis Program Pasca Sarj. Unand*, 2011.

[3] D. Novia, S. Melia, and I. Juliyarsi, "Kadar kolesterol telur asin mentah hasil pengasinan dalam larutan abu," Pros. Semin. Nas. Peran. Teknol. Pangan Dan Gizi Dalam Meningkat. Mutu Keamanan Dan Kehalalan Prod. Pangan Lokal Padang Indones., pp. 234–239, 2013.

[4] D. Novia, I. Juliyarsi, and P. Andalusia, "The evaluation of colony bacteria and organoleptic of salted duck eggs which soaked in onion skin (Allium ascalonicum) extract," *J. Peternak. Indones. Indones. J. Anim. Sci.*, vol. 13, no. 2, pp. 92–98, Jun. 2011.

[5] D. G. Wibowo, Y. A. Widanti, and A. Mustofa, "Penambahan ekstrak jahe (Zingiber officinale var Amarum) dan ekstrak kunyit putih (Curcuma zedoaria) pada pembuatan telurasin dengan variasi lama pemeraman," *J. Teknol. Pertan.*, vol. 8, no. 2, Jan. 2018.

[6] P. W. Harlina *et al.*, "Effect of Garlic Oil on Lipid Oxidation, Fatty Acid Profiles and Microstructure of Salted Duck Eggs," *J. Food Process. Preserv.*, vol. 39, no. 6, pp. 2897–2911, Dec. 2015.

[7] D. Novia, Juliyarsi, I, Sandra, Afriani, Yuherman, and Muhammad, R, "Soaking Salted Eggs in Gambier Liquid Waste Inhibit bacterial Growth - SciAlert Responsive Version," *Pakistan Journal of Biological Sciences Volume 17 (3): 424-428, 2014*, 2014. [Online]. Available: https://scialert.net/fulltextmobile/?doi=pjbs.2014.424.428. [Accessed: 20-Jul-2018].

[8] D. Novia, E. Vebriyanti, and H. F. Hakim, "Evaluation of Heating the Gambier Liquid Waste on the Quality of Raw Salted Eggs," *Int. J. Poult. Sci.*, vol. 16, no. 9, pp. 369–373, Aug.

2017.

[9] A. Allismawita, D. Novia, and I. Putra, "Evaluation of bacterial colony forming and shelf life salted eggs with soaked in Aloe vera (Aloevera barbadensis Miller) solution," *J. Peternak. Indones. Indones. J. Anim. Sci.*, vol. 16, no. 2, pp. 71–77, Jun. 2014.

[10] D. Metro, V. Cernaro, M. Papa, and S. Benvenga, "Marked improvement of thyroid function and autoimmunity by Aloe barbadensis miller juice in patients with subclinical hypothyroidism," *J. Clin. Transl. Endocrinol.*, vol. 11, pp. 18–25, Mar. 2018.

[11] M. Shahrezaee, S. Soleimanian-Zad, N. Soltanizadeh, and S. Akbari-Alavijeh, "Use of Aloe vera gel powder to enhance the shelf life of chicken nugget during refrigeration storage," *LWT*, vol. 95, pp. 380–386, Sep. 2018.

[12] I. Dammak *et al.*, "In vitro antifungal and anti-ochratoxigenic activities of Aloe vera gel against Aspergillus carbonarius isolated from grapes," *Ind. Crops Prod.*, vol. 123, pp. 416–423, Nov. 2018.

[13] Y. Sari, I. Purnawan, D. W. Kurniawan, and E. Sutrisna, "A Comparative Study of the Effects of Nigella sativa Oil Gel and Aloe Vera Gel on Wound Healing in Diabetic Rats," *J. Evid.-Based Integr. Med.*, vol. 23, p. 2515690X18772804, Jan. 2018.

[14] F. Salah, Y. E. Ghoul, A. Mahdhi, H. Majdoub, N. Jarroux, and F. Sakli, "Effect of the deacetylation degree on the antibacterial and antibiofilm activity of acemannan from Aloe vera," *Ind. Crops Prod.*, vol. 103, pp. 13–18, Sep. 2017.

[15] A. A. Maan *et al.*, "The therapeutic properties and applications of Aloe vera: A review," *J. Herb. Med.*, vol. 12, pp. 1–10, Jun. 2018.

[16] Narsih, S. Kumalaningsih, Wignyanto, and S. Wijana, "Identification of aloin and saponin and chemical composition of volatile constituents from Aloe vera (L) Peel pdf - Penelusuran Google," *J Agric Food Tech*, vol. 2, no. 5, pp. 79–84, 2012.

[17] M. Hadidi, A. Ibarz, J. Conde, and J. Pagan, "Optimisation of steam blanching on enzymatic activity, color and protein degradation of alfalfa (Medicago sativa) to improve some quality characteristics of its edible protein," *Food Chem.*, vol. 276, pp. 591–598, Mar. 2019.

Acknowledgment

Awards and acknowledgments were addressed to Ristek Dikti as funding through Research Grant Number: 020/SP2H/LT/DRPM/II/2016, dated February 17, 2016.

Characteristics based of edible film made from whey with isolated lactic acid bacteria from tempoyak as probiotics packaging

Indri Juliyarsi[1], Arief [2], Akmal Djamaan[3], and Endang Purwati[4]

[1,2,4] Lecturer at the Faculty of Animal Science, Andalas University, Padang, West Sumatera, Indonesia.
Email: indrijuliyarsi@ansci.unand.ac.id.
Corresponding author: purwati17@ansci.unand.ac.id.
[3] Lecturer at the Faculty of Pharmacy, Andalas University, Padang, West Sumatera, Indonesia.

Abstract. Edible film made from whey, a waste obtained from milk and used as primary packaging. This research aimed at isolated and identification of lactic acid bacteria derived from tempoyak, subsequently applied in the manufacture of edible film and have the properties of probiotics. Physical properties measured from the edible film are moisture content, pH, solubility and the thickness of the time. This research uses material from waste milk whey and tempoyak. The method used is an experiment using a random design group that consists of four isolates tempoyak addition, treatment A (0%), B (4%), C (8%), into a solution of whey and six replicates. Treatment results show to a thickness ranging from 0.20-0.25 mm and already meet the standards JIS 1975.
Keywords – edible film, whey, isolated lactic acid bacteria, tempoyak, probiotics packanging.

1. Introduction

Tempoyak is a type of traditional Malaysian food made by Durian that is undergoing the fermentation process. The fermentation process is spontaneous tempoyak involving microorganisms that exist in nature. The production of tempoyak is quite simply where the meat Durian fruit that comes from old, anaerobic conditions are left in during the 3-7 days at room temperature. Sometimes someone makes you add salt 1-3% [1].

In the process of fermentation of lactic acid bacteria producing tempoyak, this is due to the synthesis of Durian, which consists of about 15-20% [2], which is the source of energy in the development of lactic acid bacteria. Species *Lactobacillus* It is the predominant lactic acid bacteria found in isolated tempoyak from Indonesia and Malaysia. However, lactic acid bacteria in tempoyak also depend on the location where such products exist. *Lb. plantarum, Lb. brevis, Lb. mali, Lb. fermentum* found in tempoyak, that comes from the Malaysia [3], While in Indonesia, according to a study [4] and [5] There Isolate *Lb. plantarum, Lb. casei, Lb. corynebacterium* and *Lb. fersantum*, the latest research [6], found *Fructobacillus durionis* and *Lb. plantarum* as dominant lactic acid bacteria in tempoyak from Malaysia with total lactic acid bacteria 8.88-10, 42 log CFU/g.

One way of keeping food was put (*packaging*) and the right materials. According to [7], also known as packaging, and is one way of maintaining agricultural materials, because the packaging can extend the life of the material. Together with human consciousness will this problem than the type of packaging of biological materials and derived from renewable materials (*renewable*) and financially. One of the types of packaging that is environmentally friendly packaging is edible (*edible packaging*). The advantages of *edible packaging* It is able to protect the original appearance of food products, the product can be maintained and can be eaten as well as safe for the environment and is the packaging of the future [8].

Edible film is a thin layer made of edible materials, formed over the food ingredient which serves as an obstacle to mass transport (e.g., moisture, oxygen, dissolved substances and fat) and or as *carrier* Food ingredients or additives and or to improve food handling [9].

In this study the author will use the basic ingredients whey. Whey generally a waste of casein or cheese. Possible food and energy whey will be lost if not used, given the whey contains about 55% of the total diet from milk [10]. In addition to hydrocolloid, other construction materials, edible It is a polysaccharide of cellulose and its derivatives. While the lipid used can be derived from natural wax (*beeswax, Carnauba wax, paraffin wax*), glycerin, fatty acids, and emulsifier.

Based on research [11] whey by using the carboxymethylcellulose (CMC) with as much as 1% and 3% glycerol significantly impacts against the thickness of the film but had no effect on moisture content, pH and solubility. In review research on the edible Film and *coating* [8], stating the edible film is very active food packaging is required as the future, some biopolymer It shows excellent performance as a carrier for the active compounds that are exported so taken food packaging is worth consuming for now and the future. Further research by [6], found *Fructobacillus durionis* and *Lb. plantarum* as dominant lactic acid bacteria in tempoyak from Malaysia with total lactic acid bacteria 8.88-10, 42 Log CFU/g, which can be used as active compounds antimicrobial in the production edible film.

1.1. The purpose of the research

The research purpose is to determine lactic acid bacteria from tempoyak on edible film from waste of cheese as probiotic packaging.

2. Material and methods

2.1. Material of research

This research material is Durian Tempoyak derived from the region of Pariaman, whey from the manufacture cheese "Lasi" in District Agam, West Sumatera, acid acetate, CMC, glycerine, acetic acid, ethanol, *Man Rogossa Sharpe Agar* (MRSA) (Merck), MRSB (Merck), *Muller Hilton Agar* (MHA) (Biolife), *Nutrient Agar (NA)* (Merck), *buffer peptone water,* aquadest, alcohol, spirits, HCl 0,1N, NaOH 0,1N.

The equipment used is *cool box, t*hermometer, pH meters, bowls, petri dish, incubators, needles, counting cup, analytical scales, Erlenmeyer, Bunsen, Test tubes, *beaker glass*, the dropper drops, hockey stick, *autoclave, centrifuge, vortex,* Spectrophotometer and small pipette.

2.2. Isolate lactic acid bacteria conventional

2.2.1. Sampling

Tempoyak samples taken from the area of Pariaman.

2.2.2. The measurement of physical and chemical properties and total lactic acid bacteria Tempoyak

- Physical and chemical properties of tempoyak i.e.: pH, Moisture content [12]
- Total Lactic acid bacteria [13]

2.2.3. The isolation of lactic acid bacteria [13]

2.2.4. Characteristics of lactic acid bacteria [13]

- Morphology
- Biochemistry
  - Gram pigmentation
  - Catalase Test

2.3. Procedure Edible Film [7], [14]

The realization of edible film whey with the addition of treatment isolate lactic acid bacteria A (0%), B (4%), C (8%), with 6 replications as a group, then carried out measurements of physical properties edible film whey

2.3.1. Moisture content [12]

2.3.2. pH [12]

2.3.3. Thickness of the film [15]

2.3.4. Solubility [16]

Data Analyses Statistics with variance analysis and if treatment shows different results for Real ($P < 0.05$) then conducted further tests using *Duncan's Multiple Range Test (DMRT).*

3. Result and discussion

3.1. Isolate lactic acid bacteria from tempoyak

The value of chemistry tempoyak the value of pH, moisture, protein, fat content can be treated in Table 1.

Table 1. The chemical composition of tempoyak.

The value of chemistry	Quantity
pH	3.89
Moisture content (%)	70.21
Protein (%)	5.04
Fat content (%)	6.11

Tempoyak is composed of proteins, fats and carbohydrates, and has a sour taste because of pH low. pH this low moisture content producing 3.89, moisture content 70.21%, protein 5.04% 6.11% fat content, tempoyak, associated with the age of tempoyak is greater. This is due to the accumulation of lactic acid produced during the development of lactic acid bacteria. [17] In traditional fermentation occurs spontaneously, germs that are naturally present in food will increase and lead to a succession of the dominant microbe or substitution in the next process. Lactic acid bacteria were changed by producing more acid, so that pH down about 3.89 even lower.

3.1.1. Isolation and cleaning of balancing

The calculation of the total tempoyak by Padang Pariaman District as many 16 x 10^8 CFU/g. Some bacteria are taken randomly by the moisture content in order. Further marked for the nature of the classification. After the isolated LAB found with a characteristic rounded, creamy-white and colored convex. After that, proceed with the determination of the morphology of the LAB is selected. The identification made in this investigation is microscopic and macroscopic.

Compared to some previous reports on the types of processed milk food [17]. This may indicate that the viability of the tempoyak, which in the product is a basic requirement for the development of functional food more. It is therefore necessary to the characterization microbial diversity and synthesis of conventional dairy products.

3.1.2. Characterization of isolate elected

Macroscopic (Shape, size and color) LAB Colony found white-beige, round, and Convex for MRS Agar Merck. Macroscopic observation, after cleansing is done up to two times smoother than the census of bacteria to be used for further testing, i.e. the identification of microscopic. Microscopic determination in this research is to adhere to physiological properties Isolate LAB. Observations made from test grams. Gram test conducted to know the gram positive or gram-negative bacteria is characterized by the absorption of color reagents by the bacteria. If the bacteria absorb the color reagent violet colored crystal violet, gram-positive bacteria (+), while Gram-negative bacteria (-) is a reagent color safranin which is red. The results of the laboratory isolated gram pigmentation of curd from Padang Pariaman recovered formed the morphology of the bacteria, i.e. rods (Bacillus).

Gram pigmentation results in each Isolate LAB obtained in this study is Gram positive (+) bacteria. Characterized by bacteria that absorb the color purple violet and has the form Bacillus round rods. This is in line with the declaration by [18] stating that Gram-positive bacteria will get violet purple color. Even if it has been washed with alcohol and when given safranin red, the bacteria will still be purple, while the red shows gram-negative bacteria.

Biochemical test conducted with catalase. Catalase test is done in an Isolate LAB to learn the potential of bacteria produce catalase enzymes and isolate tolerance to oxygen. Biochemical test made with hydrogen peroxide (H_2O_2) on the review of bacteria for glass material. Observations made were the occurrence or absence of gas bubbles in the study of bacteria. In the event of a gas bubble showed that the bacteria is catalase positive, while the absence of gas is catalase negative. Because of the characteristics of the parcels are usually Gram-positive pigmentation, reacts negatively to catalase and not to the spore. The results of the experiments carried out in the Isolate They have a negative catalase which is characterized by the absence of gas bubbles in the bacteria exhausted in H_2O_2. The results of this study are like [19] who gets a negative catalase isolate laboratory of the intestines of shrimps. And also received a negative catalase isolate Mango [20].

In addition to biochemical testing, catalase test is a test of the type of fermentation. The purpose of this test is to classify the BAL belonged to the groups homofermentative or heterofermentative. Observation is done by examining the formation of air bubbles in the tube Durham. Research results isolate tempoyak Padang Pariaman are bacteria homofermentative in the treatment of tempoyak fermentation without the addition.

There are two types of lactic acid fermentation bacteria, namely homofermentative and heterofermentative [21]. Homofermentative lactic acid bacteria produce lactic acid as the main fermentation products, while lactic acid bacteria lactic acid additionally heterofermentative also produce ethanol, other acids such as acetic acid and Gas $Co._2$, so when lactic acid bacteria tested to produce accommodation gas. In the Tube Durham lactic acid bacteria, said heterofermentative, while isolate which does not produce or produces a gas called homofermentative. Results of research [22], antibacterial activity research of lactic acid bacteria *Lactobacillus delbreukii* Also obtained BAL formula homofermentative of 6 isolates isolated from yogurt on the commercial market.

3.1.3. The Activity of Antimicrobial

To select and get isolate the most likely LAB became a candidate probiotic. In addition, the ability to know candidates' probiotics suppress the growth of pathogenic bacteria *Listeria monocytogenes, Staphylococcus aureus* and *Escherichia coli O157*. The screening process is conducted using qualitative diffusion methods well. In this study do not use positive controls such as antibiotics, because tests are done to learn the potential of Isolate LAB was found to block the growth of pathogenic bacteria. The diameter of the clear belt is formed in the Isolate as shown in Table 2 below.

Table 2. Diameter clear band test antimicrobial (mm).

Isolate	Clear Zone (mm)		
	E. coli O_{157}	*S. aureus*	*L. monocytogenes*
Tempoyak	12,3	19,3	17,3

Larger obstruction zones were taken from isolated tempoyak (19.3 mm) in *Staphylococcus aureus*, (17.3 mm) in *Listeria monocytogene* and suspension of low *Escheria Coli* (12.3 mm). This is the third-test bacteria that have a different character. And the opinion of [23], each test bacteria has resilience separately against many kinds of organic acids. *L. monocytogenes* They have a greater susceptibility to lactic acid than acetic acid. *E. coli* high sensitivity against lactic acid and acetic acid, while *S. aureus.* They have the highest resistance to acids compared to other bacteria. As seen by [24] that the indicator showing the ability of packages to inhibit the growth of pathogenic bacteria can be seen from the clear zone produced during the test Antimicrobial and related to the ability of packages to produce lactic and secondary metabolites bacteriocin.

The resulting results show that the Isolate Padang Pariaman's LAB were able to block the growth of pathogenic bacteria that can harm humans, especially bacteria *Staphylococcus aureus*. Probiotics can be used as a treatment at the sight of inhibitory zones tested for pathogenic bacteria.

3.2. Characteristics of Physic Edible Film Whey

Track Moisture content, pH, the thickness, solubility and total lactic acid bacteria colony edible film whey with the addition of lactic acid bacteria from tempoyak can be seen in Table 3.

Table 3. Characteristics of physic and total LAB edible film whey.

Treatments	Moisture content (%)	pH	Thickness (mm)	Time Solubility (seconds)	Total Colony LAB (x 10^6CFU/ml)
A	$19{,}5^a$	$7{,}07^a$	0,21	$32{,}89^a$	0^a
B	$15{,}1^b$	$7{,}23^b$	0,23	$33{,}48^a$	7^b
C	$15{,}9^b$	$7{,}31^b$	0,24	$34{,}68^b$	15^c

On the table, be considered that the average value of the moisture content, pH actual different shows, with the addition of lactic acid bacteria than without treatment, is A (0%). The moisture content has decreased with the addition of lactic acid bacteria this is because isolate LAB free water binds to edible film in addition to the use of other ingredients like glycerin that is hydrophilic [11]. While increasing the pH, caused by alkali from glycerol that cannot be lowered by adding lactic acid bacteria. Based on international standards (Japan JIS Z 1707:1997) Maximum moisture content is 13%, the average moisture content of the edible film whey this exceeds the standard, this is due to the fact that the long drying in the oven for 18 hours, with temperatures 50°C was not able to reduce moisture content.

Treatment with the addition of lactic acid bacteria has no effect on the thickness due to the volume of each edible film that is poured into petri dish the same 50 ml, based on JIS Z 1707:1997, the standard thickness is 0.25 mm. Thus, in this study, the thickness of the edible film whey already meets the standards. Lactic acid bacteria, including safe microorganisms, when added to food: for nature is not toxic and does not produce the toxin, it is then called food grade microorganism Otherwise known as microorganisms *Generally Recognized as Safe* (GRAS) that is not at risk in health, even various types of basil the nutritious health [25].

Time is the actual effect of C solubility treatment (8%) from A and B, this is due to an increasing number of additions of lactic acid bacteria can increase your connective strength of lactic acid bacteria with water connected that is edible film whey time, thus adding solubility. It is also in line with previous research [14], that certain ingredients like glycerin do not dissolve in water because they are non-polar.

Concerning the parameters of all lactic acid bacteria colonies, the treatment effect is really against the treatment, in which the number of lactic acid bacteria also increased, the average temperature in drying does not cause bacteria to die. Optimum growth temperature is ± 37-45 °C, in general Distract, are anaerobic, catalase negative and oxidase positive, the fermentation produces lactic acid. Lactic acid bacteria can grow in sugars, alcohol, and high levels of salt, able to ferment the monosaccharides and monosaccharides. Most of the parcels can be well developed in environments that have or do not have O_2 (Unconscious in O_2), including *Anaerobic aerotolerant*.

4. Conclucion

Based the results of research concluded the more many additions of LAB affected the in moisture content, pH, time soluble, as well as the total number of LAB from edible film whey. However, no effect of thickness of edible film.

5. References

[1] Mardalena. 2016. Fase pertumbuhan isolasi bakteri asam laktat tempoyak asal Jambi yang disimpan pada suhu kamar. Jurnal Sains Peternakan Indonesia Vol 11. No.1: 59-66.

[2] Ketsa, S and T. Daengkranit. 1998. Physiological changes during postharvest ripening of durian fruit (Durio zibethinus). J.Hortic.Sci.Biotechnol. 73: 575-577.

[3] Leisner J.J., M. Vancameyt., B. Rusul. Pott, K. Levebre and L. K. Tee. 2001. Identification of lactic acid bacteria constituting the predominant microflora in an-acid fermented condiment (tempoyak) popular in Malaysia. Int. J.Food. Microbiol. 63: 149-157.

[4] Wirawati C. U. 2002. Potensi bakteri asam laktat yang diisolasi dari tempoyak sebagai probiotik. Tesis. Institut Pertanian Bogor. Bogor.

[5] Han, H. J. 2005. Innovations in Food Packaging. Department of Food Science University of Manitoba Winnipeg, Manitoba Canada.

[6] Chuah, L. O., A. K. S. Syuhada, M. T, Liong., A. Rosma., K. L. Thong and G. Rusul. 2016. Physio-chemical, microbiological properties of tempoyak and molecular characteristic of lactic acid bacteria. Food Microbiology (58); 95-104.

[7] Syarief, R., S. Santausa., dan St. Isyana. 2001. Buku Monograf Teknologi Pengemasan Pangan. Lab. Rekayasa Proses Pangan, PAU Pangan Dan Gizi IPB. Bogor.

[8] Mellinas C., A. Valdes., M. Ramos., N. Burgos., D. C. Garigos and A. Jimensz. 2015. Review: Active *edible films*: current state and future trends. J.Appl. Polym. Sci: DOI: 10.1002/APP 42631.

[9] Krochta, J. M., Baldwin, E. A. and M. O. N. Carriedo. 1994. Edible Coatings and Film to Improve Food Quality. Echnomic Publication Company., Inc., USA.

[10] Vinderola, C. G, Gueimonde, M, Delgado, T, Reinheimer J. A. and de los Reyes – Gavilan, C. G. 2000. Characteristics carbonated fermented milk and survival of probiotic bacteria. Internasional Dairy Journal, 10-213-220.

[11] Juliyarsi, I., S. Melia., and A. Sukma. 2011. The quality of edible film by using glycerol as plastisizer. Pak. J. Nutr., 10: 884-887.

[12] Apriyantono, A., D. Fardiaz., N. L. Puspitasari., Sedarnawati dan S. Budiyanto. 2006. Analisis Pangan. PAU Pangan dan Gizi. Institut Pertanian Bogor. Bogor.

[13] Purwati, E., S. N. Aritonang., S. Melia., I. Juliyarsi., dan H. Purwanto. 2016. Manfaat Probiotik Bakteri Asam Laktat Dadiah Menunjang Kesehatan Masyarakat. Penerbit Lembaga Literasi Dayak dan Universitas Andalas. Padang.

[14] Juliyarsi, I., S. Melia dan Tiara Novita. 2009. Pengaruh penambahan carboxylmetylcellulose dan sorbitol dalam pembuatan edible film berbahan dasar whey. Laporan Penelitian Dosen. Universitas Andalas. Padang.

[15] Mc. Hugh, T. H., J. F. Aujard and J. M. Krochta. 1994. *Plasticized whey protein edible films : water vapor permeability.* J. Food Sci, 59: 416 – 419.

[16] Gontard, N., S. Guilbert, and J. L. Cuq. 1993. Water and glycerol as plasticizer affect mechanical and water vapor barrier properties of an edible wheat gluten film. J. Food Sci, 58: 206 – 211.

[17] Taufik, E. 2004. Dadih susu sapi hasil fermentasi berbagai bakteri probiotik yang disimpan pada suhu rendah. Laporan Penelitian Dosen. Institut Pertanian Bogor.

[18] Surono, I. S. 2016. Probiotik, Mikrobiome dan Pangan Fungsional. Deepublish. Yogyakarta.

[19] Romadhon, Subagyo dan S. Margino. 2012. Isolasi dan karakteristik bakteri asam laktat dari usus udang penghasil bakteriosin sebagai agen antibakteri pada produk-produk hasil peternakan. Jurnal Saintek Perikanan. Vol. 8 No. 1 Hal 59-64.

[20] Koswara, S. 2006. Proses dan produk fermentasi pangan. Ebook Pangan.com.

[21] Mawarwati, S. B. Widjanarko dan T. Susanto. 2001. Mempelajari Karakteristik Edible Film Berantioksidan dari Germ Gandum (Tetricum Aestivum. L) dan Pengaruh Dalam Pengendalian Pencoklatan Pada Irisan Apel (Malus Sylvesrtris). Jurnal Biosain. Vol 1 No. 1 . Pp 61-7.

[22] Magalhaes, Madigan, M. T., Martinko, J. M., and Parker, J. 2011. *Brock Biology of Microorganism 10th ed.* Pearson Education, Inc. New Jersey. xxv + 1019 p.. 2001.

[23] Rodriguez, M, J. Oses, K. Ziani, and J. I. Mate. 2006. Combined effect of plasticizers and surfactants on the physical properties of starch-based edible films. Food Res. Int. 39: 840-846.

[24] Prihatiningsih, N. 2000. Pengaruh penambahan sorbitol dan asam palmitat terhadap ketebalan film dan sifat mekanik edible film dari zein. Jurusan Teknologi Hasil Pertanian. Universitas Gadjah Mada. Jogjakarta.

[25] Kusmiati dan Malik, A. 2002. Aktivitas bakteriosin dari basil Leuconostoc mesenteroides Pbac1 pada banyak media. Kaprikornus Kesehatan.

Acknowledgements

Thanks to Head of Laboratory Technology and Processing Animal Products, Dean of Faculty of Animal Science, Andalas University, LPPM Andalas University and Kementerian Riset Teknologi dan Pendidikan Tinggi.

The characteristics of the pericarp of *garcinia mangostana* (mangosteen) extract as natural antioxidants in *rendang*

S Melia, D Novia, I Juliyarsi, and E Purwati

Department of Animal Science, Andalas University, Padang, West Sumatera, Indonesia
Email: srimelia75@ansci.unand.ac.id

Abstract. The purpose of this study is to determine the role of pericarp of *Garcinia mangostana* (commonly known as *Mangosteen*) extract as natural antioxidants in *rendang*. The materials used in this study were obtained from *Mangosteen* of Pariaman, West Sumatera, Indonesia. Xhantones were extracted from the pericarp of *Garcinia mangostana by* using ethanol. The *rendang* were made with additional *Mangosteen* extract: 0%, 0.25%, 0.5%, 0.75% and 1%. The data obtained were processed statistically. If treatment showed significant results (P <0.05), it then further tested by using Duncan's Multiple Range Test (DMRT) with SPSS method. Results indicated that pericarp of *Mangosteen* produces the antioxidants. Xanthones content of *Mangosteen* extract was 37.5 mg/100ml. Another component were found: total phenolic content 90 mg gallic acid equivalents (GAE/g dry weight) mg GAE/g, tannins 3.02% and anthocyanins 1.85%. The organoleptic test showed that addition of the extract in rendang was not significant (P>0.05) and addition of *mangosteen* extract could extend the storage life of rendang.
Keywords – pericarp, *garcinia mangostana* extract, antioxidants, rendang.

1. Introduction

The fruit of Garcinia mangostana L Rind, also known as the 'queen of tropical fruits', has been used for hundreds of years around the world, mostly in Southeast Asia as medicine. Over the past decades, it was shown that G. mangostana (commonly called Mangosteen) contains of xanthones, a class of polyphenolic compounds [1].

Various kinds of xanthones in Mangosteen had been proven to have strong antioxidant activity including alpha mangostin [2], [3]. Besides containing antioxidant activities, Mangosteen has anti-inflamatory, antiallergy, antibacterial, antiviral and antifungal activities [4], [5]. According to [6], antioxidant activity of G. mangostana rind can be extracted with 50% ethanol, which concentration is IC50 wa 5.94 μg/ml.

Rendang is one of the famous traditional food from Indonesia which has unique characteristic in ingredients and it's processing. It is made of beef and coconut milk as the main ingredients; it is the reason of Rendang contains high percentage of fat. Those ingredients tend to flaw during the storing which lead to the alteration in favor; the taste will be rancid due to fats oxidized. The breakage process on fat will recall the free radical which is dangerous if the contamination is consumed in large amount. Free radical tends to cause severe illness, but it can be avoided by adding antioxidant to the food. Lipid oxidation due to raical reactions is one of the major problem. The antioxidant systems have evolved for protection, and are generally a prerequisite for the survival of aerobic organisms.

The antioxidants are defined as any substance that present in low concentrations compare to those of an oxidisable substance which significantly delay or prevent oxidation of the substrate. The antioxidants present in all plants and all parts of the plant. A natural antioxidant which can be used as a treatment is xhantones from the pericarp of Mangosteen.

1.1. The purpose of the research
The research purpose is to determine the characteristics of pericarp of Garcinia mangostana (*Mangosteen*) extract as natural antioxidants in rendang.

2. Materials and methods
2.1. Material of the research
The material used in this research was beef which was sold in *abattoir*, *Garcinia mangostana* (*Mangosteen*) from Pariaman Distric. The chemical compound used to analyze the proximate was taken from the Laboratory of the Husbandry Production Technology Department, Faculty of Husbandry, Andalas University. They included etil asetat, ethanol, folin-ciocalteu (Sigma chemical Co., St., Lois, Mo., U.S.A.), DPPH/ 2,2-*diphenyl*-1-*picrylhydrazil* or 1,1-*diphenyl*-2-*picrylhydrazitl* (Merck), NaOH 0,1 N, and gallic acid. The instrumentations used during the research were spectrophotometer UV-Vis 1800, shimadzdu/UFLC, filter paper Whatman, *water bath* and rotary evaporator

2.2. Mangoteen extract **[7]**
Mangosteen extraction by maceration using 96 % ethanol solvent. The macerates evaporated with a rotary evaporator.

2.3. Rendang processing [8]
Preparing the formulation of ingredients:1 kg beef, 3 mature coconuts, 150 gr red chili, 65 gr of red onion, 35 gr of garlic, 20 gr of galangal, 20 gr of ginger, 5 gr of turmeric, 4 pieces bay leaf, 2 pieces turmeric leaf, lemongrass and salt. Beef was cut into 2 x 4 x 5 cm dimension and put into the boiled mixture of coconut milk and the spices. The cooking process took around 3 hours. Then, catechin was added after the meat curry became oily; and the treatment was done randomly. It needs to be heated until the curry becomes brownish. The rendang sample is ready to be observed based on its observation variable.

This experimental research was applying randomized complete block design (RCBD) with 5 treatments and 4 repetitions. Rendang was made with *Mangosteen* extract with levels ranging between 0%, 0.25%, 0.5%, 0.75% and 1%. The data obtained were processed statistically, if treatment showed significant results ($P < 0.05$) then it was further tested by using Duncan's Multiple Range Test (DMRT) with SPSS method.

2.4. Analysis
- Compounds of pericarp of Mangosteen extact
  1. Xanthone [9]
  2. Antosianin[9]
  3. Tannin [9]
  4. Antioxidant activity/ DPPH assay[10]
  5. Total Phenol [11]
- Organoleptic test [12]
- Shelf life

3. Result and Discussion
3.1. Compounds of pericarp of mangosteen extract
Mangosteen fruit (*Garcinia mangostana* L Rind), also called queen of fruit (fruit queen), because its skin is blackish-red, white meat and sweet as well as its superior compounds are xanthones as the substance of natural chemistry, class of polyphenolic secondary metabolites [13].

Mangosteen benefits health because of the xhantones content as antioxidant, anti-inflammatory and antimicrobial. Antioxidant properties are vitamin E and vitamin C xanthone and its derivatives containing compounds: 3 - isomangoestein, Alpha - mangostin, Gama - mangostin, Garcinone A, Garcinone B, C, D and E garcinone, Macrulin and Mangostenol [3]. The compounds of *Mangosteen* extract can be seen in Table 1.

Table 1. The compounds of mangosteen extract.

Compounds	Quantity	Unit
Alfa-mangostin (Xanthones)	37.5	mg/100g
Anthocyanin	1.85	mg/100g
Tannin	3.20	%
Antioxidant activity (IC_{50})	210.45	µg/ml
Total Phenolic	90	mgGAE/g

The pericarp of *Mangosteen* extract contained alpha - mangostin 0.59 mg/g, anthocyanin 1.13 mg/g, the phenolic content of 8.49 mg/g and the antioxidant capacity of 19.72 mg/g [14]. While the results of the study indicates that the value of alpha - mangostin is quite high at 37.5 mg/g and anthocyanin content of 1.85 %.

The antioxidant activity was measured by using the method of DPPH free radical scavenging (2,2 - diphenyl - 1 – picrylhydrazil). [6] reported that the antioxidant activity (IC50) of extract from peel of mangosteen was 0.023 µg/ml, which was less than the IC_{50} found in this experiment (210 µg/ml). This implies that different extraction methods could effect on the activity. The antioxidant activity was correlated with phenolic content in extracs [15].

Figure 1, showed the linier regression curves of concentration of mangosteen extract plotted against percent inhibition of DPPH (y = 0.152x + 17.80). Furthermore, it was compared to a standard curve of ascorbic acid in order to obtain the value of DPPH with IC_{50} was 210.45µg / ml.

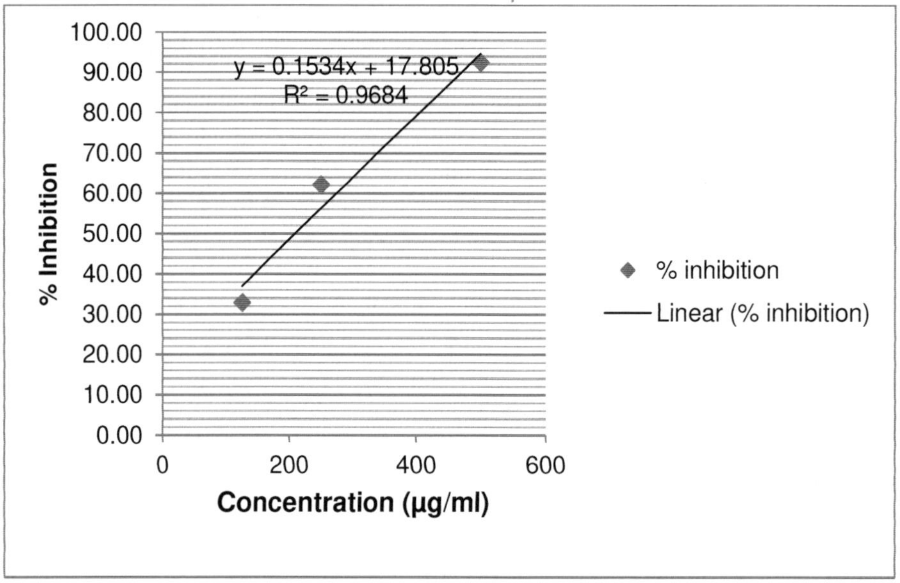

Figure 1. Percent inhibition of DPPH against *Mangosteen* extract

Xanthones compounds were extracted by using ethanol. It was consistent with the research [10], the results from the antioxidant activity of Mangosteen extract extracted with ethanol has the greatest activity. In Table 2, it shows the results of phytochemicals in Gambir extract containing flavonoids, phenolic and saponin. While the Mangosteen extract contained flavonoids, phenolics, saponins, alkaloids and steroids.

Table 2. The phytochemicals in pericarp of *Mangosteen* extracs.

Compounds	*Mangosteen* extract
Flavonoid	(+)
Phenolice	(+)
Saponin	(+)
Alkaloid	(+)
Steroid	(+)

One of the most well-known groups of antioxidant compounds is the phenolic. Phenolics and polyphenolics can provide relief from certain physical ailments and degenerative disease in human (reduction cardiovascular disiase and cancer) [16]. Some research has shown that the sources of phenolics (fruits, vegetables, grains) can lead to the increasing quantity of antioxidants in human body [16].

Flavonoids are one of the common and widely distributed group of phenolic compounds in plant, which can be closed as flavons, flavonols and anthocyanidins). Tannins (Phenolic Polymer) has capacity to bind proteins in the transformation of animal hides to leather. Tannins can be subdivided into two classes: hydrolysable and condensed tannins [16].

3.2. Organoleptic test

Based on the results of the research shows that adding the pericarp of Mangosteen extract at several levels presented the organoleptic test of rendang. For more details, the scores of each treatment can be seen in Table 3.

Table 3. the average score of taste and flavor's rendang with the addition of the pericarp of *mangosteen* extract.

Treatment	Taste	Flavor
A	3.75±0.99	3.60±1.25
B	3.50±0.94	3.40±1.19
C	3.50±0.83	3.50±1.10
D	4.10±1.04	3.65±0.72
E	3.75±0.83	3.80±0.97

From Table 3, it can be seen that the average score of taste rendang with the addition of the pericarp of Mangosteen extract ranging from 3.50 to 4.10. In this case, the level of preference for the rendang with the addition of pericarp of Mangosteen extract until 1.00 % level is really like. Results of analysis of variance shows that the treatment effect is not significant ($P > 0.05$) to the value of rendang taste with the addition of Mangosteen extract. The addition of mangosteen extract at rendang is not significant ($P>0.05$) to the value rendang aroma. It causes the rendang with the mangosteen extract did not produce a distinctive aroma. So it does not affect the rendang's flavor. From Table 3, it shows that the score of rendang's flavor from 3.40 to 3.80.

3.3. Shelf life

Table 4. shows that rendang containing the Mangosteen extract (1%) has a storage life of 12.50 days, longer than the rendang without the addition of Mangosteen extract (6.75 days). It was because the mangosteen extract contains chemical compounds called polyphenols which fuction as antioxidants. In accordance with the results of research conducted that xanthones Mangosteen extract

IOP Conf. Series: Earth and Environmental Science **287** (2019) 012028 doi:10.1088/1755-1315/287/1/012028

contains 37.5 mg / 100ml, antioxidants activity (210.45 µg/ml). Another components were analysed was polyphenol 58.28 mg GAE/g, tannins 3.02% and anthocyanins 1.85%.

Table 4. Shelf-life rendang with addition gambir extract.

Treatment	Self-life (Days)
A	6.75 ± 0.50[a]
B	12.00 ± 0.00[b]
C	12.00 ± 0.00[b]
D	12.25± 0.50[b]
E	12.25 ± 0.57[b]

The important application of antioxidant was their inclusion in food products as preservatives storage-life and to maintain quality of food. Antioxidant in food depends on many factors, such as: the antioxidant location in food, interactions with other food constituents and condition of environment (pH, ionic strength, hydrophilic and lipophilic balance) [17]. In the previous research, [18] rendang with addition of gambir extract (1%), could extend the storage-life until 18 days.

4. Conclusions

The results indicate that the pericarp of *Mangosteen* provides antioxidants (11.70 % inhibition). Xanthone content of *Mangosteen* extract was 37.5 mg/100ml. Another components analysed were polyphenol 58.28 mg GAE/g, tannins 3.02% and anthocyanins 1.85%. The organoleptic test shows that addition of the extract in rendang was not significant (P>0.05) and addition of *mangosteen* extract could expand the storage life of rendang.

5. References

[1] Obolskiy, D, Ivo P, Nisaral S and Michael H., 2009. Garcinia mangostaan L.: Phytochemical and Pharmacological Review. Phytother 23: 1047-1065.

[2] Tjahjani, S., Wahyu W., Khie. K., Adrian S., and Rita T., 2013. Antioxidant of *Garcinia mangostana* L (*Mangosteen*n) Rind. Procedia Chemistry 13, pp: 198-203.

[3] Jung H.A, Su BN, Keller WJ., Melna, RG., Kinghorn AD., 2006. Antioxidant Xanthones from the Pericarp of Gacinia mangostana (*Mangosteen*). J. Agric. Food. Chem. 54: 2077-82.

[4] Aisha A.F.A, Khalid M., Zhari I., And amin M.S.A.M., 2013. Determination of Total Xanthones in *Garcinia mangostana* Friut rind Extract by Ultaviolet (UV) Spectrophotometry. Journal of Medical Plant Research. Vol 7(1), pp: 29-35.

[5] Pedraza-Chaverri J., Cardenaz Rodriguez N., Orozco-Ibarra M., Perez-Rojaz JM., 2008. Medical Properties of *Mangosteen* (*Garcinia mangostana*). Food Chem Toxicology. 46 3227-3239.

[6] Palawakong C., SophanodoraP., PisuchpenS., Pongpaichit S., 2010. Antioxidant and Antimicrobial Activity of Crude Extract from *Mangosteen* (*Garcinia mangostana* L) Parts and Some Essensial Oils. Intn Food Res J., 17: 583-9.

[7] Depkes RI., 2000. Parameter Standar Umum Ekstrak Tanaman Obat. Jakarta, Departemen Kesehatan RI. Page: 8-11.

[8] Novelina and Nurhaida H., 1997. Studi Mengenai Cita Rasa Rendang yang Diawetkan Dengan Cara Sterilisasi dan Pemberian Antioksidan Setelah Dua Bulan Penyimpanan. Prosiding Seminar Tek. Pangan. Pp. 169-174.

[9] Sudarmadji, S., B. Haryono dan Suhardi. 1996. Analisis Bahan Makanan Dan Pertanian. Liberty Yogyakarta Bekerja sama dengan Pusat Antar Universitas Pangan dan Gizi Universitas Gadjah Mada, Yogyakarta.

[10] Weecharangsan W., Opanasopit, P, Sukma, M., Ngawhirumpat, T., Sotanaphun, U., and Siripong, P., 2006. Antioxidative and Neuroprotective activities of Extract from the Fruit Hull of Mangosteen (*Garcinia mangostana. L)* Medical Pronciples and Practise 15. 281-287.

[11] Singh, R, R.Kumar., Ramesh, V., Bimlesh M., and Sudhir K.T. 2013. Studies of Physicochemical and antioxodant properties of strawberry polyphenol extract-fotified stirred dahi. International Journal of dairy Technology Vol 60 (1).
[12] Rahayu, W.P. 2001. Penuntun Praktikum Penilaian Organoleptik. Fakultas Teknologi Pangan dan Gizi. Institut Pertanian Bogor. Bogor.
[13] Yatman , E, 2012. Kulit buah manggis mengandung xanthone yang berkhasiat tinggi. Widya, 29, no. 324.
[14] Permana, A., Widayanti SM., Prabawati, S., dan Setyabudi DA., 2012. Sifat antioksidan bubuk kulit buah manggis (Garcinia mangostana L) instan dan aplikasinya untuk minuman fungsional berkabonasi. J. Pascapanen, 92(2): 88-95.
[15] Liu, R.H. 2004. Potensial Synergy of Phytochemicals to cancer Prevention Mechanism of Action. J. Nutr 134: 3479S-85S.
[16] Liu X, Zhao, M, Wang J, Yang, B and Jiang, Y. 2008. Antioxidant Activity of Methanolic Extract of emblica Furit from sic Region in China. Journal of food Composition and analisys 21 : 219-228.
[17] Craft, B.D., Adrian L., Ryszard A. and Ronald B.P., 2012. Phenol-Based Antioxidants and In Vitro Methods Used For Their Assesment. Cmprehensive Reviews in Food Science and Food Safety. Vol 11.
[18] Melia, S., D. Novia and I. Juliyarsi, 2015. Antioxidant and Antimicrobial Activities of Gambir (Uncaria gambir Roxb) Extracts and Their Application in Rendang. Pakistan Journal of Nutrition 14(12): 938-941.

Aknowledgements

The financial support from Direktorat Jenderal Pendidikan Tinggi Kementrian Pendidikan dan Kebudayaan, Research Grant Implementation Agreement Number: 030/SP2H/PL/DIT.LITABMAS/II/ 2015, Date February 15, 2015 was greatly appriciated.

The effect of whippy cream adding on the quality of frozen soyghurt as symbiotic ice cream

Salam N Aritonang[1], Elly Roza[2], and Evy Rossi[3]

[1,2] Department of Animal Production, Faculty of Animal Science -Universitas Andalas Padang, Indonesia.
[3] Department of Agricultural Technology, Faculty of Agricultural, Universitas Riau, Indonesia.
Email: snaritonang@ansci.unand.ac.id.

Abstract. The presence of Lactic Acid Bacteria (LAB) in food is very important, as most of these LAB have beneficial effects in human body, so most LAB are generally potential as probiotics. This study aims to determine the quality of frozen soyghurt/synbiotic ice cream with the addition of whipping cream. Synbiotics is an ingredient and foods that contain both probiotics and prebiotics that beneficially affects the host by improving the survival and implantation of live microbial dietary supplement in the gastro intestinal tract. Prebiotics is a non-digestible food ingredient such as oligosaccharides substrates that match the enzymatic glycosidies hydrolase capacity of health promotion bacteria, that elicits the selective stimulation of growth and/or activity of one or a limited number of microbial genus/species in the gut microbiota that convers health benefits to the host. Soyghurt is one of the ingredients that includes synbiotics. This study began by making a synbiotic drink/soyghurt by inoculating *Streptococcus thermophillus* and probiotics isolated from okara is *Lactobacillus plantarum* SRCM 1 004 34 strain into a mixture of cow's milk and soy milk containing prebiotic, and then freezing it so that it is expected to produce a healthy Synbiotic Ice cream. This research was conducted experimentally using Completely Randomized Design with 4 treatments and 5 replications. The treatment is the addition of whipping cream with Soyghurt with the ratio: A = Whipping cream:Soyghurt (10:90), B = Whippy cream:Soyghurt (20:80) C = Whippy cream:Soyghurt (30:70) and D =Whippy cream:Soyghurt (40:60). The variable was observed the quality of frozen soyghurt that consisted of protein and fat content, pH, Lactic Acid Bacteria count, overrun, melting point and sensoryc evaluation. The result of the research indicated that the adding of whipping cream in frozen soyghurt making was significantly increased fat content, pH, overrun, melting point, texture and decreased protein content, Lactic Acid Bacteria count and flavour of frozen soyghurt. The conclusion is the addition of whipping cream up to 40% (D) has produce frozen soyghurt with good quality yet.
Keywords – frozen soyghurt, probiotic, prebiotic, synbiotic, whipping cream.

1. Introduction

Synbiotics is an ingredient and foods that contain both probiotics and prebiotics [13] that beneficially affects the host by improving the survival and implantation of live microbial dietary supplement in the gastro intestinal tract by selectively stimulating the growth and/or by activating the metabolism of one or a limited number of health promoting bacteria and thus improving host welfare. The prebiotic presence in synbiotic mixture are generally considered to have one of two possible function. First, they stimulate the number and activity of beneficial microbes endogenous to the gut of the host, independent of the activity of the probiotic. Secondly, they are considered a lunch box for the probiotics and should increase their survival while the beneficial microbes travel through the hostile environment of the gastro intestinal tract [12].

Probiotics are live microorganism which classified as capable of improving the intestinal microbial balance beneficial effect on the health of the individual when ingested regularly and in adequate concentration. Probiotic microorganisms generally produces lactic acid as a product of carbohydrates breakdown, beside produce bacteriocin and hydrogen peroxide form will suppress the growth of pathogenic bacteria [10] [16]. These bacteria called lactic acid bacteria [1]. Lactic acid and bacteriocin produced by lactic acid bacteria such as *Lactobacillus lactis* and *Streptococcus thermophillus*, act as biopreservatives because of their inhibitory effect against pathogenic microorganisms that endanger health [20].

Prebiotics is a non-digestible food ingredient that elicits the selective stimulation of growth and/or activity of one or a limited number of microbial genus/species in the gut microbiota that convers health benefits to the host [19]. Dietary fibre is classified as a non-digestible food ingredient, but not all fibre is considered a prebiotic. Owing to the fact that not all dietary fibre can be selectively fermented by one or a few members of the microbiota and thereby confer healt benefit. Saccharolytic bacteria which are able to break down and ferment complex oligosaccharides contain glucosyl, galactocyl, and xylosyl are generally considered to be beneficial due to the metabolic by-product that the produce [6]. Therefore, food ingredients that contain the non-digestible oligosaccharides substrates that match the enzymatic glycosidies hydrolase capacity of health promotion bacteria are primary prebiotics candidates. The main bacteria genera that are targeted by prebiotics are Bifidobacterium and Lactobacillus which most prebiotics that have been developed so far are carbohydrates or fibres [21].

Soyghurt is one of the ingredients that includes synbiotics. Soygurt is a fermented product such as yoghurt made from soy milk with or without the addition of cow milk using probiotic bacteria such as *Lactobacillus acidophilus* and *Streptococcus thermophillus* [22]. In this study these probiotic bacteria can be replaced by the isolation from Okara that is *Lactobacillus plantarum* SRCM 1 004 34 strain [2]. Soyghurt has advantages compared with other milk fermentation products that are beneficial to health such as can prevent degenerative diseases such as cancer, osteoporosis, coronernal heart [24]. This is because soy milk contains isoflavones as natural antioxidants [25]. In addition, soy milk also contains carbohydrates, fats, calcium, phosphorus, ferrum, vitamins A, B1, B2 [3].

Consumption of soyghurt is also beneficial to the balance of ecosystems in the intestinal tract with increasing probiotic population and decreased population of pathogenic bacteria, thus reducing abdominal bloating, increasing flavor, reducing taste, giving new textures, preventing intestinal infections, and assisting intestines flora [14]. This is because soy milk contains nutrients that are prebiotic, which contains dietary fibre which includes non-digestible oligosaccharide type, in two main components of 20% rafinosa trisakarida with α-galactosida bond between galactose- glucose and stakiose tetrasaccharides of 71% having two α-galactoside bonds between galactose-galactose and galactose-glucose [23].

The quality of soyghurt is usually assessed based on the acidity (pH), total lactic acid bacteria (LAB) and viscosity. The taste of soyghurt is still less desirable and favored by the consumer. One effort that can be done is to process soybean into soyghurt ice cream (frozen soyghurt). The making of soyghurt ice cream can also further increase the diversification of soybean process and support the marketing value in the community, because ice cream much favored by all age groups. In addition, ice cream is a frozen dish with semi-solid texture and high nutritional value.

Soyghurt is fermented milk which contains low fat content, so it needs to be added fat derived from vegetable fats, one of which is whippy cream. The function of fat in making ice cream soyghurt is as a delicate texture, flavor producer and gives a synergistic effect on the additional flavor used [9]. If the fat content of ice cream is less than 10%, then the ice texture is more coarse and cooler [8]. Therefore, fat is the most important part in making ice cream so it should get attention in every formulation. The source of whippy cream fat comes from vegetable fat that the amount of fat content is not less than 12%. The use of isolated probiotics from okara along with prebiotics from soy milk is then applied in the manufacture of synbiotic milk products which then freeze it, it is expected to obtain a healthy Frozen Soyghurt as Synbiotic Ice Cream.

2. Material and methods

This research began by making a sinbiotic drink by inoculating probiotics isolated from okara into a mixture of cow's milk with soy milk containing prebiotic then freezing it so that it is expected to produce a healthy Frozen Soyghurt as Synbiotic Ice cream.

2.1. Making soyghurt

A mixture of cow's milk with soy milk each of 500 ml added 5% sugar. It is then pasteurized at 72°C for 15 minutes, then cooled rapidly to 42°C. After that inoculated starter as much as 3% [2] that consisted of 1% *Streptococcus thermophillus* and 1% *Lactobacillus plantarum* SRCM 1 004 34 strain then mix evenly. Furthermore, it was incubated for 8 hours at 37°C [2]. The soyghurt produced is made into the ingredients of frozen soyghurt.

2.2. Making frozen soyghurt

The process of making frozen soygurt includes: mixing ingredients, pasteurization, homogenization, aging, freezing, hardening and storage. Mixing ingredients for ice cream is done gradually. The mixture of ingredients to be frozen into ice cream is called Ice cream Mix (ICM). The manufacturing process is: mix whippy cream into liquid ingredients (soyghurt) at temperature < 3°C). Liquid whippy cream is shaken at high speed for ± 3 minutes until it expands, then added soybean and sugar. Homogenizing for 10 minutes at speed 1500 rpm. Put into ICM with temperature 5°C for 45 minutes. Then stored at freezing temperature -10°C for 24 hours. This research was conducted experimentally using Completely Randomized Design (RAL) with 4 treatments and 5 replications. The treatment is the addition of Whippy cream with Soyghurt with the ratio:

A = Whippy cream : Soyghurt (10:90)
B = Whippy cream : Soyghurt (20:80)
C = Whippy cream : Soyghurt (30:70)
D =Whippy cream : Soyghurt (40:60)

2.3. Variable observed

2.3.1. Protein Content (Kjeldahl Method)

2.3.2. Fat Content (Soxhlet Method)

2.3.3. pH (pH meter)

2.3.4. Lactic Acid Bacteria

Microbiological tests use the spread surface plate method: Take 1 ml sample of frozen soyghurt using a micro pipette and then adding 9 ml of 0.85% physiological salt for a 10^{-1} dilution and continuing until 10^{-7} dilution. Then pipette 0.1 ml starting from 10^{-5} dilution to 10^{-7} dilution to be inoculated on MRS-agar medium by dripping a sample on a petri dish containing MRS-agar and the samples were flattened on the surface of the medium with shiny steek / curved glass rods that had been sterilized by burning over a bunsen fire. The process of inoculation is carried out in a sterile room, namely a laminar flow cabinet. The inoculated petri dish is then incubated in the incubator for 48 hours at 370C in order to avoid dripping water that may be attached to the inner wall of the cap. The LAB colonies that grow are counting using a colony counter. Calculation of LAB colonies can be calculated by the formula (log cfu/ml).

$$\text{LAB total per ml} = \text{Colony count} \times \frac{1}{\text{Pengenceran}} \text{ x } 10$$

2.3.5. Overrun [15]

Overrun is the development of the volume of ice cream which is calculated based on the difference in the ice cream volume with the volume of dough in the same mass or based on the difference of ice cream mass and the dough mass on the same volume. The weight of the dough is weighed on the same volume before the cooling process and heavy weighting of ice cream is put into the freezer.

Overrun value is calculated by formula:

$$\text{Overrun} = \frac{W1 - W2}{W2} \times 100\%$$

Where, W1: Dough weight /volume unit
W2: Ice cream weight/ volume unit

2.3.6. Melting point [7]

Meelting points were measured by: put 50 grams of ice cream on filter paper clamped between Petridish at room temperature (± 26°C) and incubator temperature (± 37°C). The time required to melt the ice cream melts complete measured in minute.

2.3.7. Sensoryc evaluation of frozen soyghurt

Sensoryc Evaluation uses hedonic and descriptive tests. The panelists who tested were semi-trained panelists with 25 people. Sensoryc evaluation aims to determine the level of panelist preference which includes flavour and texture with a range from very like up to not very like. Descriptive test aims to determine the characteristics of frozen soyghurt in each treatment that was tested for flavour and texture. The sample is placed in a clean container with random number code. Panelists are asked to provide an assessment of each sample on the questionnaire form presented.

3. Results and discussion

3.1. Protein

Statistic analysis shows that protein content of Frozen Soyghurt significantly ($P<0,5$) influenced by whippy cream adding (Table 1). The addition of whippy cream will reduce the use of soyghurt and reduce the protein content of Frozen Soyghurt. The protein content of Frozen Soyghurt added whippy cream up to 40% (D) was lowest (5.69%) significantly followed by treatment C (6.10%), B (6.23%), and the highest in treatment A is 6.45%. In which the protein content of Frozen Yoghurt treatment B to treatment C did not significantly ($P > 0.05$). The decrease of protein content of Frozen Soyghurt along with the increase of whippy cream adding caused of that soyghurt as the source of protein, so that with the increase in whippy cream the use of soyghurt decreases, which is followed also by the decrease protein content of Frozen Soyghurt. The high protein content in the Frozen Soyghurt was produced images illustrates that Frozen Soyghurt has a higher nutritional value and can provide benefits to consumers. The results of this study are in accordance with [5] that the protein content of Frozen Soyghurt is 4.5 - 5.7%. According to the ice cream quality standards (SNI 01-3713-1995), the protein content is at least 2.7%, while the protein content in this study ranges from 2.06 to 3.73%, which means that the ice cream in this study has in line with the quality standards of ice cream.

Table 1. The effect of whippy cream adding on the quality of frozen soyghurt.

Variable	A	B	C	D
Protein (%)	6.45^{a}	6.23^{b}	6.10^{b}	5.69^{c}
Fat (%)	6.78^{a}	7.21^{a}	7.40^{a}	8.50^{b}
pH	4.83^{a}	4.90^{b}	5.22^{c}	5.40^{d}
LAB Count (log cfu/ml)	9.57^{a}	9.57^{a}	9.06^{a}	8.26^{b}
Overrun (%)	0.87^{a}	1.10^{b}	1.27^{c}	1.40^{c}
Melting Point (hour)	1.06^{a}	1.10^{a}	1.19^{b}	1.31^{c}
Flavour	4.03^{a}	4.20^{a}	3.10^{b}	3.23^{b}
Texture	3.67^{a}	3.60^{a}	3.73^{a}	4.30^{b}

Values in the same row followed by different letter are significantly different ($P<0.05$)

3.2. Fat

The adding of whippy cream significantly (P < 0.05) increased fat content of Frozen Soyghurt. Statistic analysis on Table 1. shows that fat content of Frozen Soyghurt on treatment D is highest (8.50%) followed by treatment C (7,40%), B (7,21) and the lowest of fat content of Frozen Soyghurt in treatment A is 6.78%. The highest fat content of Frozen Soyghurt in treatment D caused in this treatment the highest addition of whippy cream while the lowest use of soyghurt is lowest. Whippy cream contains high fat, which is 17%, while soyghurt is a low-fat food ingredient. The soyghurt is a product of processing milk mixture with soy milk and lactic acid bacteria. Lactic acid bacteria in addition to decomposing lactose into lactic acid simultaneously utilize crude fiber in soy milk to increase its activity [12] in producing lipolytic enzymes to decompose fat into fatty acids. So the higher of soyghurt added in Frozen Soyghurt making, the higher the lipolitic enzymes concentration to breaking fat into fatty acids so that the fat content of the Frozen Soyghurt is the lowest. Fat is needed in the formation of ice cream structures where large globular fat forming granules determines the size of the air cavity in the cell. This fat also cause a soft texture on ice cream where evenly distributed fat with a homogeneous and small size can help the expected melting point of ice cream

3.3. pH

The higher the addition of whippy cream, the higher the pH of Frozen Soyghurt. Statistical analysis showed that the highest addition of whippy cream in treatment D the highest Frozen Soyghurt pH significantly (P<0.05) which was 5.40, followed by treatment C (5.22), B (4.90) and the lowest in treatment A (4.83). Increasing the pH of Frozen Soyghurt along with the increasing whipping cream addition is influenced by the presence of lactic acid bacteria and carbohydrate content. Soyghurt is a kind of yoghurt made from soy milk in fermentation process using a mixture of lactic acid bacteria such as *Lactobacillus bulgaricus* and *Streptococcus thermophillus* [11]. LAB in soyghurt in this ice cream mixture will degrade polysaccharides and ferment lactose, sucrose, and galactose to produce more lactic acid, consequently the pH decreased. The higher the addition of whippy cream, the lower the use of soyghurt so that the LAB count in frozen soyghurt which degrade polysaccharides into lactic acid was also lower followed by the low lactic acid produced. As a result, frozen soyghurt pH was highest.

3.4. Lactic Acid Bacteria count

Statistic analysis showed that Lacto Acid Bacteria Count of Frozen Soyghurt significantly (P<0,5) influenced by whippy cream adding (Table 1). The addition of whippy cream will reduce the lactic acid bacteria count of Frozen Soyghurt. The Lactic Acid Bacteria of Frozen Soyghurt added whippy cream up to 40% (D) was lowest (8.26 x10^6 cfu/ml) significantly followed by treatment C (9.06 x10^6 cfu/ml, treatment B (9.57 x10^6 cfu/ml), and the highest in treatment A is 9.57 x10^6 cfu/ml. In which Lactic Acid Bacteria count of Frozen Soyghurt treatment C to treatment B and to treatment A did not significantly (P > 0.05). The low of lactic acid bacteria count in Frozen Soyghurt added with whippy cream in treatment D was caused by the lowest soyghurt used in this treatment. The low level of soyghurt in the ice cream mixture means the low number of lactic acid bacteria in ice cream sinbiotis because in soyghurt contains lactic acid bacteria. In accordance with the opinion of [22] that soyghurt is a fermented product of soy milk using probiotic bacteria such as *Lactobacillus delbruecki*, *Lactobacillus achidopilus*, and *Lactobacillus bulgaicus.*

The decrease of LAB count during the process of Frozen Soyghurt making also occurs because the cooling process of the dough so that LAB cannot grow optimally. The activity of mixed culture of *Streptococcus thermophillus* and *Lactobacillus plantarum* in Frozen Soyghurt has an optimum growth temperature of 15-37°C and is able to produce antimicrobial plantaricin which is able to suppress the growth of pathogenic microbes so that it is safe for consumers' health. The decrease in LAB count is caused bacterial cell damage at freezing temperatures, but not as fast as at high temperatures. If bacteria survive early in the freezing temperature, then the bacteria can stay alive for a long time in a frozen state, where the function of the cell halt temporarily and if the surrounding media is thawed again the metabolism can last again [4].

The lactic acid bacteria count in all treatments ranged from 8.26 to 9.57 x 10^6 cfu/ml. This has in line to the standard number of probiotics contained in food ingredients. Similar with [26] which states that the minimum of lactic acid bacteria count in probiotic food products is 6 log10 cfu/ml, so the greater of BAL count contained in ice cream the better. [22] stated that one of the requirements of probiotic products in fermented milk is that it contains live microbial cells of 6 log10 up to 8 log10 cfu/ml.

3.5. Overrun

The adding of whippy cream significantly ($P < 0.05$) increased overrun of Frozen Soyghurt. Statistic analysis on Table 1. showed that overrun of Frozen Soyghurt on treatment D is highest (1,40%) followed by treatment C (1,27%), B (1,10%) and the lowest of synbiotic ice cream overrun in treatment A is 0,87%. Overrun of Frozen Soyghurt treatment D to treatment C did not significantly ($P > 0.05$). Increasing the overrun of Frozen Soyghurt along with the addition of whippy cream caused in the manufacture of ice cream in addition with functioning as a texture softener, whippy cream also functions as a swell.

Whippy cream use to help the development process and the formation of cream and the texture and shape of a product [8]. Thus, the more whippy cream is used, the more the frozen soyghurt overrun value will increase. Whippy cream has an important role in ice cream such as increasing flavor, producing a smooth texture, giving the body and helping melt because it plays a role in fat destabilization [18].

3.6. Melting point

Statistic analysis shows that the melting point of Frozen Yoghurt significantly ($P<0,5$) influenced by whippy cream adding (Table 1). The addition of whippy cream will reduce the use of soyghurt and increase the melting point of frozen yoghurt. The addition of whippy cream to 40% (D) significantly resulted in the highest melting point for sinbiotic ice cream, which was 1.31 hours followed by treatment C, B, and the lowest in treatment A, which was 1.06 hours.

The increasing melting point of frozen yoghurt along with the increase of whipping cream added due to whipping cream plays a role in increasing the volume of synbiotic ice cream products [8]. As a result, with the increase in ice cream symbiotic volume, the air in the dough will form air spaces which are released immediately along with melting ice cream. This is in line with the state of [17] that the speed of melting ice cream is influenced by the amount of air trapped in mixed ingredients. The higher the amount of fat aggregate, the higher the resistance to melting ice cream. As in this study, the addition of the highest whippy cream (D) in the manufacture of Frozen Soyghurt has increased the fat content of Frozen Soyghurt followed by increasing melting points of Frozen Soyghurt in the same treatment.

3.7. Flavour

The adding of whippy cream significantly ($P < 0.05$) decrease flavour value of Frozen Soyghurt. Statistic analysis on Table 1. showed that flavour value of Frozen Soyghurt on treatment D is lowest (3.23) followed by treatment C (3.1), A (4.2) and the highest of flavour value of Frozen Soyghurt in treatment B is 4.2 in which flavour value of Frozen Soyghurt treatment A to treatment B and treatment C to treatment D did not significantly ($P > 0.05$) each other. The decrease in the value of the Frozen Soyghurt flavor added with higher whippy cream in treatment C and treatment D caused of the lower composition of the soyghurt in Frozen Soyghurt so that the flavour was less acidic and the panelists did not liked it.

When associated with the pH value in this study, Frozen Soyghurt with a low pH has a higher preference score. This is due to the flavour of the acid produced by fermentation activity of Lactic Acid Bacteria in the soyghurt. This is in line with the state of [22] that soyghurt is a fermented product such as yogurt made from soy milk with or without the addition of cow milk using lactic acid bacteria such as *Lactobacillus acidophilus* and *Streptococcus thermophillus.*

3.8. Texture

The adding of whippy cream significantly ($P < 0.05$) increased texture value of Frozen Soyghurt. Statistic analysis on Table 1. showed that texture value of Frozen Soyghurt on treatment D is highest (4.30) followed by treatment C (3.73), A (3.67) and B (4.2) in which texture value of Frozen Soyghurt treatment A to treatment B and treatment C did not significantly ($P > 0.05$) each other. The increasing value of the Frozen Soyghurt texture added with the highest whipping cream (D) is caused by the increase in whipping cream followed by increasing the fat content of Frozen Soyghurt. This fat content inhibits the formation of large ice crystals during freezing so that the Frozen Soyghurt produced has a slightly soft texture and is preferred by the panelists. In accordance with [9] stated that the function of fat in whipping cream in the Frozen Soyghurt making is to a smooth texture and improve the flavour.

4. Conclussion

The addition of whippy cream in the Frozen Soyghurt making significantly ($P < 0.05$) increased fat content, pH, overrun and melting point and reduced protein content, Lactic Acid Bacteria count and the flavour of Frozen Soyghurt. The adding whippy cream up to 40% (D) has produce Frozen Soyghurt with good quality yet.

5. Reference

[1] Aly, S., C. A. T. Ouattara, I. H. N. Bassole, and S. A. Traore. 2006. Bacteriocins and lactic acid bacteria – a mini review. African Journal of Biotechnology Volume: 5, Pp: 678-6.
[2] Aritonang, S. N., E. Roza and E. Rossi., E. Purwati., Husmaini. 2017. Isolation and identification of lactic acid bacteria from Okara and evaluation of their potential as candiadate probiotics. Pakistan Journal of Nutrition Vol. 16 (8): 618-628.
[3] Bhathena, S. J. and M. T. Velasquez. 2002. Beneficial role of dietary phytoestrogens in obesity and diabetes. Am J Clin Nutr 76:1191–1201.
[4] Buckle, K. A., R. A. Edwards., G. H. Fleet and M. Wootton. 2007. Food Microbiology Universitas Indonesia. Jakarta.
[5] Criscio, T. D., A. Fatrianni., ar. Amignogna., L. Cinquanta and G. Panfii. 2010. Production of functional probiotic, prebiotic and symbiotic ice cream. J. Dairy Sci. 93 (10): 4555-4564.
[6] Delzene, N. M. 2003. Oligosaccharides, state of the art.Proc. Nutr. Soc. 62:177-182.
[7] Goff, H. D. 2002. Finding Science in Ice Cream. Departement of Food Science. University of Guelp. http://www.Foodsci.uoguelph.Ca/dairyedu/home.html.
[8] Goff, H. D. and R. W. Hartel. 2014. Ice Cream and Frozen Desserts. Dalam: Handbook of Frozen Foods. Springer. Willey Publisher.
[9] Goff, H. D. 2006. Quality and Safety of Frozen Food Processing and Packaging. Editor: E. Sun. CRC Press, Boca Raton.
[10] Husmaini., Sabrina., F. Arlina., E. Purwati., S. N. Aritonang and H. Abbas. 2017. Impact of administration age of probiotic *Lactococcus plantaraum* on the intestinal microflora and performance of broilers. Pakistan Journal of Nutrition Vol. 16 (5): 359-363.
[11] Irkin, R. and Eren, V. U. 2008. A Research about Viable *Lactobacillus bulgaricus* and *Streptococcus thermophilus* Numbers in the Market *Yoghurts.* World Journal of Dairy & Food Sciences 3 (1): 25-28.
[12] Kolida, S. and G. R. Gibson. 2011. Synbiotics in health and disease. Annu, Rev. Food. Sci. Technology 2: 373-393.
[13] Lee, Y. K. and S. Salminen. 2009. Handbook of Probiotics and Prebiotics. Second edition. John Wiley & Sons Inc. Hoboken. New Jersey.

[14] Liu, K. 2004. Fermented soy foods. In: Sinha, N. K., Y. H. Hui., E. O. Evranuz., M. Siddiq and J. Ahmed (Eds). 2011. Handbook of Vegetables and Vegetable Processing. Wiley Blackwell. USA.

[15] Masykuri, Y. B. Pramono, D. Ardilia. 2012. Resistance of melting, overrun and preference levelof vanilla ice cream made from main material combination of milk cream and milk coconut. Journal of Food Technology Application Vol.1 No. 3.

[16] Melia, S., E. Purwati., Yuherman., Jaswandi., S. N. Aritonang and M. Silaen. Characterization of the antimicrobial activity of lactic acid bacteria isolated from buffalo milk in West Sumatera (Indonesia) against *Listeria monocytogenes*. Pakistan Journal of Nutrition Vol. 16 (8): 645-650.

[17] Muse, M. R. dan R. W. Hartel. 2004. Ice cream structural elements that affect melting rate and hardness. J. Dairy Sci. 87:1-10.

[18] Rajah K. K. 2014. Fats in Food Technology. Willey Publisher.

[19] Roberfroid, M., G. R. Gibson., L. Hoyles., A. L. Mc Cartnel., R. Rastall., I. Rowland., D. Wolvers., B. Watzl., H. Szajewska., B. Stahl. 2010. Prebiotic effect: metabolic ans health benefits. Br. J. Nutr. 104 Supp. 2: 61-63.

[20] Savadigo, A., C. A. T. Quattara., I. H. N. Bassole and A. S. Traore. 2004. Antimicrobial activities of lactic acid bacteria strains isolated from Burkin Faso fermented milk.Pakistan Journal Nutrition 3 : 174-179.

[21] Slavin, J. 2013. Fibre and prebiotics mechanisms and health benefits. Nutrients 5: 1417-1435.

[22] Tamime, A. Y. and R. K. Robinson. 2007. Tamime *and* Robinson's Yoghurt. CRC Press. New York.

[23] Trindade, C. S. F., S. C. Terzi, L. C. Trugo, R. C. Della, Modesta, dan S. Couri. 2001. Development and sensory evaluation of soy milk based *yoghurt*. Arch Latinoam Nutr 51, 100-104.

[24] Wilson, T and Temple, N.J. 2004. Beverages in Nutrition and Health. Humana Press. New Jersey.

[25] Woodside, J. V. and M. S. Morton. 2004. Are Soy-Milk Products Viable Alternatives to Cow's Milk? In: Wilson, T. and N. J. Temple (Eds). Beverages in Nutrition and Health. Humana Press. New Jersey. 9.

[26] Vrese, M. D., and P. R. Marteau. 2007. Probiotics and Prebiotics: Effects on Diarrhea. The Journal of Nutrition 0022-3166/70: 803S-811S.

Acknowledgment

This research was supported by BOPTN Research Funding from Andalas University, Ministry of Research Technology and Higher Education Republic of Indonesia through contract for 2018. Author are very grateful to the Rector of Andalas University and the Minister of Research Technology and Higher Education, who have given us a chance to conduct this research.

Effect of mega floral booster addition on carcass characteristics of quail meat

U H Sharif[1], A S Kamarudin[1], and N Huda[2]

[1] School of Food Industry, Faculty of Bioresources and Food Industry, Universiti Sultan Zainal Abidin, Besut Cámpus, 22200, Besut, Terengganu Darul Iman, Malaysia.
[2] Faculty of Food Science and Nutrition, Universiti Malaysia Sabah, 88400, Kota Kinabalu, Sabah, Malaysia.

Abstract. This study evaluated the effect of Mega Floral Booster (MFB) as probiotic supplement on the carcass yield and physicochemical properties of quail meat. The treatment is divided to four treatments based on percentage of MFB include MFB-0.00 (control), MFB-0.05 (0.05% of MFB), MFB-0.20 (0.20% of MFB) and MFB-0.35 (0.35% of MFB). The 18 of quail species *Cortunix* japonica (Japanese quail) used as the main material in the studyfor each treatments. For the first until fourteen days, the quails are supplied with commercial feed only. After fifteen days, the quails begin provided with the MFB-0.00, MFB-0.05, MFB-0.20 and MFB-0.35 mixed with commercial feed. The quails for 42nd days slaughtered and deboned of breast and whole leg separated from the bone for analysis. Probiotic supplement (Mega Floral Booster) had no influence on body weight, and carcass characteristics (breast, whole leg muscle and carcass yield) ($p>0.05$). MFB-0.20 had influence on color parameter (L*, a* and b*), pH and WHC of quail meats compared to MFB-0.00 ($p<0.05$). MFB show cooking loss, juiciness and Warner-Bratzler shear force not affected by treatment. Thus, Mega Floral Booster is not affecting the quality of quail meats.
Keywords – mega floral, carcass, quail meat.

1. Introduction

The quail is not an established branch, but occupies a relevant place in poultry breeding and contributes to the variety in poultry meat production according to [1]. Along the increase of the chicken meat production, there is also increase of the production meat from quails and other non-commercial bird species, reared under industrial conditions. The highest consumption of quail meats tends to encourage the quail farming thus increase of the feed additive or supplement usage.

Dietary supplementations of probiotic have been proposed to antibiotics in the poultry industry that can improve animal production performance, regulate intestinal micro-flora imbalance, and enhance immune function that stated by [2]. In a limited study, the meat quality attributes of probiotic- fed have been evaluated to ensure natural and wholesome characteristics. Some of studies reported that probiotic supplements to poultry able to improve meat quality attributes such as water holding capacity (WHC), tenderness, lipid oxidation stability, sensory properties, and microbial safety [3, 4, 5]. On the other studies were stated that no synergistic effect of probiotics on meat quality [2, 6]. Postmortem aging leads to considerable biochemical and physical changes in skeletal muscles through endogenous proteolytic system, results in an improvement in meat quality attributes such as tenderness, WHC, juiciness and flavor [2].

IOP Conf. Series: Earth and Environmental Science **287** (2019) 012030 doi:10.1088/1755-1315/287/1/012030

The objectives of this study was to determine the carcass yield of quail, evaluate physicochemical properties include color, pH, water holding capacity, cooking loss, juiciness, texture profile analysis and Warner-Bratzler shear force between four treatments with different concentration of Mega flora booster in feed intake.

2. Materials and methods

The *Coturnicx* japonica quails (1st day) were obtained from commercial hatchery in Terengganu, Malaysia (Surada Jaya Kemaman). For quail growth process compose to two phase, brooding process (1st until 14th days) and growing process (15th until 42nd days). During brooding process, the quails were weighed and divided for four group, and supplied with commercial feed (Permula Ayam Daging 201C by Gold Coin Feedmills (M) Sdn Bhd) in brooding room. At growing process, pens were randomly assigned to four treatment of Mega Floral Booster (MFB) with different percentage 0.0% (MFB-0.00), 0.05% (MFB-0.05), 0.20% (MFB-0.20) and 0.35% (MFB-0.35) mixed with 100%, 99.95%, 99.98% and 99.65% of commercial feed. MFB was the commercial advanced probiotic formula, fortified with multi strain of microbes (*Bacillus* natto, *Lactobacillus* and yeast) and methionine, organic acid, enzyme protease and amylase manufacture by CPEX (Japan) Group. The proper design of building area required to avoid any infection that may occur along the experiment. The quails (in large amounts) was separate to three different part (in small amount) for each treatment; part A, part B and part C with twelve of quails for each part to minimize the rate of mortality. The 20g of feed was given two times morning and evening.

At the end of the feeding (42nd days), the six of quails per pen were randomly taken for each treatments (totally 18 for each treatments). The quails were slaughtered, scalded (60-65 °C for 15 seconds), de-feathered, eviscerated and packed in blast freezer (overnight) according [1]. Then, the quails were deboned for breast and whole leg for carried out yield analysis, physicochemical properties analysis include pH, color, juiciness water holding capacity, texture profile analysis and shear force and the nutritional composition includes chemical composition, and amino acid composition.

The colour of raw meat was measured by the Konica Minolta CR-400 Chroma meter to determine the colorimetric index of chromaticity. The color assessed by L* (lightness), a*(redness) and b*(yellowness) system. According to [1] show L*, a* and b* colorimetric coordinates as followed: L* - a value 100 corresponded to absolute white; a*, a+ corresponded to red spectrum; a-corresponded to green spectrum and b*, b+ corresponded to yellow spectrum;b- corresponded to blue spectrum.

The value of pH was examined by using electrode pH meter. The pH meter calibrated as follow the instruction in the manufacturer manual before used with 6.88 or 4.00 buffers every 2 or 3 hours. A sample (5 g) was homogenized with distilled water (50ml), transferred into beaker and along electrode with temperature electrode. The constant value appeared on screen of pH meter base and recorded of different samples refers to [7] with modifications. After each of the measurement, the electrode was rinsed with distilled water.

Water holding capacity was determined as follow the methods interpreted by [8] with modifications. 20 g sample was homogenized with 40 ml of distilled water in a waring blender at low speed setting for 30 s. 20 g of aliquot of the homogenate was weighed into 50ml centrifuge tube and centrifuged by setting at 5°C, 1000xg for 15min (low speed) of Sigma 3-16K - Mid bench centrifuges. Water holding capacity was calculated by using the following equation:

$$\text{WHC}(\%) = \frac{\text{before centrifuged-after centrifuged}}{\text{before centrifuged}} \text{x } 100$$

Cooked quail meat was conducted according to [9] Method to prepare cooked quail meat was a piece breast muscles then placed on the beaker and cooked in an 80°C water bath until achieve the core temperature 72°C. The cooked samples were cooled to room temperature for 3 hours, before undergo analysis.

Method of quail meat sausage preparation with modification on ingredient as follows [10]. Quail meat sausage was prepared by mixing 6.5% cold water, 6.5% ice, 3% salt and 84% quail meat for 2 min by using wiring blender to homogenize sample. The samples were stuffed into 2.5cm diameter sausage casings. The stuffed samples were cooked at 36°C for 30 min, followed by heating at 90°C for 10 min in a Memmert water bath. Quail meat gel were cooled in ice water for 30 min and stored in a chiller overnight prior analysis.

Cooking loss was determined by taken a piece of meat from the breast muscle, weighed, placed in beaker and cooked in 80°C of water bath for one hour until core temperature achieve above 72°C. The cooked samples were cooled to room temperature for 3 hours and re-weighed. Percentage of cooking loss was calculated according to [2] with modifications as follow:

$$\text{Cooking loss (\%)} = \frac{\text{weight after cook - weight before cook}}{\text{weight before cook}} \text{x } 100$$

Juiciness measurement was conducted according to [11] with modifications. The sample was cut into 3 mm pieces with a knife. The sample was placed between a pair of pre-weighed Whatman No. 40 filter paper covered with aluminium foil and press for 2 minute by 5 kg force. The residue was weighed and recorded. Percentage of juiciness was determined as follows:

$$\text{Juiciness (\%)} = \frac{\text{weight of filter after pressing-weight of filter paper}}{\text{weight of sample before processing}} \times 100$$

The folding test was conducted to analyze the gel strength of the cooked fish sausages and was determined according to [12] with modifications. Cooked samples were cut into three-millimeter thick portions. The slices were held between the thumb and the forefinger and folded to observe the way that they broke. The scale used was as follows: (1 = breaks by finger pressure, 2 = cracks immediately when folded in half, 3 = cracks gradually when folded in half, 4 = no cracks showing after folding in half and 5 = no cracks showing after folding.

Shear force of cooked meat and quail meat sausage were examined according to [2] with slightly modifications by using Warner-Bratzler (WB) shear attachment on texture analyzer (Double Arm TA.XT Plus Texture Analyser, Stable Micro System Ltd., Survey, UK) with the blade set (HDP/BS*) comprises a Warner Bratzler blade with blade holder by attach with 30 kg load cell. According to [13], Warner-Bratzler method was better when samples are made on cooked meat, although it is not very accurate for the prediction of meat texture. The sample was cut to 1 cm×1 cm× 2.5 cm-strips. The test cell consist of 3 mm thick steel blade which fit through a 4mm wide slit in a small table, the blade move down with a constant speed-pre-test speed 3.0 mm s^{-1}, test speed 1.0 mm s^{-1}, post-speed 3.0 mm s^{-1} and down stroke distance 30.0 mm. For every 0.01 second, the resistance of the meat sample to shearing was recorded and plotted by a computer in a force-deformation plot.

Texture profile analysis (TPA) of quail meat sausage was test for hardness (tenderness), cohesiveness, springiness and chewiness by using Double Arm TA.XT Plus Texture Analyser, Stable Micro System Ltd., Survey, UK with 30 kg load cell. TPA was measure the compression force (Newtons) when compressing a piece of meat by texturometer. The cooked sample of meat sliced for 1 cm x 1 cm x 1 cm strips was placed under the cylindrical probe that move downwards at a constant speed of pre-test 3.0 mm s^{-1}, test 1.0 mm s^{-1} and post-test 3.0 mm s^{-1}. According to [13] when the probe first came in contact with the sample, the thickness of the sample was automatically recorded by the software. The probe was continued downwards a pre-fixed percentage of the sample thickness (75%), return to the initial point of contact with the sample and stop for a set period of time (2 second), before the second compression cycle is start. The resistance of the sample was recorded every 0.01 second and plot in a force-time (grams-seconds).

The statistical analysis from this experiment was conducted by using SPSS statistic 17.0 software. Data was analyzed by using analysis of variance (ANOVA). The software was used to calculate the overall mean and standard deviation of the data and to determine the significant different of the carcass yield and physicochemical analysis. Besides that, SPSS is use to form the graph and find the liner equation for certain graph. Turkey test was used to determine significance of difference (p<0.05) of four samples.

3. Results and discussions

Table 1 shows the slaughter characteristics of 42- day old of Japanese quails on different treatments. No significant differences (p>0.05) were observed for the treatments of Mega Floral Booster (MFB) with body weight, carcass yield, breast with bone, breast meat, whole leg with bone and whole leg meat the treatments.

Table 1. Slaughter characteristics of 42 day Japanese quails for male-M and female-F.

Indications		Treatments			
		MFB-0.00	MFB-0.05	MFB-0.20	MFB-0.35
Live weight (g)	M	254.17^a±19.85	254.00^a±19.24	258.33^a±9.31	256.67^a±22.29
	F	259.17^a±21.08	256.67^a±15.71	260.00^a±22.14	259.17^a±13.57
Carcass yield (%)	M	64.93^a±1.62	64.85^a±0.90	65.48^a±1.60	64.94^a±2.97
	F	66.28^a±1.43	66.10^a±1.21	67.83^a±2.49	66.98^a±2.08
Breast with bone (%)	M	27.61^a±0.67	27.55^a±0.76	28.30^a±0.62	27.65^a ±0.34
	F	28.75^a±0.40	28.68^a±0.68	29.04^a±2.10	28.90^a±1.17
Breast meat (%)	M	23.92^a±0.70	23.64^a±0.56	24.04^a±0.66	23.93^a±0.48
	F	24.74^a±0.53	24.68^a±0.45	24.94^a±1.92	24.70^a±2.45
Whole leg with bone (%)	M	16.11^a±0.21	15.92^a±0.51	16.13^a±0.50	15.87^{a}0.68
	F	16.50^a±0.39	16.90^a±1.43	16.92^a±1.22	16.50^a±0.86
Whole leg meat (%)	M	13.87^a±0.60	13.38^a±0.46	13.97^a±0.12	13.20^a±0.10
	F	14.24^a±0.27	14.63^a±1.71	14.74^a±1.02	13.29^a±0.77

Each value is expressed as mean ± SD.
[a]Means with same small letters within same row are significantly different (p<0.05).

Carcass characteristics and body weights are major determinants of profitability and sustainability of any animal enterprise [1]. Although the treatments were not shown the significance difference but on the treatment MFB-0.20 (Mega floral booster 0.20 %) shown the body weight, carcass yield, breast, breast meat, whole leg and whole leg meat were the highest value than others treatment (MFB-0.00, MFB-0.05 and MFB-0.35). Slaughter characteristics of treatment 0.20% were highest of both male and female 258.33 g and 260.00 g of body weight and 65.48 % and 67.83 % of carcass yield.

After six weeks, weight of female bird had a higher mean than male had been interpreted [14] [15]. Female quail birds had a higher mean value than the male counterpart (163.0 g and 148.0 g) was stated by [15]. The growth pattern and carcass weight in male and female quails could be due to feed metabolism or onset of fattening and differences in hormonal profile, aggressiveness and dominance especially when both sexes are reared together respectively [16]. Carcass yield of quails in range 64-71% after six weeks old and Genhev et al. (2008) reported carcass yield without skin of Japanese quails of 35-day for male was 60.13% and female 58.97% [17].

The breast, which was economically important and valuable part of the carcass, was determined by [18] to be 27.45 % of the body weight. It was reported breast with bones had 25.35-25.42 %, breast meat in range 20.3-20.80 %, thigh with bones were 15.97-16.63 % and leg meat 12.06-12.62 % [1]. Breast with bone yield and breast meat were higher on the treatment MFB-0.20 for male and female (28.30 %, 29.04 %, 24.04 % and 24.94 %) than MFB-0.35 (27.65 %, 28.90 %, 23.93 % and 24.70 %), MFB-0.00 (27.61 %, 28.75 %, 23.92 % and 24.74 %) and MFB-0.05 (27.55 %, 28.6 %, 23.64 % and 24.68 %).

On the treatment of probiotic tend to improve the health of broiler chickens referred to [19] and [20]. Probiotic like *Bacillus* natto, *Lactobacillus* and *Bacillus subtilis*, spore forming probiotic, were metabolically dormant to face extreme environments, such as lowering pH through acid fermentation, extreme high and temperatures in intestine tend to improve intestinal health and body health. Reference [5] illustrated that probiotics could improve broiler performance by enhancing the immunomodulation capacity of broilers. By treating with 0.20 % and 0.35 % were showed a little difference with the control treatment of body weight and carcass characteristics.

The averages of L*, a* and b* values of breast meat and whole leg meat were presented in Table 2. The color parameters L*, a* and b* of breast meat on treatment MFB-0.20 had high significant than treatments MFB-0.00, MFB-0.05 and MFB-0.35. For whole leg muscle, L* value on treatment MFB-0.20 had significant with other treatments while a* value for MFB-0.20 and MFB-0.35 high significant than treatment MFB-0.00. But for b* value, there are no significant different between the treatments. L* value was show the lightness and darkness of breast and whole leg muscles on post slaughter after 3 days. From the research, the mean values of L* on breast muscles was lower and no significant (p>0.05) on treatment MFB-0.20 (46.46) followed by MFB-0.35 (47.36), MFB-0.05 (48.34) and had significant (p<0.05) with MFB-0.00 (49.93).

Table 2. Color parameters of meat from 42nd day old Japanese quails.

Parameters	Parts	Treatments			
		MFB-0.00	MFB-0.05	MFB-0.20	MFB-0.35
L*	Breast	49.93^{a}±2.12	48.34^{ab}±1.42	46.46^{b}±2.45	47.36^{ab}±1.54
	Whole leg	44.92^{a}±0.84	44.23^{a}±0.85	42.43^{b}±0.92	43.89^{a}±0.73
a*	Breast	6.33^{a}±0.44	7.26^{ab}±1.87	8.10^{b}±0.65	7.36^{ab}±0.78
	Whole leg	7.36^{a}±0.41	7.86^{ab}±0.33	8.23^{b}±0.40	7.98^{b}±0.37
b*	Breast	9.73^{a}±0.98	8.96^{a}±0.96	7.52^{b}±0.38	8.78^{a}±0.57
	Whole leg	7.15^{a}±0.45	7.02^{a}±0.51	6.73^{a}±0.16	6.84^{a}±0.21

Each value expressed as mean ± SD.
a-bMeans with different small letters within same row are significantly different (p<0.05).

For whole leg muscles, the mean of L* value on the treatment MFB-0.20 (42.43) had significant with other treatments but MFB-0.00 (44.92) not significance with treatments MFB-0.05 (44.23) and MFB-0.35 (43.89). L* value of whole leg meat lower than breast meat for all treatments means more dark. For a* value (redness), MFB-0.00 (6.33) shows no significant with MFB-0.05 (7.26) and MFB-0.35 (7.36) while had significant with MFB-0.20 (8.10) but a* value for whole leg on treatment MFB-0.00 also hadn't difference with MFB-0.05 (7.86) but it shows significance with MFB-0.20 (8.23) and MFB-0.35 (7.98). Whole leg muscle showed a* value higher than breast muscle, shows the redness was higher. Reference [11] was state that dark meat contains myoglobin, an oxygen binding protein that related to hemoglobin in red blood cells, which makes dark muscle appear red. Dark muscles are suited for durability activity and slow to fatigue because they use oxygen to generate energy. The color parameter for b* value (yellowness) in this study show MFB-0.00 had similar with treatments MFB-0.05 and MFB-0.35 but highly significance on treatment MFB-0.20 for breast muscle of Japanese quails. Whole leg muscles show b* value of the treatments no have significant.

According to [1], L*, a* and b* values of muscle pectoralis of Japanese quails were 45.67, 11.68 and 14.48, respectively after 7th days storage after slaughter. They also stated in their studies, the pectoral muscle composed of muscle, connective tissue and fat. The muscle bundles of the pectoral muscle were composed of dark muscle fibers and the ratio between dark and light fibers was 95.1-96.7 % to 3.7-4.9 %. In the study by [21] that carried out on broiler chicken meat quality, the ideal L* must be between 46 and 53, and meats with L* value below 46 were called to be dark, firm, dry, which shows they had a dark color, high water holding capacity and short shelf life. Reference [22] reported the meat quality was significantly improved (p<0.05) in *Bacillus subtilis* group supplement compared with that in control group during storage. At 8 days storage, a* (7.9-5.0) value was significantly decreased (p<0.05), while, L* (43-49) and b* (14-19) value were significantly increased (p<0.05) in *Bacillus subtilis* group supplement compared with the control group.

Table 3 presented the pH values and water holding capacity of breast muscles and whole leg muscles. The pH value of breast MFB-0.20 (5.98) showed significantly higher (p<0.05) than MFB-0.05 (5.83) and MFB-0.35 (5.87) but higher than MFB-0.00 (5.78). For pH value of whole leg meat, MFB-0.20 (6.50) had highly significant difference (p<0.05) with MFB-0.00 (6.39). Comparison of water holding capacity under treatment of Mega Floral Booster supplement was show MFB-0.20 had significant difference compared to others treatments for breast muscle while WHC of whole leg muscle hadn't difference between the treatments. The pH values of meat depended on glycogen content in muscles, and glycogen stores were highly influenced by the locomotors activity and the presence of stress factors in the pre-slaughter period. The glycogen stores of pectoral muscles and whole leg muscles were gradually depleted, leading to lactic acid accumulation and thus, pH value decrease after slaughter. Reference [23] reported pH value of breast 6.12 (female) and 6.17 (male) and for legs 6.23 (female) and 6.31 (male) of Japanese quail. Reference [4] found that pH of breast muscle form chickens 5.78-5.98. Reference [9] was stated microbial probiotics supplementation could increase ultimate pH of chicken breast muscle. In general, broiler meats with pH between 5.7 and 6.1 were called normal and do not reveal any quality problems.

Table 3. pH and % WHC values of breast meat (B) and whole leg meat (W) on different treatments.

Characteristics		Treatments			
		MFB-0.00	MFB-0.05	MFB-0.20	MFB-0.35
pH	B	5.78^{a}±0.05	5.83^{b}±0.04	5.98^{c}±0.05	5.87^{b}±0.04
pH	W	6.39^{a}±0.05	6.44^{ab}±0.04	6.50^{b}±0.04	6.46^{ab}±0.07
WHC (%)	B	32.22^{a} ±2.31	31.89^{a} ±2.14	40.49^{b}±6.20	31.89^{a}±2.14
WHC (%)	W	117.25^{a}±17.79	106.42^{a}±12.09	109.06^{a}±17.53	110.28^{a}±6.67

Each value is expressed as mean ± SD.
$^{a-c}$Means with different small letters within same row are significantly different (p<0.05).

For whole leg meat of sample MFB-0.20 (6.50) showed higher compared to sample MFB-0.00 (6.39). The pH value of breast meat shows lower than whole leg meat because of the different morphological structure that determines the functional affiliation of both muscles. The oxidative muscle (dark muscle) fibers in thigh muscles was higher compared to that in pectoral muscle and thus, a precondition for slower depletion of glycogen stores. According to [9], decreased ultimate pH, could reduce protein functionalities of chicken breast muscle such as WHC, protein solubility and gel formation ability. WHC was the ability of meat to hold all or part of its own water. Reference [1] reported water retention capacity of pectoral muscles for male 22.39 % and female 21.68 % while for legs muscles of male were 26.91 % and for female 25.08 %. Water holding capacity of whole leg muscle was higher than breast meat for all treatments in range 106.25-117.42 %. The mechanism is influenced by both the pH of the tissue and by the amount of space in the muscle cell and particularly the myofibril that exists for water to reside according to [24]. The percentage of water holding capacity is also influenced by the different pressure or speed of centrifuge machine.

By treatment with probiotics like *Bacillus subtillis* and *Bacillus natto*, the connective tissues among muscle fibers were significantly less than without treatment with probiotics but breast muscle fibers under the treatment showed neat muscle arrangement. The cross-section of breast muscle fibers was significantly larger than without probiotic treatment. The water holding capacity of breast muscle under treatments MFB-0.20 was higher than MFB-0.00 similar to research by [6] that reported WHC of breast muscle under the treatment probiotic (49.15-52.35 %) higher than without treatment (43.30 %). The increase of water-holding capacity and improve meat color, so it is a good alternative for antibiotics in their feed. On the physicochemical characteristics of quail leg and breast meat from five week show leg meat had higher moisture content, ether extract, collagen level, water-holding capacity, pH and shear force values, with lower myofibrillar and total protein than breast meat [25]. Connective tissue in whole leg was more than in breast muscle. The components in connective tissue were collagen and elastin able to hold the water [6]

Quail breast meats were cooked in an 80 °C water bath to measure the cooking loss, juiciness and shear force of cooked meat like in Table 4. Cooking loss, juiciness and shear force were shows no significant difference between the treatments. By treatment with MFB supplement was show the decrease of cooking loss and juiciness but higher the shear force. In the treatments, MFB-0.20 presented the lower percentage of cooking loss (19.16 %) and juiciness (16.78 %) but higher the shear force (11.92) compared to MFB-0.00, MFB-0.05 and MFB-0.35.

Table 4. Cooking loss, juiciness and shear force of cooked quail meat.

Treatment	MFB-0.00	MFB-0.05	MFB-0.20	MFB-0.35
Cooking loss (%)	19.86^a±0.04	19.52^a±0.16	19.16^a±0.35	19.30^a±0.49
Juiciness (%)	17.62^a±0.38	17.42^a±0.30	16.78^a±0.07	17.25^a±0.08
Shear force (N)	10.62^a ±0.44	11.09^a ±0.06	11.92^a ±0.79	10.67^a ±0.50

Each value is expressed as mean ± SD.
[a]Means with the same letter in same row is not significantly different (p>0.05).

In a few studies, the meat quality attributes of probiotic-fed chicken have been evaluated to ensure natural and wholesome characteristics [2]. In a similar result it was reported no difference in percentage of cooking loss from broilers fed probiotics (19.41 %) compared to non-probiotics fed chicken counterpart (19.48 %). Reference [9] was found that the cooking loss also no difference of cooking loss (18.2-18.7 %) in chicken breast muscles of probiotic feeding levels. Reference [20] reported cooking loss by treatment *Bacillus subtilis* lower than control along 8 days of storage.

Shear force by Warner-Bratzler shearing knife was used to determine the toughness of breast meat but some studies was run by identify the tenderness of meat. The reading of shear force was similar with percentage cooking loss that no significant from the treatment with probiotics. Reference [6] were analyzed the shear force of breast quail meat by treat with probiotics (*Bacillus subtilis, Bacillus natto* and *Bacillus licheniformis*) were 1.49-1.58 kg (14.61-15.49 N) hadn't difference with no treated with probiotics (1.69 kg or 16.57 N). They explained that the different results of shear force compared to other studies could be related to different probiotics feeding. References [20] and [9] were reported shear force (tenderness) not influence among the treatments with the value 1.41-1.57 kg (13.83-15.40 N) and 14.4-16.7 N (1.47-1.70 kg). The lower of tenderness value of meat means the meat was tough that were similar with this study. The fast growing of probiotic-fed chicken could influence meat tenderness due to the presence of large diameter fibers and an increase in glycolytic fibers.

Table 5 was presented about the percentage of cooking loss, juiciness, folding test and shear force of quail meat sausage. No significant different (p>0.05) of the parameters with the treatment of Mega Floral Booster (MFB). By application of treated quail meat in sausage making could determine the quality of quail meat. Cooking loss and juiciness percentage of MFB-0.20 were 4.82% and 15.79 %, receptively showed lower than other treatment (MFB-0.00, MFB-0.05 and MFB-0.35). According to [26] reported percentage of cooking loss of chicken sausage was 4.32 % lower than quail meat sausage it is because the variety of ingredients was improve the structure of sausage. Folding test is a simple method used to measure the textural quality of sausages. Generally, all commercial chicken sausages samples showed good gel strength because their folding test scores were more than 4.0. The folding test of the sausages had no significant difference between the treatments had 3.00 means the slices of sausage crack gradually when folded in half. Folding test upper 3.00 was declared as good texture of sausage.

By the lower cooking loss and juiciness were tend to high the shear force of breast quail meat of MFB-0.20 was 4.12 N. Shear force of commercial chicken sausage and duck chicken sausage were 0.38-1.31 kg (3.73-12.84 N) and 0.09-2.89kg (0.88-28.34 N) while quail meat sausage in range 2.72-4.12 N. Moisture, water holding capacity and fat content were influenced the texture of sausage. The lower moisture content tends to high shear force. Chicken sausage added with more water tends to reduce fat content and had low shear force.

Table 5. Cooking loss, juiciness and folding test and shear force of quail meat sausage.

Treatment	MFB-0.00	MFB-0.05	MFB-0.20	MFB-0.35
Cooking loss (%)	5.97^a±0.07	5.75^a ±0.08	4.82^a ±0.61	5.07^a ±0.28
Juiciness (%)	16.99^a ±0.35	17.72^a ±2.83	15.79^a ±2.62	17.33^a±1.43
Folding test	3.00^a±0.00	3.00^a±0.00	3.00^a±0.00	3.00^a±0.00
Shear force (N)	3.12^a ±0.02	2.72^a ±0.42	4.28^a ±0.17	3.44^a ±0.78

Each value is expressed as mean ± SD.
^{a}Means with the same letter within same row is not significantly different (p>0.05).

Five parameters (hardness, springiness, cohesiveness, gumminess and chewiness) by TPA recorded in Table 6 to analyze texture of quail meat sausage. Values for hardness, cohesiveness, springiness, gumminess and chewiness of quail meats influenced by MFB treatment. Treatments of MFB-0.20 (2981.14 g) and MFB-0.35 (2590.84 g) had high significant changes on hardness parameter than MFB-0.00 (2045.99 g) and MFB-0.05 (1847.35 g). Hardness of fish surimi sausage analyzed by [10] was 1103.69 g lower than quail meat sausage (1847.35-2981.14 g), while hardness of commercial chicken in range 3840-5000 g.

Table 6. Texture profile analysis (TPA) of quail meat sausage.

Parameters	MFB-0.00	MFB-0.05	MFB-0.20	MFB-0.35
Hardness (g)	2045.99^a±56.19	1847.35^a±110.28	2981.14^b±186.31	2590.84^b±70.22
Springiness (mm)	0.74^a±0.02	0.66^{ab}±0.03	0.78^b±0.01	0.70^{ab}±0.03
Cohesiveness	0.32^a±0.02	0.30^a±0.01	0.35^a±0.03	0.31^a±0.02
Gumminess (g)	621.84^a±14.33	544.31^a±14.46	1043.96^b±60.91	806.05^c±36.39
Chewiness (gmm)	459.64^b±23.34	333.60^a±14.72	865.66^c±36.81	621.42^d±17.2

Each value is expressed as mean ± SD.
$^{a-d}$Means with different small letters within same row are significantly different (p<0.05).

The higher hardness value within the sample was related to the higher protein content and lower content of fat. Meat type (young or old meat), meat part (breast or thigh), deboning method (hand or mechanically), water added and additives could also influence the hardness of the samples sausage. Among texture attributes, hardness was the most important to the consumers as it determine the commercial value of a meat according to [27].

Second parameters, springiness, gumminess, cohesiveness and chewiness behave similarly on which depend on the hardness parameters. Springiness of quail meats were in range (0.66-0.78 mm) and gumminess (544.31-1043.96 g). Quail meat from MFB-0.20 had higher springiness, gumminess, cohesiveness and chewiness were 0.78 mm, 1043.96 g, 0.35 and 865.66 gmm, respectively than other treatments. Commercial chicken sausage was show springiness, gumminess, cohesiveness and chewiness in range 12-15 mm, 1200-2300g, 0.2-0.4 and 1600-3300 gmm, respectively while fish surimi sausage show the springiness was 0.59 mm, cohesiveness 0.20 and chewiness 129.45 gmm lower than chicken sausage.

The differences could be due to the ingredients used for example salt. Salt significantly increased meat product cohesiveness. Mixture of duck and chicken meat sausage show the hardness 1200-4000 g, springiness 7.3-11 mm, gumminess 200-1000 g, cohesiveness 0.23-0.32 and chewiness in range 2910-9290 gmm. In term of protein content, the variety and high protein content resulted hard of texture and low fat content tend to increase the harder, gummier, cohesive, high chewiness and lower springiness of meat [28].

4. Conclusions

In conclusion, this study indicated that Mega Floral Booster (dietary probiotic supplement) had little or noticeable on carcass characteristics include live bird weight, carcass yield, breast with bone yield, breast meat yield, whole leg with bone yield and whole leg meat yield but MFB-0.20 was show a little impact on carcass characteristics. Mega Floral Booster-supplemented-0.20% diets show a little differ in color of breast muscle and whole leg muscle (L*, a* and b*), pH and WHC, cooking loss, juiciness and shear force compared with MFB-0.00, MFB-0.05 and MFB-0.35. In addition, usage of quail meat in sausage production not shows a large differs in sausage quality but still acceptable as good quality. Although the treatments not show the large improvement on meat quality but it not gives a bad effect.

5. Reference

[1] Genchev A and Mihaylov R 2008 *Trakia J. Sci.* **6(4)** 66-71.
[2] Kim H W, Yan F F, Hu J Y, Cheng H W and Kim Y H B 2016 *Poult. Sci.* **95** 1457–64.
[3] Aksu M I, Karaoğlu M, Esenbuğa N, Kaya M, Macit M and Ockerman H W 2005 *J. Muscle Foods **16***(**4**) 306-17.
[4] Zhang A W, Lee B D, Lee S K, Lee K W, An G H, Song K B and Lee C H 2005 *Poult. Sci.* **84(7)** 1015-21.
[5] Yang X, Zhang B, Guo Y, Jiao P and Long F 2010 *Poult. Sci.* **89(2)** 254-60.
[6] Zhou X, Jin E, Li S, Wang C, Qiao E and Wu G 2015 *Turkish J. Vet. Anim. Sci.* **39(2)** 203–10.
[7] Ockerman H W 1985 *Quality control of post-mortem muscle tissue.* (Columbus: The Ohio State University) p 1000.
[8] Ismail I, Huda N, Ariffin F and Ismail N 2010 *Int. J. Poult. Sci.* **9(6)** 556-61.
[9] Kim H W, Cramer T, Ogbeifun O O, Seo J K, Yan F, Cheng H W and Brad K Y.H 2017 *Meat Muscle Biol.* **1(1)** 81-89.
[10] Muhammad N A N, Hashim H and Huda, N 2017 *J. Agrobiotech.* **8(1)** 25-32.
[11] Boni I, Nurul H and Noryati I 2010 *Int.Food Res; J.* **17** 661-66.
[12] Lanier, T. C., 1992. Measurement of surimi composition and functional properties *Surimi Technology* eds. Lanier T C and Lee C M (New York: Marcel Dekker Inc.) pp 123-166.
[13] Huidobro F R D, Miguel E, Blázquez B and Onega E 2005 *Meat Sci.* **69(3)** 527-36.
[14] Bonos E M, Christaki E V and Florou-Paneri, P C 2010 *Int. J. Poult. Sci.* **9** 264-72.
[15] Walita K Z, Tanganyika J and Mussah S R 2017 *Int. J. Avian Wildlife Biol.* **2(2)** 50–53.
[16] Ilori B M, Peters S O, Ikeobi C O N, Bamgbose A M, Isidahomen C E and Ozoje M O 2010 *Int. J. Poult. Sci.* 2010 **9(4)** 368–75.
[17] Inci H, Sogut B, Sengul T, Sengul A Y and Taysi M R 2015 *Rev. Bras. Zoo.* **44(11)** 390-96.
[18] Narinc D, Aksoy T, Karaman E, Aygun A, Firat M Z and Uslu M K 2013 *Poult. Sci.* **92(7)** 1735-44.
[19] Tabidi M H, Mukhtar A M and Hassan I M 2013 *Global J. Medic. Plants Res.* **1(1)** 136-42.
[20] Bai W K, Zhang F J, He T J, Su P W, Ying X Z, Zhang L L and Wang T 2016 *PloS one*, **11(12)** 1-17.
[21] Zhang L and Barbut S 2005 *British Poult. Sci.* **46** 687–93.
[22] Bai K, Huang Q, Zhang J, He J, Zhang L and Wang T 2016 *Poult. Sci.* **96(1)** 74-82.
[23] Genchev A, Mihaylova G, Ribarski S, Pavlov A and Kabakchiev M 2010 *Trakia J Sci.* **6(4)** 72-82.
[24] Huff-Lonergan E 2002 *Fact Sheet* PIG 12-04-05.
[25] Shanaway M M 1994. *Quail production systems: a review.* (Rome: FAO Animal Production Health Division and Engineering) p 145.
[26] Grizotto R K, Andrade J C, Miyagusku L and Yamada E A 2012 *Ciência Tecn. Alimen.* **32(3)** 538-46.
[27] Huda N, Wei L H, Jean, A T and Ismail I 2010 *Int. J. Poult. Sci.* **9(10)** 954-58.
[28] Huda N, Lin O J, Ping Y C and Nurkhoeriyati T 2010 *Int. J. Poult. Sci.* **9(6)** 550-55.

Acknowledgement

The authors acknowledge with gratitude the support provided by Universiti Sultan Zainal Abidin to conduct research in the area of Food Science and Technology through research grant UniSZA/2017/DPU/11.

IOP Conf. Series: Earth and Environmental Science **287** (2019) 012031 doi:10.1088/1755-1315/287/1/012031

Proximate and amino acid composition of quail meat treated with mega floral booster addition

U H Sharif[1], A S Kamarudin[1] and N Huda[2]

[1]School of Food Industry, Faculty of Bioresources and Food Industry, Universiti Sultan Zainal Abidin, Besut Campus, 22200, Besut, Terengganu Darul Iman, Malaysia
[2]Faculty of Food Science and Nutrition, Universiti Malaysia Sabah, 88400, Kota Kinabalu, Sabah, Malaysia

Abstract. This study evaluated the effect of Mega Floral Booster (MFB) as probiotic supplement on the quality properties such as the carcass yield, physicochemical properties and nutritional composition of quail meat. Mega flora booster is the multi- probiotic animal feed additive. The treatment is divided to four treatments based on percentage of MFB include MFB-0.00 (control), MFB-0.05 (0.05% of MFB), MFB-0.20 (0.20% of MFB) and MFB-0.35 (0.35% of MFB). The 18 of quail species *Cortunix* japonica (Japanese quail) used as the main material in the study for each treatment. For the first until fourteen days, the quails are supplied with commercial feed only. After fifteen days, the quails begin provided with the MFB-0.00, MFB-0.05, MFB-0.20 and MFB-0.35 mixed with commercial feed. The quails for 42nd days slaughtered and deboned of breast and whole leg separated from the bone for analysis. Probiotic supplement (Mega Floral Booster) had no influence on proximate and amino acid composition. Chemical score and amino acid score was show tryptophan as amino acid limited. Essential Amino Acid index of breast meats presented MFB-0.20 (61.66%) higher than MFB-0.00 (60.26%). Thus, Mega Floral Booster is not affecting the nutritional quality of quail meats.
Keywords proximate, amino acid, quail meat, mega floral booster.

1. Introduction

The quail is not an established branch, but occupies a relevant place in poultry breeding and contributes to the variety in poultry meat production according to [1]. Along the increase of the chicken meat production, there is also increase of the production meat from quails and other non-commercial bird species, reared under industrial conditions. The highest consumption of quail meats tends to encourage the quail farming thus increase of the feed additive or supplement usage.

Dietary supplementations of probiotic have been proposed to antibiotics in the poultry industry that can improve animal production performance, regulate intestinal micro-flora imbalance, and enhance immune function that stated by [2]. In a limited study, the meat quality attributes of probiotic- fed have been evaluated to ensure natural and wholesome characteristics. Some of studies reported that probiotic supplements to poultry able to improve meat quality attributes such as water holding capacity (WHC), tenderness, lipid oxidation stability, sensory properties, and microbial safety [3, 4, 5]. On the other studies were stated that no synergistic effect of probiotics on meat quality [2, 6]. Postmortem aging leads to considerable biochemical and physical changes in skeletal muscles through endogenous proteolytic system, results in an improvement in meat quality attributes such as tenderness, WHC, juiciness and flavor [2]. Therefore, the objectives of this study was to determine the nutritional composition include chemical composition (moisture content, ash, protein and fat), and amino acid composition between four treatments with different concentration of Mega flora booster in feed intake.

2. Materials and methods

Coturnicx japonica species of quails (after 42nd days) were getting from collaboration with Animal Science final year students. The quails (1st day) were obtained from commercial hatchery in Terengganu (Surada Jaya Kemaman). For quail growth process compose to two phase, brooding process (1st until 14th days) and growing process (15th until 42nd days). During brooding process, the quails were weighed and divided for some group, and supplied with commercial feed (Permula Ayam Daging 201C by Gold Coin Feedmills (M) Sdn Bhd) in brooding room. At growing process, pens were randomly assigned to four treatment of Mega Floral Booster (MFB) with different percentage 0.0% (MFB-0.00), 0.05% (MFB-0.05), 0.20% (MFB-0.20) and 0.35% (MFB-0.35) mixed with 100%, 99.95%, 99.98% and 99.65% of commercial feed. MFB was the advanced probiotic formula, fortified with multi strain of microbes (*Bacillus* natto, *Lactobacillus* and yeast) and methionine, organic acid, enzyme protease and amylase manufacture by CPEX (Japan) Group. The proper design of building area required to avoid any infection that may occur along the experiment. The quails (in large amounts) was separate to three different part (in small amount) for each treatment; part A, part B and part C with twelve of quails for each part to minimize the rate of mortality. The 20g of feed was given two times morning and evening.

At the end of the feeding (42nd days), the six of quails per pen were randomly taken for each treatments (totally 18 for each treatments). The quails were slaughtered, scalded (60-65 °C for 15 seconds), de-feathered, eviscerated and packed in blast freezer (overnight) according Genhev (2008). Then, the quails were deboned nutritional composition includes proximate composition, and amino acid composition.

Proximate composition was determined according to [7] standard methods. The determination of moisture (oven drying method), total protein, fat (ether extract method) and ash (muffle furnace technique) content were performed for breast quail meats.

Amino acid composition was determined according to [7] the analysis preparation of amino acid composition, meat sample was dried by using oven drye at 70-80°C. meat sample was analyze in triplicates. Amino acid analysis of meat was determined by using High Performance Liquid Chromatography (HPLC) based on Waters auto Tag^{TM} OPA Pre Column Derivation method. The test samples was analyzed with three hydrolyses in three (6N HCl, peromic acid + 6N HCl and 4.2N NaOH). 6N HCl at 110°C for 24 hours and derivatized with AccQ reagent (6-aminoquinolyl-N-hydroxysuccinimdyl carbamite) before conducting the chromatographic separation by using an AccQ Tag^{TM} reversed phase (3.9 x 150mm) analytical column (Waters®). The amino acid analysis will be perform on a HPLC system consist of Waters 1525 Binary HPLC Pump, 717 Plus Autosampler (Waters®) and Water 2475 Multi Fluorescence detector (wavelength excitation 250nm, emission 395nm). Chromatographic peaks was integrate, identify and quantify using Breeze^{TM} software version 3.20 by comparing it to identify standards (Amino acid standard H, Pierce, Rockford,Illinois). Methionine and cysteine was determined from the same method of acid hydrolysis after treatment using performic acid oxidation. Samples hydrolyzed with 4.2N NaOH for determination of tryptophan.

Chemical score: chemical score was conducted by comparing the essential amino acid content amino acid of egg [8]). Essential amino acid content of egg according to [9] were lysine 6.98, methionine + cysteine 5.79, isoleucine 6.29, leusine 8.82, valine 6.85, fenilalanine + tyrosine 9.89, threonine 5.12 and tryptophan 1.49/100g protein. Amino acid score: amino acid score was determined by comparing the essential amino acid content of sample and type of amino acid content that was suggested for human needed [10]. The type of amino acid content suggested for 2-5 years old used for determining this score (FAO, 1985) was histidine 1.9, lysine 5.8, methionine + cysteine 2.5, threonine 3.4, isoleucine 2.8, leusine 6.6, valine 3.5, fenilalanine + tyrosine 6.3and tyrtophan 1.1g/100g. Essential amino acid (EAA) index: essential amino acid (EAA) index was used to determine the chemical score. The score for every amino acid was converted to log10. The average score was given to get the index score [8].

The statistical analysis from this experiment was conducted by using SPSS statistic 17.0 software. Data was analyzed by using analysis of variance (ANOVA). The software was used to calculate the overall mean and standard deviation of the data and to determine the significant different of the proximate analysis and amino acid analysis. Besides that, SPSS is use to form the graph and find the liner equation for certain graph. Turkey test was used to determine significance of difference ($p<0.05$) of four samples.

3. Results and discussions

Japanese quail composed of chemical composition include moisture content, protein, carbohydrate, fat and ash in wet weight basis (%) had been presented in Table 1. Treatment of Mega Floral Booster (MFB) on chemical composition of breast quail meats hadn't significant difference. Although the treatments not have significant difference but MFB-0.20 was presented the high level of moisture content (75.00 %), protein (22.86 %), ash (1.25 %) and low level of fat (1.70 %) compared to chemical compositions of MFB-0.00, MFB-0.05 and MFB-0.35.

Table 1. Chemical composition of 42 days old of Japanese quail.

Chemical composition	Treatments			
	MFB-0.00	MFB-0.05	MFB-0.20	MFB-0.35
Moisture content (%)	$74.59^a\pm0.41$	$74.38^a\pm0.50$	$75.00^a\pm0.35$	$74.51^a\pm0.42$
Protein (%)	$22.82^a\pm0.13$	$22.78^a\pm0.17$	$22.86^a\pm0.16$	$22.80^a\pm0.30$
Ash (%)	$1.24^a\pm0.06$	$1.24^a\pm0.04$	$1.25^a\pm0.06$	$1.14^a\pm0.10$
Fat (%)	$1.77^a\pm0.13$	$1.78^a\pm0.12$	$1.70^a\pm0.14$	$1.72^a\pm0.34$

Each value is expressed as mean ± SD.
[a]Means with the same letter within same row is not significantly different ($p>0.05$).

According to [12], the moisture content and protein of hand deboned meat of young quail were 73.01 % and 20.13% and spent quails were 70.28 % and 22.33 % slightly similar with moisture content in the [1] while [13] reported moisture content (75.63 %) and protein (20.18 %) with high fat content (5.02) and low ash (0.94) of quail meats. Reference [1] reported protein content, fat content and ash content were 23.38 to 22.23 %, 2.21 to 2.75 % and 1.51 to 1.61 % of 35 days old Japanese quails while chemical composition of quails reported by [14] was show the moisture, protein, fat and ash were 72.35 %, 21.65 %, 3.57 % and 2.47 % of 6 weeks of breast quails. Fat content in breast muscle in this result was lower than fat content that stated by [13] and [14] especially fat content of breast quail muscles treated with probiotic was lower.

Reference [15] was run the research on breast meat quality with different probiotic level and reported chemical composition was unaffected by probiotic feeding levels ($p>0.05$) in which moisture, protein, fat and ash contents were 75.5 to 75.9 g/100 g, 23.3 to 23.4 g/100 g, 1.5 to 2.1 g/100 g and 1.2 g/100 g, respectively. They were stated that probiotic supplementation could improve protein solubility (sarcoplasmic protein) of breast muscle from chicken in term on protein functionality. Reference [6] was found by treatment with (*Bacillus subtilis, Bacillus natto* and *Bacillus licheniformis*) tend to improve the muscle fiber (myofibrillar protein) and decreased of connective tissue in breast muscle cause protein content in breast muscle was high by treatment with probiotics. In increase of protein and moisture content tend to lower of fat content.

The dietary value of meat defined the composition and ratio among the different group of nutrients. Reference [1] stated the analysis of amino acids showed that quail meat was very rich in essential amino acids. According to most dieticians, the normally functioning non-trained human organism needs 8 essential amino acids includes lysine, valine, leucine, isoleucine, threonine, tryptophan, methionine and phenylalanine. Table 2 and 3 was showed amount of amino acid for quail meats under treatment and the amino acid composition (g/100 g protein) for different ages of quail meat analyses.

Breast quail meats of MFB-0.20 was higher than MFB-0.00 but lower than young and spent quails. If amino acid type was almost similar with amino acid type in human body determine the quality of protein will be higher [16]

Table 2. Amino acid compositions (Essential Amino Acid) in quail meats treated with MFB, young and spent quails.

Essential Amino acid composition	MFB-0.00	MFB-0.20	Young quail [12]	Spent quail [12]
Histidine	2.13	2.34	4.07	3.93
Methionine	2.38	2.37	3.18	3.12
Cysteine	0.87	0.89	0.25	1.57
Isoleucine	3.51	3.66	5.37	5.81
Leusine	5.74	5.94	7.44	8.66
Lysine	6.09	6.20	7.12	7.70
Phenylalanine	2.96	3.09	5.48	5.25
Threonine	3.18	3.37	6.02	5.86
Tyrosine	2.74	3.19	2.85	2.68
Valine	3.67	3.88	7.12	7.77
Tryptophan	0.82	0.71	ND	ND
Total EAA	34.09	35.64	48.9	52.35

Table 3 Amino acid compositions (Non-Essential Amino Acid) in quail meats treated with MFB, young and spent quails.

Non Essential Amino acid composition	MFB-0.00	MFB-0.20	Young quail [12]	Spent quail [12]
Alanine	3.95	4.16	4.70	5.04
Arginine	4.96	5.06	10.76	8.04
Aspartic Acid	6.69	7.07	6.57	6.80
Glutamic Acid	10.28	10.73	12.31	12.54
Glysine	3.13	3.38	7.77	6.65
Proline	2.54	2.77	4.14	3.83
Serine	2.83	3.05	4.87	4.75
Total non-EAA	34.38	36.22	51.12	47.65

In essential fraction in breast meat treated by MFB, leucine (5.74-5.9 %) lysine (6.09-6.20 %) similar in breast meat of young and spent quails lysine (7.12-7.70) and leucine (7.44-8.66) such results in [14] Moreover, the data indicated that among all amino acid glutamic acid recorded the highest value in MFB-0.00 and MFB-0.20. On other hand presented that tryptophan was the lower compared with other content. Reference [12] also found the highest amino acid in young and spent quails was glutamic acid (12.31-12.54), respectively. However, the relatively higher content glutamic in the breast meat had a great importance when it comes to the taste and flavor of meat indicated by [17].

The quality of protein in foods was determined by the type of its amino acid and total of its essential amino acid. The quality of protein will be higher if its amino acid type almost similar with the amino acid type of human body. Chemical score, amino acid score and EAA index was using to identify the quality of quail meat. Reference [12] was reported chemical was a standard for protein quality which already done for long time before.

Amino acid content was compare with egg protein as a standard to get the chemistry score in range 0-100. As shown in Table 4, chemical score indicated that the first limiting amino acid of breast meat quail treating with MFB-0.00 and MFB-0.20 was tryptophan (54.97 and 47.45) followed by valine (53.64 and 56.66 %) in six weeks breast quail meat. The first limiting amino acid of spent quail was methionine + cysteine (81.00%) was more higher followed by young quail (59.24 %), MFB-0.20 (53.54 %) and MFB-0.00 (51.64 %) as shown in Table 4.

Table 4. Chemical score of amino acids from breast quail meat.

No	Type Of Amino Acid	MFB-0.00	MFB-0.20	Young quail [12]	Spent quail [12]
1	Methionine + cysteine	56.15	56.37	59.24	81.00
2	Isoleucine	55.85	58.12	85.37	92.37
3	Leusine	65.07	67.29	84.35	98.19
4	Lysine	87.29	88.88	102.01	110.32
5	Threonine	62.09	65.86	117.58	114.45
6	Valine	53.64	56.66	103.94	113.43
7	Phenylalanine + Tyrosine	57.59	63.55	84.23	80.18
8	Tryptophan	54.97	47.45	ND	ND
	Score	51.64	53.54	59.24	81.00

It is important to know more about the role of essential amino acid in source of protein for human need. Amino acid score was above 0 to 100 by compared the essential amino acid content of sample and type of amino acid content that is recommended for 2-5 years old. If the amino acid score counted above 100, it is calculated as 100. Amino acid score of MFB-0.00 (74.45 %) was higher than MFB-0.20 (64.27 %) while spent and young quail were the higher of amino acid score 100. First of amino acid limiting from quail meat by treat with MFB-0.00 and MFB-0.20 was trytophan and the second amino acid limiting was leusine as shown in Table 5.

Table 5. Amino acid score of amino acid from breast quail meat.

No	Type Of Amino Acid	MFB-0.00	MFB-0.20	Young quail [12]	Spent quail [12]
1	Histidine	112.26	123.16	214.21	206.84
2	Methionine + cysteine	130.04	130.56	137.20	187.60
3	Isoleucine	125.46	130.57	191.79	207.50
4	Leusine	86.95	89.92	112.73	131.21
5	Lysine	105.05	106.97	122.76	132.76
6	Threonine	93.50	99.18	177.06	172.35
7	Valine	104.97	110.89	203.43	222.00
8	Phenylalanine + Tyrosine	90.41	99.76	132.22	125.87
9	Tryptophan	74.45	64.27	ND	ND
	Score	74.45	64.27	100	100

Essential amino acid index is the geometrical mean of the ratio of all essential amino acid in the evaluated protein relative to their content in a highly nutritive reference protein such as whole egg. Essential amino acid (EAA) index were identifying for evaluating proteins in diets according to [19]. The data presented in Table 6 indicated the EAA index of quail meats MFB-0.00 was lower than quail meats of MFB-0.20. EAA index of spent quail had higher (97.60 %) followed by young quail (89.16 %), MFB-0.20 (61.66%) and MFB-0.00 (60.26%). Spent quail meat was most of a valuable source of protein with very good amino acid profile and EAA index than young quail, age of quail was affected the quality of protein and amino acid associated with protein degradation [12] The breed of quail also influenced the composition of protein content and amino acid. From this analysis, quail meat from the treatment of MFB-0.20 was improving the amino acid profile and EAA index than MFB-0.00.

Table 6. EAA index from breast quail meat.

No	Type Of Amino Acid	MFB-0.00	MFB-0.20	Young quail [12]	Spent quail [12]
1.	Methionine + cysteine	1.75	1.75	1.77	1.91
2.	Isoleucine	1.75	1.76	1.93	1.97
3.	Leusine	1.81	1.83	1.93	1.99
4.	Lysine	1.94	1.95	2.01	2.04
5.	Threonine	1.79	1.82	2.07	2.06
6.	Valine	1.73	1.75	2.02	2.05
7.	Phenylalanine + Tyrosine	1.76	1.80	1.93	1.90
8.	Tryptophan	1.74	1.68	ND	ND
	Average (N=7)	1.78	1.79	1.95	1.99
	Score	60.26	61.66	89.16	97.60

4. Conclusions

In conclusion, this study indicated that Mega Floral Booster (dietary probiotic supplement) had noticeable effect on nutritional characteristics of quail meat. Proximate composition (moisture content, protein, fat and ash) of quail meat not affected by difference percentage of MFB while MFB-0.20 show a high value in amino acid analysis include essential amino acid and non-essential amino acid.

5. Reference

[1] Genchev A and Mihaylov R. 2008. *Trakia J Sci.* **6(4)**: 66-71.
[2] Kim H W, Yan F F, Hu J Y, Cheng H W and Kim Y H B. 2016. *Poult. Sci.* **95**: 1457–64.
[3] Aksu M I, Karaoğlu M, Esenbuğa N, Kaya M, Macit M and Ockerman H W. 2005. *J. Muscle Foods* ***16*(4)**: 306-17.
[4] Zhang A W, Lee B D, Lee S K, Lee K W, An G H, Song K B and Lee C H. 2005. *Poult. Sci.* **84(7)**: 1015-21.
[5] Yang X, Zhang B, Guo Y, Jiao P and Long F. 2010. *Poult. Sci.* **89(2)**: 254-60.
[6] Zhou X, Jin E, Li S, Wang C, Qiao E and Wu G. 2015. *Turkish J. Vet. Anim. Sci.* **39(2)**: 203–10.
[7] Association of Official Analytical Chemists International. 2005. *Officials methods of analysis. 18th ed.* (Washington: Association of Analytical Communities International).
[8] Acton J C and Rudd C L. 1987. Protein quality method for seafood *Seafood quality determination* eds. Kramer D E and Liston J (Amsterdam: Elsevier) pp 453-73.
[9] Food Agricultural Organization. 1970. *Amino acid content of foods and biological data on proteins* (Geneva: Report of a joint FAO/WHO Ad hoc expert committee) Tech. Rept. Series 522.
[10] Sawar G and McDonough F E. 1990. *J. Assoc. Anal. Chem.* **73**(3): 347-56.
[11] Food Agricultural Organization. 1985. *Energy and protein requirements* (Geneva Report of a joint FAO/WHO/UNU expert committee). Tech. Rept. Series 724.
[12] Boni I. 2010. *Quality characteristics of quail meatballs* (Penang: Master Thesis. University of Science Malaysia) p 165.
[13] Karakök S G, Ozogul Y, Saler M and Ozogul F. 2010. *J. Muscle Foods* **21(2)**: 210-23.
[14] Khalifa A H, Omar M B, Hussein S M and Abdel H E. 2016. *Assiut J.Agric.Sci.* **47**: 58–71.

[15] Kim H W, Cramer T, Ogbeifun O O, Seo J K, Yan F, Cheng H W and Kim Y H B. 2017. *Meat Muscle Biol.* **1(1)**: 81-89.
[16] Friedman M. 1996. *J. Agric. Food Chem.* **44(1)**: 6-29
[17] Brudnicki A, Kulakowska A, Pietruszynska D, Lozyca-Kaplon M and Wach J. 2012. *Czech J. Food Sci.* **30(4)**: 309-13.
[18] Bunda M G, Tumbokon B L and Serrano Jr A E. 2015. *AACL Bioflux* **8(3)**: 411-20.

Acknowledgement

The authors acknowledge with gratitude the support provided by Universiti Sultan Zainal Abidin to conduct research in the area of Food Science and Technology through research grant UniSZA/2017/DPU/11.

Analysis of prospects application of integrated farming system models for lowland rice with beef cattle to strengthen farmers' economy in West Sumatera

Asdi Agustar[1], Juniarti[2], and Ida Indrayani[3]

[1] Professor at Livestock Development and Business, Andalas University, Padang, West Sumatera, Indonesia.
Corresponding email: asdiAgustar @ yahoo.com.
[2] Lecturer at Social Science Department, Andalas University, Padang, West Sumatera, Indonesia.
[3] Lecturer at Livestock Development and Business, Andalas University, Padang, West Sumatera, Indonesia.

Abstract. Crop-Livestock-System (CLS) is an agricultural system approach that can improve business efficiency, maintain land fertility and realize sustainable agriculture, especially on small-scale agriculture. This research was conducted in Gunung Talang Subdistrict, Solok Regency, West Sumatra, aimed to find out the habits of farmers and the potential to integrate and calculate farmers' income if they only do one commodity farming, namely rice, or rice and cattle. Data collection was conducted from June to August 2018 using survey methods. Data collection was carried out through interviews with 30 lowland rice farmers as respondents, randomly selected. Data collected includes crop and livestock farming data such as: farmer characteristics, average rice field area, average productivity and production value, average number and value of inputs (seeds, inorganic fertilizers and organic fertilizers, and organic and inorganic pesticides, the average supply of laboratory in an integrated farming system between plants and cattle. The results showed that average rice yield was 5.74 tons/ha of dried ground rice. The use of inorganic fertilizer urea 100 kg/ha (N 71.43%), SP-36 50 kg/ha (50%), KCl 50 kg/ha (50%).The income of farmers with only rice IDR 11,816,807 for ones hectare land. Total income from paddy (1 ha) and livestock (2 heads) integrated farming system about IDR 17, 612,405/season and provide R/C value about 1.44, meanwhile R/C value of non integrated farming system about 1.33. It means, paddy and livestock integrated farming system could increase farmer's income about 49,45% per season compared to non integrated farming.
Keywords – integrated farming system, lowland rice, beef cattle, economy.

1. Introduction

Fragmentation and converted land use are the causes of the increasingly small land tenure in lowland rice farmers. Therefore to increase production, intensification with the use of inorganic fertilizers cannot be avoided. In the long run (approximately 30 years) can reduce land productivity [1], due to decreased nutrients in the soil. Besides that it will also reduce business efficiency, thereby reducing farmers' income.

The integration of lowland rice with beef cattle maintenance is conceptually seen as an ideal solution. Livestock waste will produce organic fertilizer while rice waste can be used as a source of cattle feed. This model application will be able to maintain soil fertility and at the same time reduce

IOP Conf. Series: Earth and Environmental Science **287** (2019) 012032 doi:10.1088/1755-1315/287/1/012032

production input costs. According to data from the West Sumatra Central Bureau of Statistics (BPS), rice production in 2015 reached 2.55 million tons of milled dry grain with a harvested area of 440,000Ha. If the straw is produced as much as 1.44 x of MPD production, then 3.6 million tons of rice straw will be produced each year. If each cow needs 10 kg of hay / day for feed, then wet rice in West Sumatra has the potential to provide feed to increase the cattle population of ± 800,000, - tail. This amount means almost 3 times the number of cattle population in West Sumatra. Thus rice-beef cattle integration will improve business efficiency, improve the quality of land use, and produce a clean and comfortable environment. Such conditions will lead to the strengthening of the farmers' economy, especially the destruction of rice fields.

This study aims to describe the habits of farming patterns and the potential integration of cattle-rice in the area of rice farming centers in Solok Regency, West Sumatra. Furthermore, it is also to find out the farmers' income if they only do rice farming in the same area.

2. Materials and methods

The study was conducted in *Kenagarian* Jawi-Jawi, Gunung Talang Subdistrict, Solok Regency which was chosen purposively. The consideration was that nigari almost all of its inhabitants had intensive rice farming activities conducted intensively. Determination of sample farmers Stratified random sampling was carried out as many as 30 farmers. Data collection is done by interviewing to fill out a list of questions (questionnaires) that have been prepared for it. Data analysis was carried out with descriptive statistics. For economic analysis using cost-benefit analysis. Furthermore, financial analysis is done using the formula: Marginal Benefit Cost Ratio (MBCR) which is modified from the formula [2].

$$MBCR = \frac{\text{Revenue integrated farming} - \text{Revenue of rice farming}}{\text{Total Cost of integrated farming} - \text{Total cost rice farming}}$$

3. Result and Discussion

3.1. Characteristics of Research Locations

Kenagarian Jawi-Jawi has a rice field area that is 100% technical irrigation. Rice fields are the main source of livelihood for the farming family and can be planted 2-3 times / year because the presence of water is available throughout the year. From the socio-economic aspect, rice fields are land to fulfill basic needs and are the main employment opportunities for villagers. The dominant cropping pattern is rice-rice-rice and most farmers around the study site plant local varieties. Fertilization for good quality rice generally uses urea doses of 300-400 kg / ha, SP-36 100-150 kg / ha, and KCl 100-150 kg / ha. Manure is not used as a source of fertilizer for rice fields.

3.2. Rice Cultivation and Utilization of Rice Straw as Sources of Feed

Rice cultivation planted local varieties of *Cisokan* and *Anak Daro*, N fertilization with BWD indicators, fertilizing P and K according to recommendations. With an average yield of 5.74 tons / Ha. This result is higher than that reported [3] that the P3T program can produce rice production 5.5 t / ha. Based on rice planting area, fresh straw obtained 13.20 tons / ha, after fermented to 7.92 tons (60% yield) and can be used for feed 2 heads of cattle for a year assuming feed consumption of 10 kg / head.

Proximate analysis of straw fermented showed can improve protein quality at 4.88% from 4.01% to 9.09%, reducing crude fiber 6.32% from 24.76% to 18.44%, and increase TDN 6.95% from 41.68% to 48.63%. There is an increase in protein and very low levels of crude fiber support in its use as animal feed, because in general which is a limiting factor in the utilization of agricultural waste as Animal feed is a low nutritional value.

3.3. Beef Cattle Raising and Utilization of Manure

The number of cattle ownership in the study location is an average of 2 heads. By the product of the process of raising cattle is manure which will become a quality compost after fermentation. The results of the analysis of manure showed that fermentation could improve the quality of fertilizer and reduce the C / N ratio from 33.67 to 19.03. With 2 heads of cattle, organic fertilizer will be produced 0.24 tons / month up to 0.3 tons / month.

If the average ownership of the farmer's land area is 0.32 hectares, fine compost needs for own land of 0.6 tons / MT (if the dose is 2 tons / ha), which can be obtained during 2 months of collecting manure. The average production of compost is 1.2 tons / 2 heads / 4 months, if it is assumed that the selling price is IDR 1000/kg, an additional income of IDR 1,200,000 will be obtained. Contribution of additional revenue from fine compost is 9.70%, while the farmer pattern that was carried out at the time of research there was no revenue from manure.

3.4. Financial Analysis

Farmer's income in rice farming is influenced by the amount of production and the emphasis on the purchase of inorganic fertilizer (57.14%). The results of the analysis of the R / C ratio obtained 2.28, that each expenditure of production costs of IDR 1,000 gives revenue of IDR 2,280 and IDR 1,990 (farmer level). Some assumptions used are the price of rice IDR 6,200/kg, fine compost IDR 800/kg, urea IDR 1,800/kg, TSP IDR 2,500/kg, KCl IDR 2,500/kg.

Financial analysis of cattle fattening through utilization of fermented straw as animal feed and utilization of organic fertilizer, showed that the average income reached IDR 1,383,100/head/4 months with R / C 1.24. The MBCR value is 1.43, means that every additional cost in applying the introduction technology of IDR 1,000 can increase the revenue of IDR 1,430. This shows that the cattle raising with an integrated system approach can increase the productivity and income of farmers as well as financially feasible to be developed when compared to the pattern of agriculture only rice plants.

Livestock integration systems are significantly capable of providing added value on farm produce and on livestock productivity. Integrated farming can reduce production costs, especially for the provision of forage, as source of laboratory, and can contribute to savings purchase of fertilizer.

4. Conclusions

Integration of cattle with rice are farming which is efficient and very relevant for the condition of people's farming with ownership limited land in the countryside. This system benefits farmers because (1) Animal waste that have not been optimally used can be used as organic fertilizer to increase soil fertility and sources income and (2) agricultural waste (rice and bran straw) available used as a source of high quality food that reduces costs feed supply.

5. References

[1] Haloho, R.D, S.I. Santiso dan S. Marzuki. 2004. Profit Function Analysis of Beef Cattle Raising. J.Livest. Sci. 134: 108-115.

[2] Sudaryanto T, Ilham N. 2001. Upaya Peningkatan Efisiensi Usaha Ternak Ditinjau Dari Aspek Agribisnis Yang Berdaya Saing. Apresiasi Teknis Program Litkaji Sistem Usaha Tani Tanaman Ternak (*Crop Animal System*). Bogor: Puslitbangnak.

[3] Sumanto, Darwis M, Rafiq A, Darmawan A, Mulyawan H. Dan Khairuddin. 2002. Peningkatan produktivitas padi Jawa Barat. Prosiding Lokakarya Pelaksanaan Program Peningkatan Produktivitas Padi Terpadu. Badan Litbang Pertanian. Yogyakarta, 17-18 Desember 2002.

Basic mapping data of Jambi Province potency for a continuous livestock development strategy

D Nora

Agiculture Faculty Livestock Study Program, Muara Bungo University, Jambi, Indonesia.
Email: delvianora@yahoo.com.

Abstract. Local commodity in Jambi Province consists of 20 kinds of product for plantation area, which in its entirety have the power of the biggest labour variation amount that is in plantation area (85%) followed by farm area (66,67%) and animal husbandry area (63,64%). Technique to identify regional input contribution. As for the quantity of input resources variation becomes the measurement of : (1) the unit size of commodity area, (2) effort variation and effort specification, (3) quantity of effort contribution, (4) alternative potency and prior potency of area effort that becomes a measurement. The result above can be used as precisely basic identification program in specifying program policy that is integrated to strengthen the commodity power. Jambi Province has the ability to apply integration program between farming-plantation, animal husbandry (P3) to support reinforcement of self sufficiency in food and effort. The action of maintaining this final potency of commodity can be taken as a recommendation to keep the stability of raw and final material availability as commodity of area requirement accomplished also can be developed continuously in Jambi Province To answer the second problem in this research, it is stated that the lowest standard pattern that can be done in developing local resources program to maintain food is by P3 pattern (Plantation, Animal Husbandry and Agriculture). Basic mapping data horizontally, the plantation mapping possesses 3 area categories of commodity : green zone A , orange zone B and C yellow zone. Sub-provinces with green zone commodity area 1,2,3,4,5,6,7,8,9. Zone B in sub-province (none), 5,6,7,8,9,11 and C Yellow zone sub-province is 10. Horizontally farming mapping owns 3 (three) area categories of commodity A green zone, B orange zone (none) and C yellow zone. Sub-province with green zone commodity are 1,2,3,4,5,6,7,8,9, zone B in sub-province in none and C yellow zone in sub province are 10,11. While horizontally animal husbandry mapping owns 4 area categories commodity. For Non Poultry livestock, B (non poultry livestock). Sub province non poultry livestock 1,2,3,4,5,6, Zone B (7,8,9,10,11), zone poultry livestock (1,2,3,4,5,6,7,8), B (9,10,11). while Vertically animal husbandry mapping owns 4 zone, they are active/high potency (A1,A2) non poultry livestock and poultry livestock conservation zone B and C. According to Nora (2016) identification of rescue conservation measures for livestock stock can be seen from: 1) whether or not there is integration of potential natural resources and animal husbandry activities (in terms of waste utilization), 2) an increase in deductions which causes changes in population to exceed the existing population, 3) absence of linearity between stock and cutting numbers, 4) mapping of conservation positions starting from level 2, 3, and 4.
Keywords – mapping data, livestock development, Jambi Province.

1. Introduction

The position of agricultural commodities is a commodity that has a promising regional potential in terms of the availability of raw materials for waste, but when linked to the role of leading commodities (plantations) it turns out that agriculture still cannot support integration efforts, so that the relationship between availability and effort that can be developed requires utilization strategies. The right SDL (Local Resources) and the empowerment of the role of the intervention of the side processing technology of the plantation business as a supplier of raw materials for animal feed business need to be improved. Agricultural and plantation by-products have their own strength for livestock development businesses. Agricultural business needs to be intensified and livestock businesses can be used as other businesses that need to be developed as sustainable superior potential with integration patterns. Raw materials in the form of raw materials in fresh form and waste can be used as indicators of the potential of commodity resources. Leading commodity and business relationships as the basis for evaluating Local Resources (SDL).

1.1. The purpose of the research

The purpose of this research is : the act of maintaining this potential commodity can be used as a recommendation to maintain the balance of raw material availability, results as a commodity that meets the needs of the region, can be developed as a WASH-WASTE AND CLS (Croop Livestock System) or wild plats PROGRAM in a sustainable manner to achieve food and raw material self-sufficiency, to answer the second problem in this study, it was produced that the lowest standard pattern in the location that this program could do in developing SDL (or :local resources-indonesia) potential for food security, and P3 (Pattern Planning Program) or (Plantation, Animal Husbandry and Agriculture) or 3 in 1 field of livestock development and livestock development

2. Materials and methods

2.1. The material of the research

The data is continued by taking a 100% sample from all regencies in Jambi Province to get the strongest potential position and possible variation of additional efforts from the business. Commodities contained in a diverse area produce a variety of potential possibilities for further development and empowerment optimally to strengthen the potential that already exists in the future. The data obtained were analyzed based on the assessment data matrix measured in percentage, ranking, scoring and comparison, and the analysis used was descriptive analysis, LQ (Location Quotation) and continued with QSPM analysis to determine priority policy.

2.2. Methods

The method used is a survey method, where the data inputted is primary data and secondary data from previous research which is a continuation of the identification of the sample location mapping in one of the districts in Jambi Province. The selected location is the basic location of alternative policy making and priority policy for the development of beef cattle in Jambi Province, while the commodity data is the result of secondary data analysis of Jambi Province in the fields of agriculture, plantations and livestock. The output of the previous research was in the form of the position of the conservation area of livestock, especially in densely populated areas of livestock and the lack of utilization of natural resources which led to changes in the situation of livestock stock maps.

2.3. Basic mapping data processing [4]

The steps that must be taken in the basic mapping of data are as follows: position on the potential base map box, knowing the location boundaries and regional criteria based on the zone (A.B.C), give green marks for active / potential, yellow (empowerment) and orange (conservation), and than after knowing the zones, continue by looking at the potential ability of the overall potential that exists in a particular area (can be broad or small).

2.4. Processing organic fertilizer [4]

The prosedure of making basic mapping data that was : 1 The first step is to do commodity grouping in each district by creeping to find out Advanced Potentia Mapping with Livestock for covering determine Variative Commodities All Regencies, Overall Variable Amount, Potential Variation Percentage (As for this step is intended to determine the rank, grade, potential and potential value, 2) The results of the main potentials of each dominant sector look for their potential value in basic mapping data, with the provisions of existing commodities being given overall green color, then after the overall potential in potential boxes for regions with moderate potential criteria and slightly given the flowering line between the largest and the smallest so that there are differences in zones, for the most zones are colored green and medium yellow and slightly orange, and 3) The difference in vertical lines will be known as their potential zones while in the horizontal line it will be known the criteria for the regions that have the largest, medium to the least commodity.

3. Result and discussion

3.1 Basic mapping data the plantation potency of Jambi Province

3.1.1. Basic Jambi Province plantation mapping data

Based on the results of research on Basic Mapping Data of Jambi Province, the results were shown in Table 1.

Table 1. Mapping variation of plantation potential in Jambi Province.

Commodity Judging Criteria	Number of Similar Province Commodity Scores	Number of Commo dities	Indonesia Code	Commodity Name	Percentage of Commodities	**Ranking**	Advanced Potential Mapping with Livestock
Score A	30	3	kd,Kr,pi	Kelapa dalam, Robusta Coffee, Pinang	3,88372	1	Coffee
Score B	18	2	k,cl	Rubber, Chocolate	2,32558	2	Chocolate
Score B	6	1	ki	Candlenut	6,97674	3	
Score B	10	2	l,a	Pepper, Aren	11,62791	4	
Score B	8	2	kh,kkm	Hybrid coconut, cinnamon bark	9,30233	5	
Score B	6	2	tu,n	Tobacco, Patchouli	6,97674	6	
Score B	6	3	ka, kp,tb	Arabica coffee, kapok, sugar cane	6,97674	7	
Score B	2	2	ck,v	Cloves, Vanilla	2,32558	8	
Score C	0	3	ks,jp,th	Palm oil, Jatropha curcas, Tea	0	9	Palm oil
Variative Commodities All Regencies	86	20			Total: 20,93022		
Overall Variable Amount		220					
Potential Variation Percentage	39,09						

Source: Primary data analysis (2017)

The strengths of the highest potential in the plantation sector that can be included in the CLS Program are 3 types of commodities (15%) in the coffee and chocolate specifications which are sufficiently wasteful and not widely used and used as a source of CLS Program activities. Variations in provincial plantation commodities were found as many as 86 types (39.09%) of 220 potential if considered to cover all districts. Although the percentage based on coffee and chocolate scores is not the largest position produced, but when viewed from the source of conventional SDL utilization, this type of waste is quite abundant and can be used as the main potential supporting business for the CLS program. The potential that can be developed for the integration of the plantation and livestock sector is the use of coffee waste and chocolate waste as an alternative feed source for livestock, which of course must go through the process of making feed processing technology for livestock from the waste.

3.1.2. Mapping of plantation potential

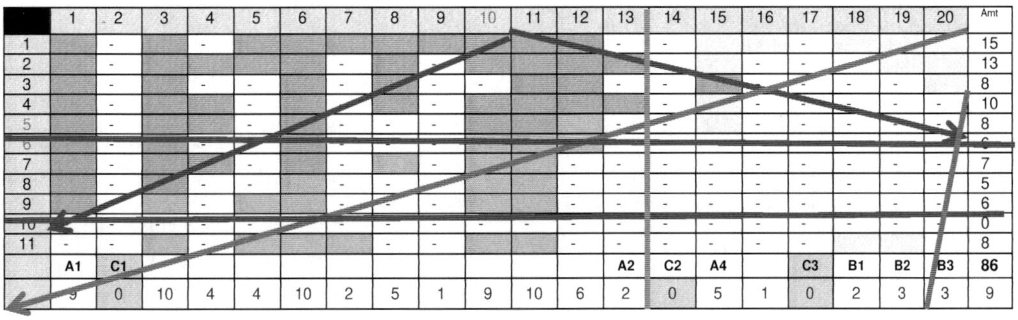

Figure 1. Mapping of plantation potential.

In Figure 1, the plantation mapping horizontally / transversely has 3 (three) areas of commodity categories for green zone A (A1, A2), zone orange B (A4, B1, B2, B3) and yellow zone C (C1, C2, C3). Regency with green zone commodity is 1, 2, 3, 4, 5, 6, 7, 8, 9. zone B in regency 5, 6, 7, 8, 9, 11 and yellow zone C in regency 10. While vertically mapping plantation has 2 zones namely Active / High Potential (A1, A2) and Passive / Medium Potential zone (A4, B1, B2, B3), and conservation zones (C1, C2, C3).

3.2 Basic mapping data the agriculture potency of Jambi Province

3.2.1. Basic Jambi Province plantation mapping data

Based on the results of research on Basic Mapping Data of Jambi Province, the results were shown in Table 2.

Variative diversity of agricultural fields is dominated by corn and rice plants. Thus integration efforts can be optimally carried out for the development of animal feed in the form of integration of agricultural crops and livestock waste in supporting alternative animal feed self-sufficiency in Jambi Province with strength variation 66.66% greater than the byproducts of waste from plantations such as rubber and coffee. Conversion of productive agricultural land and marginal land for non-agricultural activities will automatically have an impact on the reduced availability of feed base for ruminant livestock production. If the condition continues until there are no more areas left for overgrown fodder, the productivity of the national assets in the form of ruminants will also automatically decrease. To prevent this from happening, the maintenance of existing animal feed forages must certainly be a priority so that the future need for feed for the ruminant industry can be believed to be fulfilled sustainably. The intended industry will still be needed by the new generations of the coming Indonesian nation.

The amount of contribution of agricultural waste dry matter product Y (135,764,61 ton / BK / year) can be classified as the highest for the comparison of superior commodity waste supporting self-sufficiency in animal feed from waste. Mapping of the potential of Jambi Province Agriculture can be seen in Figure 2.

Based on the results of research on Basic Mapping Data of Jambi Province, the results were shown in Table 2.

Table 2. Mapping variation of agriculture potential in Jambi Province.

Kabupaten	P	TON	Ranking	Grade			Information	Potential Value
				A	B	C		
Kerinci	85,71	809,29	1		B		Potency	Normal
Merangin	100,00	524,76	3	A			Potency	Normal
Sarolangun	100,00	418,86	10	A			Potency	Normal
Batang Hari	85,71	443,95	6		B		Potency	Normal
Muaro Jambi	100,00	443,98	7	A			Potency	Normal
Tanjung Jabung Timur	85,71	409,29	11		B		Potency	Normal
Tanjung Jabung Barat	100,00	475,80	5	A			Potency	Normal
Tebo	100,00	440,92	8	A			Potency	Normal
Bungo	100,00	476,18	4	A			Potency	Normal
Kota Jambi		420,61	9				Normal Conservation	Normal
Kota Sungai Penuh		589,45	2				Normal Conservation	Normal
Amount	999,99	5.444,28		6	3	2		
Average Harvest	90,91	494,93						
District Conservation Map	0,01						Conservation Low	
Percentage				54,55	2	18,18		

Source: Primary data analysis (2017)

Districts with high harvest rates will produce high side waste and can be used as a source of alternative feed ingredients, especially cassava in the districts 2,3,4,5,6, 8,9,10 with the ability to harvest (ton) 90,91% if it is associated with the inter-population of livestock and the contribution of waste (135,775.61) then the potential map is n is the conventional raw material (food) and waste (By Product) for alternative feed. Maintaining the planted area (Ha) and harvest area (tons) of plantation crops resources will maintain the availability of alternative feed ingredients especially for grade A areas (the plantation harvest barns and waste barns indicated from the area's information have more plantation potential and lead to conservation value low (0,01).

[4] stated that, some of the problems in the development of beef cattle business in Indonesia are: (1) livestock productivity is still low, (2) the availability of local superior seeds is limited, (3) human resources are less productive and low knowledge, (4) the availability of feed is not continuous, especially in the dry season, (5) the livestock business system is not optimal, and (6) yield marketing is not yet efficient. Tawaf and Kuswaryan (2006) stated that, the low productivity of livestock and the limited availability of local superior seeds were caused by: (1) nursery sources were still dominated by folk farmers who spread widely with low ownership (1-4), (2) the existing nursery institutions (nursery business groups) have not developed towards professional business, (3) the weak reach of nursery UPT services due to the wide distribution of livestock, and (4) the high cutting of productive female livestock as a result of high demand for beef.

The results of agricultural mapping show that the division of zones is based on the same criteria with the results on the mapping of the farm, namely the green zone, orange zone and yellow zone. The yellow zone is a conservation zone, for areas that are not shaded are areas that have no supporting potential for this type of commodity variation.

3.2.2. Mapping of agriculture potential

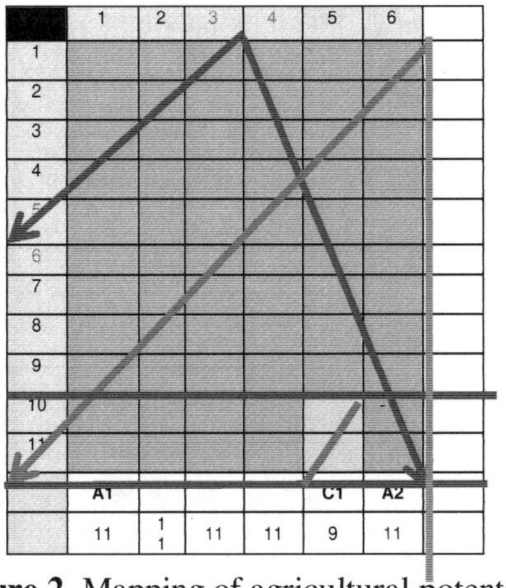

Figure 2. Mapping of agricultural potential.

In Figure 2, the horizontal / transverse mapping of agriculture has 2 (three) areas of commodity categories for green zone A (A1), zone orange B (B1) and yellow zone C (0). Regency with green zone commodity is 1,2,3,4,5,6,7,8,9. zone B in district 0 and yellow zone C in district 10, 11. While vertically mapping plantations have 1 zone (Figure 2), namely Active / High Potential (A1, A2) and Passive / Medium Potential zone (0), and conservation zone (C1). Basic mapping of agricultural potential that can be developed as a varied potential in the empowerment of local natural resources is the effort to integrate agricultural livestock as an alternative feed procurement business, especially those that are areas in the active zone. Thus the district can be developed as an agricultural-livestock integration zone.

Corn plants can be used as alternative feed ingredients for livestock through a fermentation process to increase the value of its protein content so that the value of crude fiber raw materials that previously had a high content. In addition, corn plants have not been widely used so that efforts are needed to empower local resources in the form of corn waste to be used as a source of animal feed which is widely available from agricultural activities, besides that this agricultural potential is also actually strong from commodity commodities (66, 66%) contributors to the overall source of animal feed ingredients such as rice in the form of rice bran waste, soybean in the form of soybean meal, and cassava in the form of left over waste (Indonesian : *onggok)* as the implementation of crop and livestock integration (Crop Livestock System), namely through optimizing the utilization of crop farming waste for feed and utilization of livestock manure for fertilizer plants.

The potential of a region's resources has different characteristics. These differences exist that are very visible and invisible when analyzed, as a result there is a division of regions that differ in size based on the availability of resources. To simplify the identification of an area from a simple observation angle, it is often constrained because of the parameters that are not owned to support an analysis of the things to be examined. Various formulas and techniques are used with the aim to answer about a problem that will be examined. The strengths and weaknesses encountered in an area lead to a new paradigm of technical fast observing the potential whether it leads to the development or improvement of conditions (Nora, 2016).

Mapping of the varied potential of agriculture is seen from various sources of waste, namely the harvested area of rice, corn, cassava, sweet potato, soybeans and peanuts. The biggest contribution of waste is corn plants can be seen in Table 3.

Table 3. Mapping of variative potentials of agriculture in Jambi Province.

No	District	Number Y (tons of BK / yr)	Regency Rank	Main Commodity	Supply Map of Commodity				Amount Total Commodity
		1	2	3	Biggest Commodity				
		Y	R	C	Rice	corn	Soy	Cassava	
	Y (Tons / BK / Yr)	135.764,61			3	6	1	1	71
	Percentage				27,27	54,55	9,09	9,09	
	Province Commodity Ranking				2	1	3	4	
	Commodity Category :								
	1. Most Commodities	Corn				54,55			
	2. Medium Commodity	Rice			27,27				
	3. Little Commodity	Soy, Cassava					18,18		
	Number of Commodity Diversities	4							
	Percentage of Commodity Diversity	66,66%							

Source: Results of Research Analysis (2017)

3.3 Basic mapping data the animal husbandry potency of Jambi Province

3.3.1. Basic Jambi Province plantation mapping data

In Figure 3, the horizontal / transverse mapping of farms has 4 (four) areas of commodity categories in green zone A Non Poultry Cattle (A1, A2, A3, A4), zone orange B (B1, B2, B3), and orange zone C (C1, C2, C3).

Regency with commodity of green zone of non-poultry livestock namely 1,2,3,4,5,6, zone B of non-poultry livestock (B1, B2) in regency namely 7,8,9,10,11 and green zone of poultry in district 1,2,3,4,5,6,7,8 and zone B of poultry in district 9.10.11. While vertically mapping farms have 4 zones, namely Active / High Potential (A1, A2) non-poultry livestock and poultry and the Passive / Medium Potential zone (B1) of non-poultry livestock and poultry and conservation zones (B1, B2) and for livestock non poultry and poultry (C1, C2, C3).

Based on the results of research on Basic Mapping Data of Jambi Province, the results were shown in Table 4.

Most districts have the potential for livestock farming, especially 1,5,6,8,9. With illustrations as follows:

1	2	3	4	5	6	7	8	9	10	11
11	2	9	3	11	11	4	11	11	10	7

Table 4. Mapping variation of the animal husbandry potential in Jambi Province.

No	Analysis Results	Non Poultry Farming							Poultry Farming				Total
		1	**2**	**3**	**4**	**5**	**6**	**7**	**8**	**9**	**10**	**11**	**KV**
1	Amount	11	2	9	3	1 1	1 1	4	1 1	1 1	10	7	90
2	Province Commodity Ranking	1	6	3	5	1	1	5	1	1	2	4	
3	Commodity Category												
4	1. Most Commodity	11				1 1	1 1		1 1	1 1	10		**65**
5	3. Medium Commodity			9								7	**16**
6	3. Little Commodity		2		3			4					**9**
7	**Number of Commodity Diversities**												**90**
	Percentage of Commodity Diversity	74,38											

Source: Analysis Results (2017).

3.3.2. Mapping of animal husbandry potential

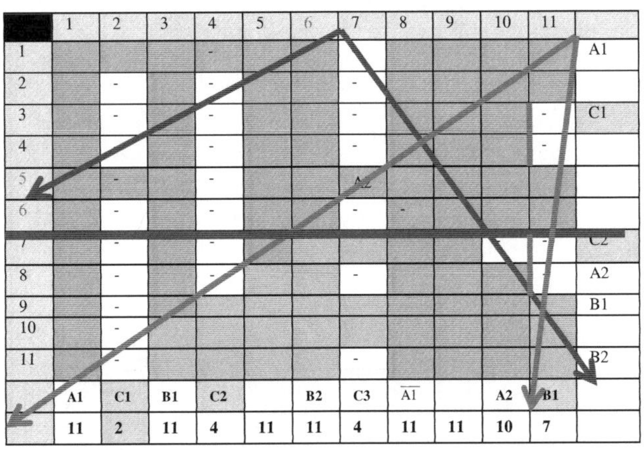

Figure 3. Mapping of animal husbandry potential.

The most commodity in the province is in the field of livestock, the highest variative commodity in the district of Kerinci. The diversity of provincial commodities is quite good and has the potential if CLS technology is developed to increase the variative diversity of commodities through the creation of alternative technologies in addition to the main waste.

Fields that can be used as a basic integration effort in Jambi Province have the potential to develop agricultural-livestock integration efforts seen from the production of animal feed commodities in the form of corn waste production output (54.55%), rice (27.27%), soybeans and cassava respectively - (9.09%). Districts that have waste output contribution from agricultural business are in Kerinci Regency with rice harvest area of 26,142 Ha or 25.58% and corn harvest area of 2,017 Ha or 23.77% of the total land area in Jambi Province.

The diversity of the percentage of livestock potential is 74.35%. The results of the study get basic mapping data information as a whole that Jambi Province can be used as a CLS program with the following data bases, namely: 1) basic data mapping obtained from the diversity of commodity potential, 2) varied combinations of commodity potential and 3) varied potential wastes for Crop programs Livestock System (CLS). Waste and livestock barns can be seen in Table 5 below.

Table 5. Mapping of variative potential of Jambi Province.

No	District	Potency			Commodity	
		PV 1	PV2	PV3	Amount	Status
1	Total	71	84	90	245	Potential
2	Global Province Commodity Ranking	3	2	1		
3	Commodity Category				Potential Ability	Commodity Total
3.1	1. Most Commodity			PV3	11	121
3.2	2. Medium Commodity		PV2		15	165
3.3	3. Little Commodity	PV1			7	77
3.4	**Number of Commodity Diversity**	245				**363**
3.5	**Percentage of Commodity Diversity**	67,49				

Source: Analysis Results (2017)

CLS policies that can be applied in Jambi Province are: 1) all potential commodities have strong integration, 2) integration of variative waste commodities based on efforts made to optimize waste, 3) total varied combinations of waste connecting CLS business with a strength of 69.70% (Table 6). The variant religious data produced from waste can be seen in Table 6.

Table 6. Mapping of Jambi Province variative waste potential.

No	Commodity	Amount		Potential Value	Status	Output
		Variative Combination	Factory Waste			Authority
1	Plantation	71	71	1	Based on the dominant kind	CLS
2	Farming	84	84	1	Based on the dominant kind	CLS
3	Animal Husbandry	90	8	0,08	Feces, Urine, Blood, Skin, Bones, Fur, Nails, Horns	CLS, Factory Waste Optimization
	Amount	245	163			
	Global Province Commodity Ranking	1	2			
	Commodity Category :					
	1. Most Commodity	Animal Husbandry	98			
	2. Medium Commodity	Farming	84			
	3. Little Commodity	Plantation	71			
	Number of Commodity Diversity		**253**			
	Percentage of Commodity Waste Diversities					
	1. Animal Husbandry	**98**	**38,74**			
	2. Agriculture	**84**	**33,20**			
	3. Plantation	**71**	**28,06**			

Source: Analysis Results (2017)

The diversity of sub-sectors in Jambi Province is 66.67% (Table 5) of all businesses managed in the field of food.

Table 7. Location Mapping and variative Potential of Jambi Province.

No	District	Potency			Commodity		Development
		PV 1	PV2	PV3	Amount	Rating	
1	Kerinci	6	15	9	30	1	Active potential
2	Merangin	7	13	8	28	2	Active potential
3	Sarolangun	7	8	7	22	5	Passive Potential
4	Batang Hari	7	10	7	24	3	Active potential
5	Muaro Jambi	7	5	10	22	5	Active potential
6	Tanjung Jabung Timur	7	5	8	20	7	Empowerment
7	Tanjung Jabung Barat	6	8	7	21	6	Passive Potential
8	Tebo	7	5	7	19	8	Empowerment
9	Bungo	7	6	10	23	4	Passive Potential
10	Kota Jambi	5	0	9	14	9	Empowerment
11	Kota Sungai Penuh	5	9	8	22	5	Passive Potential
	Total	71	84	90	245		

Source: Analysis Results (2017)

Jambi Province has 3 main commodities, namely livestock (38.74%), agriculture (33.20) and plantations (28.06) with the amount of waste diversity is 253 varied combinations of waste potential when compared with the overall variative potential of the province's potential strength is the percentage of the ratio of waste to overall commodities is 69.70%.

Location mapping and varied potential Jambi Province has 3 (three) potential development areas namely active, passive and empowerment potential. In detail LQ Potential analysis can be detailed in Table 8.

Table 8. QSPM analysis.

	Horizontal/Priority	Vertical/Alternative	Active potential	Passive Potential	Conservation	Amount
Picture 1	Horizontally / transversely the plantation mapping has 3 (three) areas of commodity categories for green zone A (A1, A2), zone orange B (A4, B1, B2, B3) and yellow zone C (C1, C2, C3). Regency with green zone commodity is 1,2,3,4,5,6,7,8,9. zone B in the district 5,6,7,8,9,11 and yellow zone C in district 10.	Vertically mapping plantations has 2 zones, namely Active / High Potential (A1, A2) and Passive / Moderate Potential zones (A4, B1, B2, B3), and conservation zones (C1, C2, C3).	2	4	3	9
Picture 2	Horizontally / transversely the plantation mapping has 3 (three) areas of commodity categories for green zone A (A1, A2), zone orange B (A4, B1, B2, B3) and yellow zone C (C1, C2, C3). Regency with green zone commodity is 1,2,3,4,5,6,7,8,9. zone B in the district 5,6,7,8,9,11 and yellow zone C in district 10.	While vertically mapping plantations have 1 zone, namely Active / High Potential (A1, A2) and Passive / Medium Potential zone (0), and conservation zone (C1).	1	0	1	2
Picturer 3	horizontally / transversely the mapping of farms has 4 (four) areas of commodity categories for green zone A Non Poultry Cattle (A1, A2, A3, A4), zone orange B (B1, B2, B3), and orange zone C (C1, C2, C3) Regency with commodity of green zone of non-poultry livestock namely 1,2,3,4,5,6, zone B of non-poultry livestock (B1, B2) in regency namely 7,8,9,10,11 and green zone of poultry in district 1,2,3,4,5,6,7,8 and zone B of poultry in district 9.10.11.	While vertically mapping farms have 4 zones, namely Active / High Potential (A1, A2) non-poultry livestock and poultry and the Passive / Medium Potential zone (B1) of non-poultry livestock and poultry and conservation zones (B1, B2) and for livestock non poultry and poultry (C1, C2, C3). Most districts have the potential for livestock farming, especially 1,5,6,8,9 (For Livestock Barn Program	2	3	3	8
Amount			5	7	7	19

Depreciation of the area and function of agricultural land if it occurs continuously will cause a decrease in waste production so that the ability to support waste is low against the availability of other food sources. In this condition, a breeder's business is needed to be more creative in procuring minimal animal feed, including the use of waste in the formulation of cattle rations. Consumption, efficiency and feed conversion affect the production and productivity of cattle. A good pastoral planting pattern when followed by human resource management and pasture management for integrity between existing farms and plantations. Although the land continues to experience extensive decline as a productive land for agriculture, plantations and forestry, land management techniques are an urgent interest in anticipating a decrease in land area and land management patterns (Nora, 2016)

The large percentage of provincial waste strength has the opportunity to develop CLS, considering that the general types of business carried out covering livestock, agriculture and plantations produce a diversity of regional potential, a varied combination of commodity potential and varied potential wastes to optimize the CLS (Crop Livestock System (CLS) program. a combination of data mapping results can be seen in Table 9.

Table 9. Mapping of global potential of Jambi Province.

No.	Sector	A/H	B/O Horizontal	C/K	Amount	A Active	P Passive	K Conservation
1	**Plantation**							
		1	1	1	3	1	1	1
		1	1	1	3	1	1	1
		1	1	1	3		1	1
			A/1		1		A/1	
	Amount	**3**	**4**	**3**	**10**	**2**	**4**	**3**
	K	**9**	**6**	**1**	**16**			
2	**Agriculture**	1	1	0	2	1		1
	Amount	**1**	**1**		**2**	**1**	**0**	**1**
	K	**9**	**0**	**2**	**11**	**1**		
3	Non Poultry Farms:	**1**	**1**	**1**		**1/NU**	**1/NU. 1.U**	**1/U**
		1	**1**	**1**		**1/NU**	**1/U**	**1/U**
		1	**1**	**1**				**1/U**
		1						
	Amount	**4**	**3**	**3**	**10**	**2**	**3**	**3**
	K	**6**	**5**		**11**			
	Poultry Farming	**1**	**1**	**0**				
		1						
	Amount	**2**	**1**	**0**	**3**			
	K	**8**	**3**		**11**			
	Amount (NU/U)	**6**	**4**	**3**	**13**			
	Amount (K)	**14**	**8**	**22**				
	Amount Number of Regencies **(T)**	**18**	**6**	**3**	**27**			
	Potential Amount **(T)**	**4**	**5**	**3**	**12**	**3**	**4**	**4**
	Amount Number of Regencies **(T)**	**32**	**15**	**25**	**72**		**11**	
	Potential Amount **(T)**	**10**	**9**	**6**	**25**			

Source: Results of Advanced Test Analysis (2017)

An alternative strategy that can be used in developing the potential as a basis for the utilization of Jambi Province's SDL is that potential mapping leads to periodic conservation, namely positions 5-7-7 = 5 where, the comparison between conservation areas is as large as active potential areas, prioritized potential utilization of locations the green zone mapping is the main priority of SDL for the CLS Program. While the passive potential area does not have a large influence on conservation conditions.

Weakness of passive potential is used as an alternative support program even though the strength of potential is not so great, but can be improved through accelerating the development of business types to support the main program in the green zone in 32 districts for all commodities with overall provincial potential strength (44.44%) for development Program CLS. According to Young (1992) that sustainable agricultural development policies are formulating environmental damage problems and resource deterioration which result in social costs and externalities in agricultural policy and internalized in economic policies through extension and relations approaches and economic incentives.

4. Conclusions

In Figure 1, the mapping of the plantation horizontally / transversely has 2 (two) areas, namely the green zone commodity category and vertically the plantation mapping in the zone, namely Active / High Potential (A1, A2). In Figure 2, the horizontal / transverse mapping of agriculture has 2 (two) areas, namely the green zone commodity category A (A1) and vertically the plantation mapping has 1 zone, namely Active / High Potential (A1, A2). In Figure 3, the horizontal / transverse mapping of farms has 2 (two) areas of commodity categories for green zone A Non Poultry Cattle (A1, A2, A3, A4) and green zone of poultry livestock in the district 1,2,3,4,5, 6,7,8. While vertically mapping farms have 2 zones, namely Active / High Potential (A1, A2) non-poultry livestock and poultry, and conservation zone (B1, B2) for non-poultry livestock and poultry (C1, C2, C3). Most districts have CLS potential for livestock, especially 1,5,6,8,9.

5. References

[1] Arfai, Potential and Strategy for The Development of Beef Cattle Bussiness in Fifty Cities, West Sumatera (Dissertation). Bogor Agricultural University Postgraduate Scool. 2009.
[2] Central Bureau Of Statistic, Jambi in Numbers.Cooperation between Bapeda and BPS. 2013.
[3] Central Bureau Of Statistic, Jambi in Numbers.Cooperation between Bapeda and BPS. 2017.
[4] Diwyanto , Priyanti, Conditions, Potential and Problems of Ruminant Animal Husbandry Agribusiness in Supporting Food Security. In the ; Proceedings of the Ntional Seminar on Empowering Livestock Communities in the Field of Agribusiness to Support Food Security Semarang. Pp. 1-11. 2006.
[5] Daryanto, Sari Production Managemnt Lecture. Publisher PT. Tutorial Tools Nurani Sejahtera, Bandung. 2012.
[6] Directorate General of Animal Husbandry, Statistics of Animal Husbandri and Animal Health. Livestock and Animal Statistic. Jakarta. 2012.
[7] Directorate General of Animal Husbandry, Statistics of Animal Husbandri and Animal Health. Livestock and Animal Statistic. Jakarta. 2013.
[8] Muhammad F, Agribusiness Management. Bumi Aksara Publisher, Jakarta. 2012.
[9] D. Nora, Mapping of Conservation Locations for Reservation of Animal Stock. National Seminar on "Urban-Inclusive Development Planning". ASPI-PERHEPI and PPs Unand 2017 collaboration. Proceeding, pp 215-224. 2016.
[10] T. Young and Burton, Agriculture Sustainability Definition and Implementation for Agricultural and Trade Policy. FAO-UN, Rome, Italy. 1992.

Acknowledgments

The award and acknowledgment were addressed to those who have supported the implementation of this research, especially to Agriculture Faculty, Muara Bungo University in my research 2017-2018 from national seminar before. By respect Rector, Dean of Agriculture Faculty.

Extension methods in adopting reproductive biotechnology innovation for cattle breeding in Dharmasraya District, West Sumatera

Ediset and Fuad Madarisa

Departement of Livestock Development and Business, The Faculty of Animal Science, Andalas University, Padang, West Sumatera, Indonesia.
Email: edisetjami80@gmail.com; fmadarisa@gmail.com.

Abstract. A study had been carried out from May to July 2016 to evaluate the extension methods utilized in adopting process of reproductive biotechnology innovation for beef cattle breeding in Dharmasraya district, West Sumatra. Following a survey method, 40 farmers of using Artificial Insemination (AI) and 14 farmers of using Embryo Transfer (ET) were interviewed. Those farmers were chosen by quota sampling method and saturated sampling method respectively. Both primary and secondary data were analyzed by descriptive and quantitative approach using Likert scale. The results showed that extension methods operated by extension workers consisted of three types; home and farm visit; demonstration and campaign.
Keywords – Extension methods, Reproductive biotechnology, Cattle breeding.

1. Introduction

The gap between supply and demand towards livestock products tend to be greater in the near future. In the one hand, the number of population and their consciousness to consume food products with a higher level of nutritional value increase progressively. On the other hand, abundant local and natural resources potential in Dharmasraya district are not yet utilized properly. This position requires the efforts to accelarate of using potential resources by introducing biotechnological innovation, particularly in beef cattle breeding.

Dharmasraya in West Sumatra province figured out that its number of cattle heads tend to decrease. Although there was an intensified effort, reduction of number of beef cattle was 0.75 % in 2013. To a large extent, this figure was caused by beef cattle slaughter, eventhough in 2014, cattle number increase by 0.8 % [1]. Then, it was different with other districts in West Sumatra, Dharmasraya accept an introduction of reproductive biotechnology innovation in form of both of artificial insemination (AI) and embryo transfer (ET) for beef cattle breeding.

As [6] [7] [8] stated out that the introduction of reproductive biotehnology shoud be supported by co-existing institutional capacity, such as extension agents. An agricultural development and transformation could also be accelarated by both institutional capacity and human resources competencies. Within these framework, as [9] stated,there is a need for extensionrole to change beef cattle farmers' behavior.

In particular, the rate of adoption of innovation by farmers is determined by the extension method. While, a number of extension methods is available, the appropriate method depends on local conditions. One should consider several points to utilize an agricultural extension method, such as target conditions, employment, availability of time, business, emotional characteristics and types of innovation.

Thus, there is a need to evaluate the method of extension operated by extension workers in Dharmasraya District. This study was aimed at understanding the methods used by extension workers to disseminate both reproductive biotechnology for beef cattle breeding; AI and ET.

2. Research methods

This research was conducted from May to July 2016 in Dharmasraya District, West Sumatera. Particular location was at a transmigration settlement with a pilot project area of syncronized application of reproductive biotechnology innovation. Program was proposed by provincial Livestock and Animal Health extension services.

Both primary and secondary data colection was made relevant to the study objectives. Primary data was gained from interviewing the beef cattle farmers and in-depth discussion with related staeholders. All beef cattle farmers adopted the AI and ET innovation became study population. Sample was counted by quota procedure to pick up 40 farmers of using AI and saturated sampling technique for 14 farmers of using ET.

Collected data analysis utilized a descriptive and quantitative analysis, using Likert scale to measure attitudes, opinions, and perceptions of a person or group about social phenomena. Measured variables were translated into research indicators, to arrange items of instruments that could be a statement or question [11]. The quantitative analysis then the answer was scored as follows:

1. Agree : Score 3
2. Doubtful : Score 2
3. Do not agree : Score 1

The collected data transformed into tabular form, and calculated based on their respective scores. Then its result was compared to "The Guidance on Identifying Technical Determinants of Animal Husbandry from [2], resulting following categories;

a. Good, the achieved percentage; 81-100%
b. Moderate, achieved ercentage; 60-80%
c. Less, the achieved percentage below 60%

3. Result and discussion

The result showed that there were three types of methods of adopting reproductive biotechnology innovation among beef cattle farmers in Dharmasraya. Each method took into account the respective biotechnology adoption; AI and ET as follow;

3.1. Extension methods in adoption of innovation of Artificial Insemination (AI)

Table 1. Extension methods in adopting the innovation of Artificial Insemination (AI).

Nu	Methods	Indicator	Amount (n)	Total Score	Average Score (%)	Category
1	Home and Farm Visit	Agree	32	96	85.71	Good
		Doubtful	8	16	14.29	
		Disagree	0	0	0.00	
2	Demonstrasion	Agree	36	108	93.10	Good
		Doubtful	4	8	6.90	
		Disagree	0	0	0.00	
3	Campaign	Agree	29	87	82.10	Good
		Doubtful	8	16	15.10	
		Disagree	3	3	2.80	

Source: Research Results 2016.

3.1.1. Method of home and farm visit

Theimplementation of home and farm visit extension methods in the process of adoptingAI innovation was supported by the related departmental policy. The authority institution put forward extension officers and inseminators at each sub-districts of centre for livestock health services. This centre was accomplished by cattle breeding Artificial Insemination Service. When beef cattle farmers inform extension workers that they cattle was being estrus, home and famr visit method was used by the related officers. [12] stated that these extension method were effective, due to the officersfamiliar with the targets. In particular, when the target has reached the stage of trying or implementing the innovation.

3.1.2. Method of demonstrasion

The usage of demonstration method in extension has been well implemented in AIbiotechnological innovation services in Dharamasraya District. The effectiveness of this method was supported by the extension process, which carried out on the farmers' group basis. In particular where the number of cow was relatively large enough to have AI demonstration. Then extension workers could provide materials and assistance for handling postAIprocess, pregnant period and calving delivery. [3] [10] explained that demonstration is one of agricultural extension methods to prove a better farming result.

3.1.3. Methods of campaign

The study showedthat the application of campaign method in the adoption of Artificial Insemination (AI) innovationwas in good category, with average score was 82.10%. The success of campaign method was caused by the hard work of extension workers in preparing the activities, including funding, human resources and informing related stakeholders. [3] [10] stated that to work with campaign method, requires an apropriate preparation.

3.2. Extension method in adopting Embryo Transfer (ET) innovation

Table 2. Extension methods in adopting of Embryo Transfer (ET) innovation.

Nu	Methods	Skor	Amount (n)	Total Score	Average Score (%)	Category
		Agree	14	42	100	
1	Home dan Farm Visit	Doubtful	0	0	0.00	Good
		Disagree	0	0	0.00	
		Agree	14	42	100	
2	Demonstrasion	Doubtful	0	0	0.00	Good
		Disagree	0	0	0.00	
		Agree	13	39	95.12	
3	Campaign	Doubtful	1	2	4.88	Good
		Disagree	0	0	0.00	

Source: Research Results 2016.

3.2.1. Methods of home adn farm visit

The implementation of home and farm visitmethod was well implemented, due to the the number of participated farmers was selected by extension workers. The objective was to achieve an optimal result and process of adopting ET innovation. Normally, before adopting an innovtion, farmers calculate its profit and benefit based on their experience. [4] revealed that the sequence of farmers'interests of innovation attributed to innovation adoption was profit, cost, complexity, suitability, energy and time savings, and description.

3.2.2. Method of demonstrasion

The demonstration method has been appliedappropriately by extension workersdue to a relatively first-hand character of ET innovation. ET was introduced to beef cattle farmers at Dharmasraya District in 2012. [5] stated that the advantages of the demonstration method wasits ability to see a new method put into practice. Thus there was no need for high trust between farmers and extension workers.

3.2.3. Methods of campaign

The method of campaign has been also implemented by extension workers in conveying ET innovation in Dharmasraya District. This was proved by the fact that 92.86% of respondents agreed the method used in ET adoption process. The figure indicated that the use of campaugn method was in good category. Based on guidance of [2] standard, the minimum limit of good category was more than 81% of respondents agreed on the statement. The highly participation rate of farmers in the extension method was the main reason why campaign method could be operated well by officers.

3.3. Extension method in adoptingthe reproductive biotechnology (Artificial Insemination/AI and Embryo Transfer/ET) innovation

The study proved that home and farm visit method of has been well implemented by extension workers in Dharmasraya District in assisting farmers to adopt reproductive biotechnology innovation, where 92.86% of respondent agree. The result also showed that the demonstrasion and campaign methods have been operated well by the extension staff. These were proved by the percentage of average score for demonstration and campaign was 96.55% and 92.67% respectively.

Table 3. Extension methods in adopting of reproductive biotechnology innovation.

Nu	Methods	Innovation of Reproductive Biotechnology		Reproductive BiotechnologyA verage Score (%)	Category
		AI / Average Score (%)	ET / Average Score (%)		
1	Home and Farm Visit	85.71	100	92.86	Good
2	Demonstrasion	93.10	100	96.55	Good
3	Campaign	82.10	95.12	86.61	Good
Average Score (%)		86.97	98.37	**92.67**	**Good**
Category		Good	Good	**Good**	

Source: Research Results 2016.

The total scores on extension methods utilized in the adoption of reproductive biotechnology innovation in Dharmasraya Regency was 92.67%. These three types of extension methods (home and famr visits, demonstrationand campaign) have been well implemented by extension workers in processingdisseminate of reproductive biotechnology innovation AI and ET in the study area.

4. Conclusion

Extension workers in Dharmasraya District of West Sumatera have utilized properly three types of extension methods; home and farm visit, demonstration and campaign to assist beef cattle farmers in adopting reproductive biotechnology innovation, namely Artificial Insemination (IB) and Embryo Transfer (ET).

5. References

[1] Dharmasraya Livestock and Fisheries. 2015. Data Base Populasi Ternak Besar Kabupaten Dharmasraya. Dinas Peternakan dan Perikanan Kabupaten Dharmasraya. Pulau Punjung.

IOP Conf. Series: Earth and Environmental Science **287** (2019) 012034 doi:10.1088/1755-1315/287/1/012034

[2] Directorate of Animal Husbandry Development. 1992. Pedoman Identifikasi Faktor Penentu Teknis Peternakan. Proyek Peningkatan Produksi Peternakan. Diklat Peternakan. Ditjen Peternakan, Kementrian Pertanian.Jakarta.

[3] Anwar, S; F. Madarisa andA. Anas. 2009. Ilmu Penyuluhan Pertanian. Fakultas Peternakan, Universitas Andalas, Padang.

[4] Crouch, B. R.and Shankariah Chamala. 1981. Extension Education and Rural Development. New York: John Wiley & Sons Ltd.

[5] Hawkins, H.S. and A.W. Van Den Ban. 1999. Penyuluhan Pertanian, Diterjemahkan oleh Agnes Dwina Herdiasti. Kanisius, Jakarta.

[6] Juma, Calestous. 1999. Biotechnology in the global economy: beyond technical advances and risks. AgBioForum, 2(3&4), 218-222. World Wide Web: http://www.agbioforum.org.

[7] Juma, Calestous. 2005. Biotechnology in a Globalizing World: The Coevolution of Technology and Social Institutions.BioScience: Vol. 55, No. 3, pp. 265-272.

[8] Juma, Calestous. 2011. The New Harvest: Agricultural Innovation in Africa. Oxford University Press. New York.

[9] Kartasapoetra, AG. 1994. Teknologi Penyuluhan Pertanian, Edisi Pertama, Cetakan Keempat. Bumi Aksara, Jakarta.

[10] Madarisa, F; M. Reza; A. Anas dan S. Anwar. 2018. Pengantar Ilmu Penyuluhan Pertanian. Andalas University Press. Padang.

[11] Sugiyono. 2014. Metode Penelitian Kuantitatif Kualitatif dan R&D. Alfabeta.Bandung.

[12] Soedarmanto, 2003. Penyuluhan dan Komunikasi Pertanian. Teori dan Penerapannya. Buku Referensi. Program Studi Penyuluhan dan Komunikasi Pertanian. Universitas Brawijaya. Malang.

Analysis of the determining factor in profit efficiency of quail farming in Payakumbuh Sub-District, Lima Puluh Kota Regency

I Indrayani, R Wati, and M I Rias

Faculty of Animal Science, Andalas University, Padang, West Sumatera, Indonesia.
Email: ida.indrayani@gmail.com, rahmi.unand@gmail.com, ikhsan.rias@yahoo.com.

Abstract. Demand for egg consumption in Indonesian society being one of the great opportunities to establish quail farms business. The goal of this research is to measure number of profit and to identify the affecting factors of profit efficiency on quail farms in Lima Puluh Kota Regency. Cross section data obtained from quail farmers were used. Thirty samples from the farm with ≥ 4000 quails selected by accidental sampling were obtained. Income analysis and Stochastic Frontier Analysis (SFA) were used for statistical analysis. This study exhibited a considerable capacity to improve quail profitability. The determining factor in profit efficiency of quail farming business in Payakumbuh sub-district, Lima Puluh Kota regency were feed, quail breed and fixed costs and also the number of productive quails. The average level of technical efficiency achieved by quail farmers was 0.527. In other word, the average productivity achieved was 52.7%. The most important factor that affected the profits inefficiency was the farmers'age. In conclusion, profit efficiency could be improved by minimize feed, quail breed, and fix costs and increase the number of productive quails.
Keywords – Quail farming, efficiency, profit, and stochastic frontier function.

1. Introduction

The high demand for egg consumption in Indonesia was considered as one of the great opportunities for the establishment of quail farming business. In 2013 and 2016, West Sumatra became the fourth largest quail egg producer in Indonesia with the total egg production of 2,785 tons and 2,614 tons, respectively and also as the largest quail meat producer in Indonesia. Although quail farms business in Lima Puluh Kota Regency was relatively low, in West Sumatra, Lima Puluh Kota Regency was considered to have the highest quail population. For the past five years, traditional quail farm business activities still continued despite the dominance of the agribusiness development of broiler chicken farming which has been intensively carried out in every region, especially in Lima Puluh Kota Regency. Although quail farming business was not as big and intensive as other farming businesses, this business had become a potential alternative business. Quail farming could be either a permanent or side business that provided additional income for the community or the farmers who worked on it.

The success of a farm business was determined by its management, maintenance management and the resources (inputs) used. In the quail farming business done by the farmers in Payakumbuh Sub-district, Lima Puluh Kota Regency, all the input variables and marketing of farming products were carried out independently. The farmers did not only provide their own capital, but they also directly manage the quail farming, including marketing the productions.

Some of the problems faced by quail farmers in Payakumbuh sub-district were the lack of availability of corn and fine bran used for concentrate mixture for quail feed. The corns were bought from Pariaman at a price of IDR 5,000/ kg and fine bran from Solok at a price of IDR 2,500/ kg. Changes in the price of feed raw material would affect the quail farming business and its production costs which ultimately would also affect the amount of profits obtained by farmers. The profits obtained were also influenced by the production while the production itself was strongly influenced by the maintenance management. Profit efficiency was determined by some factors, namely technical factors, production costs and farmers' demographic factors. Information about profit determining factors could be useful in changing the current quail farming system towards a more profitable and competitive one. This study aimed to analyze the factors that influenced profits efficiency in quail farming business in Payakumbuh sub-district, Lima Puluh Kota Regency.

2. Research method

This research was carried out in Payakumbuh Sub-district, Lima Puluh Kota Regency. The area was chosen based on the consideration that Payakumbuh Sub-district had the highest population of quail in the Lima Puluh Kota Regency. The number of samples used was 30 samples which were chosen using accidental sampling technique. The research variables were the benefits and factors that influenced the profit efficiency of quail farming business. The data were analyzed using the method of income analysis and processed using econometric model and explained descriptively. Profit efficiency was measured using *Stochastic Frontier Analysis*. Stochastic Frontier Analysis (SFA) [2] is an econometric approach that is generally used. SFA approach simultaneously estimate the frontier production function (or profit function) and inefficiency model, where the inefficiency effect is specified as a function from other variables that affect efficiency [3]. Empirically, this approach is expanded in the form of cost, profit and production functions. The model of *Cobb-Douglas* profit function can be written as follows:

$$\pi i = f(pi, Z) \exp(vi\text{-}ui) \text{ where } i = 1,2,3, n$$

The Cobb-Douglas function can be estimated as follows:

$$ln\pi = ln\,\beta 0 + \beta 1lnpi + \beta 2lnp2i + \beta 3lnp3i + \beta nlnpni + \beta 4lnZ1i + \beta 2lnZ2i + \beta 3lnZ3 + \beta nlnZn + vi - ui$$

where π is profit, p1, p2, p3, till p_n are the variable input price, whereas Z1, Z2, Z3 till Z_n are fixed input prices and other factors other than input prices that affect the profits obtained.

In this study the *Stochastic Frontier* model of quail farming business used is as follows:

$$ln\pi = ln\,\beta 0 + \beta 1lnpi + \beta 2lnp2i + \beta 3lnp3i + \beta 4lnp4i + \beta 5lnp5i + \beta 6lnp6i + \beta 7lnZ1i + \beta 8lnZ2i + \beta 9lnZi3 + \beta 10lnZ4i + vi - ui$$

Where:
Π = Profit from quail farming business (IDR/month)
P1 = Feed price (IDR/kg)
P2 = Cost of veterinary medicine (IDR/unit)
P3 = Quail breed costs (IDR)
P4 = Fixed Cost (IDR/month)
P5 = Labour Cost (hours/month)
P6 = Business scale (quail)
P7 = Number of productive quail (quail)
β0 = intercept
βi = coefficient of parameter estimation
vi – ui = error term (ui = technical inefficiency effects in the model and vi = the effect of external factors that are not modeled)

To measure the technical inefficiency effects in this study, the formula developed by Battese and Coelli (1995) in Coelli (1996) was used. The formula is as follows:

$$\mu i = \delta 0 + \delta 1Z1 + \delta 2Z2 + \delta 3Z3 + \delta 4Z4 + \delta 5Z5 + wit$$

where:
μi = technical inefficiency effect
Z1 = the farmers' age (old)
Z2 = the farmers' formal education (years)
Z3 = experience in quail farming (years)
Z4 = *Dummy* Business Status (Z41 = 1, if it is main business and Z42 = 0, if it is side business)

The hypothesis testing used for measuring the parameter of frontier production function and technical inefficiencies was the *one-sided generalized likelihood ratio test* (*LR-test*).

3. Results and discussion

The development of quail farming business in Payakumbuh sub-district was influenced by many factors. For instances, the age of the people will affect them in terms of skills and experience, as well as their capacity to absorb new technologies and other things [7]. The result of this study showed that the age of the quail farmers ranges from 27 - 60 years old, which meant that they were still in their productive age.

The education level they received was also very important and would influence every decision they made or in adopting every decision making that deemed useful for their business development. The results of this study showed that the percentage of respondents who graduated from elementary school, junior high school, senior high school and university were 3.3%, 3.3%, 70% and 23.4%, respectively. It showed that the education level of the quail farmers was relatively high. Having good knowledge, should have helped people with higher education level to develop their business, both in terms of technical and business management.

One's experience in running a business would greatly influence their decision making to achieve maximum results. The length of quail farmers' experience in running the quail farming business in Payakumbuh sub-district varied between 3-14 years. It showed that the quail farmers in Payakumbuh sub-district had the experienced in quail farming business.

The business scale of quail farming business in Payakumbuh sub-district ranged from 4000-15000 quails, in which mostly were in the range of 5000-6000 quails. Seventy percent (70%) of the respondents (out 30 respondents) had decided to make quail farming business as their main business. This was because before starting quail farming business, they actually did not have permanent job.

The average profit obtained by quail farmers in one month was IDR 10,673,782 with the average number of quails raised per month was 6850 quails.

The estimation results of *Stochastic Frontier* production function were presented in Table 22. The sigma squared ($\sigma 2$) and gamma (γ) values were 3.984 and 0.999, respectively, and real at the level of 99 percent. The gamma indicated the existence of technical efficiency in the production process, or variations in results caused by technical efficiency differences.

Table 1. Stochastic profit frontier estimates.

Variables	Parameters	Coefficients	t-ratio
Constant	β_0	26,914	26,781
Feed Price	β_1	-0,414**	-1.712
Cost of Veterinary Medicine	β_2	0,003	0,012
Breed Costs	β_3	-0,482***	-6,672
Fixed Cost	β_4	-2,199***	-6,254
Labour Cost	β_5	0,319	1,213
Business Scale	β_6	0.676	0,802
Number of productive quail	β_7	1,363***	3,634
Log-likelihood OLS			
Log-likelihood MLE	-33,133		
LR	22,556		

Notes: Statistical significance levels: ***1% ; **5%; *10%

The value of generalized likelihood ratio (LR) from the *Stochastic Frontier* production function was 22.556. It was greater than the critical value in *Kodde and Palm* Table (1986), i.e., 11.91 which was significant at $\alpha = 5\%$. It meant that there was a technical inefficiency effect in the model of certain technology.

The feed, breed and fixed costs had a significant negative effect on the profitability of quail farming business, while the number of productive quail had significant positive effect on the profits. On the other hand, the cost of veterinary medicine, labor and business scale did not significantly affect the profits. The cost of feed has a significant negative effect on the profit because there were differences between the type and the amount of feed provided by the farmers. For instance, there were farmers who provided complete feed or bought processed feed. There were also some farmers who mixed quail feed by themselves based on the predetermined formula, and other farmers who used the combination of complete and self-mixed feed. This matter had caused a difference in terms of feed expenses, in which complete feed was more expensive than self-mixed feed. Farmers who chosen complete feed reasoned that complete feed could help increasing the productivity of quail eggs. The quail breed cost had a significant effect on the profits because the quail farmers were likely obtained the quail breed in a different way. For instance, some of them would produce their own quail breed, buy the quail eggs and hatching it themselves, buy DOQ (day old quail) and others farmer would buy quail grower (30 days old quail). The different ways of getting the quail breed had caused the cost of quail breed to be varied. The method used to obtain the quail breed would also affect the quail productivity, because by choosing the quail breed on their own, the farmers could choose the breed quality and control its growth. The number of productive quails had a positive effect at the level of 0.01. It meant that the number of productive quail was directly proportional to the profit gained. The study conducted by [4] shows that feed prices have a significant negative effect on profits in all herd size categories. [1] found the major factors affecting the technical efficiency were feed, stock size,experience and farmer's educational level.

The average value of technical efficiency achieved by the quail farmers was 0.527. It meant that the average productivity achieved was 52.7. Therefore, in the short run, it is possible to increase production by an average of 47,3% by adopting efficient farming techniques. In other words, this mean value indicates that opportunity still exists for increasing productivity and income through increased efficiency in resource utilization. [1] found the mean of technical efficiency among the quail farmers in Ilorin, quara state, Nigeria is 60,5%. according to [5] proper management such as vaccination, cleanliness and sanitization ensures efficient production and good quality products (meat or eggs).

From this result, it could be concluded that the farmers were still lacking in the technical efficiency, thus the profits obtained were not yet maximum.

Table 3. Determinants of profit inefficiency among quail farms.

Variables	Coefficients	t – ratio
Constant	- 1,038	-1,006
Age (Z_1)	0,0466***	6,815
Formal education (Z_2)	0,259	-0,305
Years of experience (Z_3)	-0,675	- 0,725
Dummy Business Status (Z_4)	-0,0313	- 0,423

Notes: Statistical significance levels: ***1% ; **5%; *10%

From the frontier analysis results, it could be seen that the factors significantly affected the technical efficiency of quail farming business was the farmers' age. while the education level, experience and business status factors had no significant effect. The age factor had a significant positive effect on the technical efficiency of the quail farming business. [4] found age may also influence farm performance positively as older farmers are more experienced, and notably, can apply accumulated knowledge to the efficient use of inputs. Farmers' education level is expected to positively influence farm performance. [6] found a positive relationship between education and technical efficiency.

4. Conclusion

The determining factor in profit efficiency of quail farming business in Payakumbuh sub-district, Lima Puluh Kota regency were the feed, quail breed and fixed costs and also the number of productive quails. The average level of technical efficiency achieved by quail farmers was 0.527. In other word, the average productivity achieved was 52.7%. The factors that affected the profits inefficiency was the farmers' age.

5. References

[1] Abdoelaziz, M.L., A.K. Busola, and A.O. Abdulrasheed. 2017. Economics of Quail Production in Ilorin, Kwara State, Nigeria. Journal of Agricultural Extension. Vol. 21 (2): 44-53.

[2] Aigner, D.J., C.A.K. Lovell and P. Schmidt. 1977 Formulation and Estimation of stochastic frontier production models. Journal of Econometrics 6:21-37.

[3] Chirwa, E.W. 2007. Sources of technical efficiency among smallholder maize farmers in Southern Malawi. *AERCResearch*172. http://dspace.africaportal.org/jspui/bitstream/123456789/32355/1/RP172.pdf?1.

[4] Bahta, S., D. Baker. 2015. Determinants of Profit Efficiency among Smallholder Beef Producers in Botswana. International Food and Agribusiness Management Review Volume 18 Issue 3: 107-129.

[5] Hamra C.F. 2010. An Assessment of the Potential Profitability of Poultry Farms: A Broiler Farm Feasibility Case Study. Research Paper Submitted to the Faculty of The University of Tennessee at Martin.

[6] Otieno, D.J., L. Hubbard and E. Ruto. 2012. Determinants of technical efficiency in beef cattle production in Kenya. Paper presented at the International Association of Agricultural Economists (IAAE) Triennial Conference, Foz do Iguacu, Brazil, August.

[7] Prawirokusumo. S. 1990. *Ilmu Usaha Tani. Edisi Pertama.* Yogyakarta: Penerbit BPFE. Rasyaf. 2005. Beternak Ayam Petelur. Penerbit Erlangga. Edisi Ketiga. Jakarta.

Acknowledgement

The authors acknowledged research funding provided by the LPPM Universitas Andalas (Grant of BOPTN Universitas Andalas No. 10/UN.16.17/PP.RD/LPPM/2018).

Study of the application of technical aspects of Pesisir cattle in several regions of West Sumatera to maintain the existence of native Indonesian beef cattle

I Iskandar and W Sartika*

Social Economic of Animal Science, Andalas University, Padang, West Sumatera, Indonesia.
*E-mail address: winda.sartika@gmail.com.

Abstract. Pesisir cattle are native Indonesian germplasm living in the coastal areas of West Sumatera. Pesisir cattle population was to decline from year to year. This research aims to examine the technical aspects of the maintenance of pesisir cattle in several regions in West Sumatera to maintain the existence of this native Indonesian beef cattle. The research method used is a survey. The research location was determined based on the distribution of Pesisir cattle in several regions in West Sumatera, namely Pesisir Selatan Regency, Pasaman Barat Regency, Padang Pariaman Regency, Agam Regency and Padang City. Data analysis was done by tabulation and scoring in accordance with the Guidelines for Animal Husbandry Technical Aspects of the Directorate General of Animal Husbandry in 1992. The results showed that the technical maintenance of Pesisir cattle in Pesisir Selatan District, Pasaman Barat Regency, Padang Pariaman Regency, Agam Regency and Padang City were in the less category with scoring scores averaged 43,34%. This shows that Pesisir cattle breeding techniques in West Sumatera still need to be considered so that the native Indonesian cattle population can continue to grow.
Keywords – Pesisir cattle, tecnical aspects, West Sumatera.

1. Introduction

The population of beef cattle in West Sumatera in 2017 reached 393,481 cattle, this number decreased by 2.4% compared to beef cattle population in 2016 which reached to 403,048 cattle. The declining also occurred in the cattle slaughter which it was 120,186 cattle in 2016, the decreased by 24.5% in 2017 of 90,797 cattle [1]. The population decline will affect the amount of needs fulfilment of meat protein in West Sumatera. The beef cattle in West Sumatera consist of Simmental, Balinese, Brahman, Limmosine and Pesisir cattle. In 2011 the highest beef cattle population was in Pesisir Selatan Regency where the population reached 23% of the total beef cattle population in West Sumatera [6]. In 2017 the largest beef cattle population is still owned by Pesisir Selatan Regency of 81,786 heads or 20.8% of the total beef cattle population in West Sumatera [1].

Pesisir cattle is a native Indonesian germplasm that lives in the coastal area of West Sumatera and has been designated as a group through the Decree of Ministry of Agriculture No. 2908/Kpts/OT.140/6/2011. The Pesisir cattle population in West Sumatera can be found in the Regency of Pasaman Barat, Padang Pariaman, Agam, Padang City and Pesisir Selatan [2]. West Sumatera local cattle play an important role in the fulfillment of animal protein needs in West Sumatera. The population of beef cattle is not yet accurately known. The Pesisir cattle as the supporter of the fulfillment of animal protein needs in West Sumatera must be carefully preserved in order to maintain one of native Indonesian germplasm.

Most of the area in West Sumatera is located in hills and coastal areas. Pesisir cattle is a type of local cattle that can survive in coastal areas by utilizing the food that grows around it. Pesisir cattle is a local beef cattle that need to be considered by both regional and central government. In order to encourage the productivity and food security of animal protein especially beef, then the cultivation and productivity of local cattle should be considered as well.

The natural condition of Sumatera Barat is cool and dominated by hills, have fertile soil contours so that cattle feed grass are well-available in West Sumatera. A good raising technique can improve the productivity of the cattle being raised. Feed is an important part in the raising technique of cattle. Sufficient feed can make cattle grow healthy and productive. The cattle feed commonly used is forage and concentrate feed. Along with the increasing prices of concentrates and limited land to plant forage feed, an alternative feed is often used by farmers. One of the alternative feed used comes from agricultural waste such as straw [7]. Many regency in West Sumatera that with largest agricultural waste that can be used as beef cattle feed like Padang Pariaman, Agam and Pasaman Barat.

Pesisir cattle are usually raised extensively by the community. Cattle are released and allowed to find feed themselves around the house. Some breeders also do not have a housing for the Pesisir cattle. They believe that cattle they raised have a strong immune system and do not need a housing. Pesisir cattle are usually raised freely and receive very little attention from the breeders. In terms of other raising technique such as seed selection, raising management and disease prevention also must be considered carefully by the cattle so that their condition are healthy and can be sold at higher price. The Pesisir cattle are currently raised in several different locations in West Sumatera such as Regency of Pasaman Barat, Padang Pariaman, Agam, Padang City, and Pesisir Selatan. These different locations also have different technical aspects on the cattle raising. Based on the problems above, this research is conducted to study of the application of technical aspects of pesisir cattle in several rergions of west sumatera to maintain the existence of native Indonesian beef cattle.

2. Research methods

This research was conducted in five Regency and one City in West Sumatera including regency of Pesisir Selatan, Pasaman Barat, Padang Pariaman, Agam and City of Padang who raise Pesisir cattle, with a total sample of 250 farmers. The determination of the number of samples in these five regions was carried out by accidental sampling, where the selected of breeders (farmer) was raising Pesisir cattle.

The research variable used was technique aspect in accordance with the standard by Director General of Animal Husbandry of 1992 which includes:

1. Breeding/Reproduction: Types of seeds raised, mating system, method of breeding selection, the first time mating, calving interval, mating season knowledge
2. Feed: The amount of forage given, quality of forage, frequency of forage feeding, provision of concentrate, mineral, drinking water quality, drinking water quantity, forage feeding processing
3. Raising procedures: cleaning/bathing the cattle, housing cleaning, energy utilizing, animal waste utilizing, recording
4. Housing : Housing location, housing construction, animal waste location, housing area/efficiency, housing equipment
5. Disease/Health: Disease knowledge: Anthrax, S.E. Snoring, A.E/Oral disease, Brucellosis, other diseases and vaccination/prevention.

The observed data were analyzed descriptively with the help of data tabulation and scoring in accordance with the guidelines by the Director General of Animal Husbandry 1992. Based on the scoring results, it can be concluded that:

1. The score percentage of 81% – 100% is categorized as good
2. The score percentage of 60% - 80% is categorized as medium
3. The score percentage less than 60% is categorized as poor

3. Results and discussion

People of West Sumatera called the Pesisir cattle as its local name, such as *jawi ratuih* or *bantiang ratuih*, which means cows that produce many calves. Pesisir cattle is a native Indonesian germplasm that lives in the coastal area of West Sumatera, which has been designated as a group through the Decree of Minister of Agriculture No. 2908/Kpts/OT.140/6/2011. The populations of Pesisir cattle are found in the Regency of Pasaman Barat, Padang Pariaman, Agam, Pesisir Selatan and City of Padang [2]. Pesisir cattle act as a source of animal protein in the form of meat for people of West Sumatera. Pesisir cattle are currently marketed to various regions outside West Sumatera such as Bengkulu, Riau and Jambi. The demand of Pesisir cattle increases before the celebration of Eid Adha. In addition, the community also raises cattle as their savings for the future of their children and families. Business orientation is not really reflected in the breeders in their cattle raising business. Limited costs and knowledge makes the breeders unable to develop their cattle business to a larger direction.

Pesisir cattle have high carcasses percentage (50.5%), high fertility rate, high survival rate, able to consume coarse fiber, able to survive with poor nutrition. They can also adapt to tropical environment, resistant to tropical disease, benign temperament so that they can be easily controlled. The specific characteristics of Pesisir cattle are adult male of 4-6 years old weighted 186 kg and 99 cm height. Calves of 1.5-2.5 years old only grows 20 gr/head/day and young mother of 3-4 years old grows around 140-225 gr/head/day. The males have short head, short and big neck, wide neck backs, large humps, short and rounded rudders. The females have rather long and thin heads, tilted, short and thinrudders, small horns and pointing outwards. The first pregnancy is 30 months. The age of first birth is 40 months [2]. Pesisir cattle have a small body size, the carcasses percentage of Pesisir cattle reached 50.6%, higher than the carcasses percentage of ongole cattle (48.80%), Madurese cattle (47.20%), PO cattle (45%), and buffalo (39.30%) [3].

Table 1. Technical aspect of Pesisir cattle in several region in West Sumatera.

No	Technical Aspect	Score Standard of Director General of Animal Husbandry of '92	Pesisir Selatan Regency	North Pasaman Regency	Padang Pariaman Regency	Agam Regency	Padang City
1.	Seedlings/ Reproduction	300	65,3	115,81	192,9	133,1	137,2
2.	Feed	300	113,4	149,31	161,5	166,11	108,34
3.	Housing	100	52,3	48,72	50,1	45,67	50,8
4.	Raising Management	100	70,6	58,61	68,1	68,12	47,36
5.	Disease Prevention	200	25,2	48,26	96,6	58,67	65,18
	Total	1000	326,8	420,71	569,2	471,67	408,88

The ability to convert fibrous feed into meat makes these cattle potential as the meat producer and popular as a sacrificial animal. The body weight is one of the characteristics of Pesisir cattle. The body weight and size are affected by the genetic and environmental factor, especially feed. This is considering that their feed still relies on grass from cattle grazing and from field, without addition of concentrate or grains. Therefore, cattle only adjust the production capability with the natural resources condition. The lack of input given causes less cattle production [3].

Reproduction performance of Pesisir cattle and Balinese cattle in Bayang Regency (part or Pesisir Selatan Regency) is different. This is apparent from the lower reproduction performance of Pesisir cattle compared to Balinese cattle with the first mating & first service after calving, calving interval, pregnancy length are subsequently 68.73 days, 545.12 days, 277.39 days for Pesisir cattle; 211.36 days, 500.63 days, 278 days for Balinese cattle. Whereas the calving rate of Pesisir cattle is higher than Balinese cattle, which is 73.03% for Pesisir cattle and 50% for Balinese cattle [4]. Based on the table above, it can be seen that the technical maintenance of Pesisir cattle in the Pesisir Selatan District has the lowest score compared to other districts / cities in the

western Sumatera region. Disease control and housing management have not been an important concern for farmers because Pesisir cattle are currently maintained with extensive or released systems. Disease control is also difficult to do because livestock roam along the road or coast in the area. [5] says low feed quality because the Pesisir cattle only consume wild grass that grows in the shrubs, roadside, or in coconut field. The forages consumed by Pesisir cattle have higher water content and lack of dry material. In addition, the breeders do not provide additional feed either in the form of forage or concentrate. As a result, the cattle lack of main nutrition supply such as protein and minerals needed for normal reproduction performance.

Table 2. Thecnical aspect of Pesisir cattle in West Sumatera.

No	Technical Aspects	Score Standard of Director General of Animal Husbandry of 1992	Score of Research Results	Percentage (%)
1.	Breeding/Reproduction	300	125.34	41.78
2.	Feed	300	139.72	46.57
3.	Housing	100	49.49	49.49
4.	Raising Management	100	62.57	62.57
5.	Disease Prevention	200	56.30	28.49
	Total	1000	433.42	43.34

According to [3] the development of Pesisir cattle face the production decline in the last few years which was reflected in the declining population and productivity due to lack of environmental support and low competitiveness to the imported cattle. Limited grazing field causes grass consumption to decrease so that the weight gain decrease by 50 g/head/day. This condition causes the breeders interest to raise Pesisir cattle decline and switch to import cattle business that have higher productivity. This is feared to affect the sustainability of local cattle in the future. Therefore, serious efforts are necessary to conserve Pesisir cattle germplasm to avoid extinction. The development of modern husbandry while still paying attention to local wisdom can be one of the efforts to maintain the sustainability of these local cattle.

4. Conclusion

The technique of Pesisir cattle raising in several region in West Sumatera have different methods both in terms of seedlings/reproduction system, feeding, raising management, housing and disease prevention. The raising technique carried out to the Pesisir cattle in West Sumatera is less categorized (43.34%) which means overall improvement on the cattle raising technique are necessary to preserve the existence of local cattle as native Indonesian germplasm.

5. References

[1] Badan Pusat Statistik (BPS) Sumatera Barat. 2018. Sumatera Barat Dalam Angka 2018. Badan Pusat Statistik Sumatera Barat.

[2] Erlita, Y. 2016. Sapi Pesisir. *www.sumbarprov.go.id/details/news/708*. Accessed, Oktober 30th 2018.

[3] Hendri, Y. 2013. Dinamikan Pengembangan sapi Pesisir sebagai Sapi Lokal Sumatera Barat. Jurnal Litbang Pert. Vol.32 No.1 hal. 39-45.

[4] Yendraliza. 2005. Performans reproduksi sapi Pesisir dan sapi Bali di daerah inseminasi buatan Kecamatan Bayang Kabupaten Pesisir Selatan. Jurnal Peternakan UIN Suskan Riau, Vol.2 No.1, hal 36-40.

[5] Sarbaini, R. Anggraini, R. Suhati and A. Husni. 2007. Profil populasi sapi pesisir di kabupaten Pesisir Selatan Sumatera Barat. Jurnal peternakan Indonesia (JPI) Vol. 12 No.3 Hal. 206-212.
[6] Sartika, W and E. Rahmi. 2012. Perkembangan Populasi Ternak Besar Dan Unggas Pada Kawasan Agribisnis Peternakan Di Sumatera Barat Jurnal Peternakan Indonesia (JPI) Vol.14 No. 3. Hal. 466-472.
[7] Sartika, W and J. Hellyward. 2017. Bauran Pemasaran Susu Sapi dan Susu Kambing Segar di Kota Padang. Jurnal Peternakan Indonesia (JPI) Vol 19 No.3 Hal. 178-184.

Acknowledgement
The authors acknowledged research funding provided by the LPPM Andalas University (Grant of BOPTN Andalas University No.14/UN.16.17/PP.RD/LPPM/2018.

Profit analysis of broiler chicken business in Beringin Village, STM Hilir District, Deli Serdang Regency

J Marisa[1,] and S A Sitepu[2]

[1] Department of Agroecotechnology, Faculty of Sains and Technology, Pembangunan Panca Budi University, Medan 20122, Indonesia.
[2] Department of Animal Husbandry, Faculty of Sains and Technology, Pembangunan Panca Budi University, Medan 20122, Indonesia.

Abstract. The purpose of this study was to analyze the benefits received by broiler chicken farmers in Beringin Village, STM Hilir Subdistrict, Deli Serdang District. The type of research is descriptive quantitative research, which is the researcher who describes the condition of the variable, namely the amount of income obtained by broiler chicken businessmen. The population in this study were farmers of broiler chickens in Beringin Village, STM Hilir Subdistrict. The analysis method used is income analysis and business feasibility analysis. The results showed that broiler chickens in Beringin Village, STM Hilir Subdistrict, Deli Serdang District, were feasible to be developed economically with the highest profit generated by respondents 5 of IDR 12,300,000, - and the lowest in respondent 5 is IDR 27,320,125 per period with an average profit of IDR 101,758,063 per period / cycle.
Keywords – profit analysis, broiler chicken business, Beringin Village, STM Hilir District, Deli Serdang Regency.

1. Introduction

Deli Serdang Regency, especially the Senembah Tanjung Muda (STM) Hilir Subdistrict, is one of the autonomous regions that have a variety of resources that can be developed to achieve the main objectives of economic development, namely increasing the number and type of employment opportunities for the local community so that the welfare of the community is evener. In an effort to achieve these goals, the government must be able to develop a potential economic sector to develop as a leading sector. Based on the Central Statistics Agency of North Sumatera, broiler chickens in the Senembah Tanjung Muda Hilir Subdistrict in 2012-2015 experienced various developments and can be seen in the table 1.

Table 1. Broiler chicken population in 2012-2015 in STM Hilir Subdistrict.

Year	Total Population (Tail)
2012	516.000
2013	554.000
2014	523.169
2015	575.000

Source: BPS North Sumatera, 2016.

Based on the statistical data above, the increase in broiler chicken population is seen from 2012 - 2013 and in 2014 there was a decline then in 2015 there was an increase. Based on surveys and statistical data, in the District of STM Hilir which included 15 villages, only 9 villages that raised broiler chickens as shown in Table 2.

Table 2. Distribution of broiler chickens in the district of STM Downstream in 2012 – 2015.

No.	Village	2012	2013	2014	2015
1.	Gunung Rintih	71.000	75.000	69.000	72.000
2.	Kuta Jurung	51.000	50.000	47.169	51.500
3.	Lau Barus Baru	20.000	25.000	22.000	32.000
4.	Limau Mungkur	92.500	97.500	91.000	95.000
5.	**Negara/Beringin**	**120.000**	**130.000**	**125.000**	**133.000**
6.	Siguci	20.500	23.500	20.500	22.500
7.	Sumbul	37.00o	42.000	40.000	50.000
8.	Taduken Raga	41.500	45.500	43.500	44.000
9.	Talun Kenas	62.500	65.500	65.000	75.000

Source: BPS North Sumatera, 2016.

Based on Table 2, it can be seen that Beringin Village is the village with the highest distribution of broiler chickens compared to other villages in STM Hilir Subdistrict. Beringin Village also experienced a high population increase compared to other villages, which was 133.000 in 2015.

In business activities, broiler breeding in Beringin Village, of course, requires capital, costs, and labor. The capital used in this nursery is its own capital and partners. While costs incurred include production costs, labor costs, and equipment depreciation costs. Costs incurred will affect the profit process received by the farmer. Every business in running its business always wants to make a big profit, where the profit is a guarantee for the survival of a business or activity of a business. Based on this description, researchers are interested in conducting research on the profit analysis of broiler chicken farming in Beringin Village, STM Hilir Village, Deli Serdang District. The purpose of this study was to analyze the benefits received by broiler chicken farmers in Beringin Village, STM Hilir Subdistrict, Deli Serdang District.

2. Methods

2.1. Location selection

This research will be carried out during February 2018 to May 2018 in the STM Subdistrict, Deli Serdang Regency, North Sumatera Province, on the grounds that the area is one of the areas that have great potential in the development of broiler chicken farming, in addition to the scale of business owned by many local farmers. Requirements for respondents are broiler breeders in the study area. The research method used is a survey method with a family analysis unit that maintains broiler chickens.

2.2. Populasi dan sampel

Samples are farmers who will be used as respondents, while the determination of samples is done by purposive sampling technique by selecting several partnerships in the STM Hilir Subdistrict, Deli Serdang Regency, North Sumatera Province and making the partnership members into research samples. In this study, the sample was determined in the Beringin Village with a partnership pattern which had the highest population among other villages.

Secondary data collection by visiting the Central Statistics Agency of North Sumatera to see the number of broiler chickens in STM Hilir Subdistrict. For the study used questionnaires and interviews as a tool to retrieve primary data.

2.3. Data collection

Data collection conducted in this study, namely:

1. Observation, which is a direct observation of broiler chickens in STM District.
2. Questionnaires and interviews are data retrieval by dividing questionnaires or questionnaires to farmers and communicating directly with respondents to obtain the necessary data.

2.4. Data analysis method

Data analysis was used to determine the income of broiler chickens in Beringin village. To find out the income of beef cattle business, the general formula is used:

$$\pi = TR - TC$$

π = Total profits earned by farmers (IDR / year)
TR = Total Revenue earned by farmers (IDR / year)
TC = Total Cost (IDR / year)

The general form of revenue from sales is

$$TR = P \times Q$$

TR = total revenue,
P = the selling price per unit of product
Q = the number of products

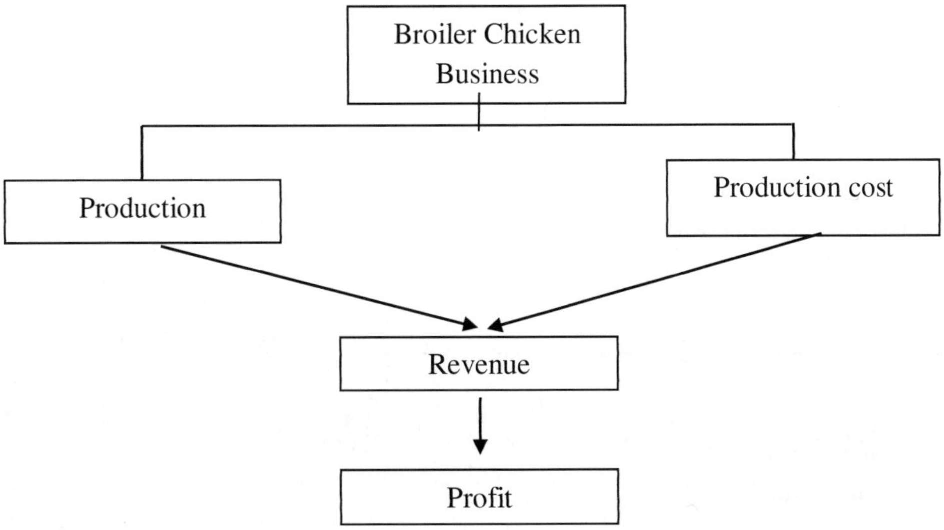

Figure 1. Scheme of research procedure.

3. Results and discussion

3.1. Analysis of costs and revenues

3.1.1. Cost analysis

Costs incurred by farmers in managing broiler chicken business consist of fixed costs and variable costs. Fixed costs are costs that cannot change (constant) for each level of the number of results produced or the costs for which they are used are not used up in one production period and are still issued even if they do not produce, among others, depreciation costs. The fixed costs incurred in this broiler chicken business are the cost of shrinkage of the cage and shrinkage of tools. One way to calculate depreciation costs is the difference between the initial value of the item and the final value of the item divided by the length of usage.

Variable costs are costs that can change at any time depending on the size of the production volume or the costs spent during production. The variable costs of broiler chicken business include the cost of broiler chickens, feed costs, costs of medicines, vitamins and labor costs. For more details, the average costs incurred in this business can be seen in Table 3.

Table 3. Total Production Costs During Research.

No	Business Scale (tail)	Total Fixed Cost (IDR)	Total Variable Cost (IDR)	Total Cost (IDR)
1.	10.800	5.130.193	212.430.000	217.560.193
2.	24.770	9.566.250	480.440.000	490.006.250
3.	11.500	5.441.693	239.046.000	244.487.693
4.	16.340	6.517.561	341.404.000	347.921.561
5.	31.500	10.748.050	547.618.000	558.366.050
6.	8.500	3.510.875	170.344.000	173.854.875

Source: Primary data, 2018.

Table 3 shows that the average fixed costs incurred in broiler chicken business at the study location amounted to IDR 6,819,104 consisting of depreciation costs and depreciation of livestock equipment. While for variable costs consist of the cost of production facilities and labor costs with an average cost of IDR 331,880,333. Based on the results of the study it can be seen that variable costs are greater than the fixed costs. This is in line with Soumena's (2012) study which states that expenditure on fixed costs does not affect the number of products produced but affects the level of profit obtained by broiler chicken farmers. While the expenditure on variable costs affects the number of products produced.

3.2. Profit business

The size of the broiler chicken business income is largely determined by the size of the production and price at harvest time. Prices of broiler chickens range from IDR 14.000 and IDR 15.000 per kilogram. The average production of broiler chickens is 17.235 birds per period. The average income and profit of broiler chicken farming can be seen in Table 4.

Table 4. Average Acceptance and Benefits of Broiler Chicken Farming Business Per Period.

No	Description	Business Value (IDR)
1	*Fix Cost*	6.819.104
2	*Variable Cost*	331.880.333.
3	**Total cost (1+2)**	**338.699.437**
4	**Total Revenues (P*Q)**	**440.457.500**
5	**Profit (4 - 3)**	**101.758.063**

Source: Primary data, 2018.

Table 4 shows the average production value (revenue) generated from broiler chicken business in this study area of IDR 440.457.500, - per period. The average total cost which is the result of the sum of fixed costs and variable costs is IDR 338.699.437. The results of the data from Table 8 show that the average profit received by broiler chicken farmers obtained from the total revenue is reduced by the total cost / total cost of IDR 101.758.063 per period. The assessment criteria for a business is profitable if a price level is multiplied by the amount of broiler chickens production exceeding all costs, then it can be ascertained that broiler chicken business is feasible to be developed. At the level of efforts to maximize profits are usually realized through increasing technical efficiency.

3.3. Economic analysis and business feasibility

After completing the analysis of production costs and income analysis, then the feasibility analysis of broiler chickens in Beringin Village, STM Downstream District can be seen in Table 3, while the average business feasibility analysis can be seen in Table 5 below.

Table 5. Average Feasibility Analysis of Broiler chicken livestock business in Beringin Village, STM Hilir Subdistrict.

No	Description	Unit	Business Value
1	Total Cost	Rupiah	338.699.437
2	Total Revenue	Rupiah	440.457.500
3	Profit/Net Income (2 - 1)	Rupiah	101.758.063
4	Revenue Cost Ratio (2/1)	-	1,30

Source: Primary data, 2018.

Table 5 shows that the average R/C ratio is 1,30. The R/C ratio is obtained from the ratio of the total average revenue of IDR 44.457.500,- with an average total cost of IDR 338.699.437. Economically, broiler chicken farming in the study area is feasible to be cultivated (developed) as indicated by the R / C Ratio > 1 value which is an average value of 1,30. This means with sacrifice (production costs) of IDR 1,00 then the broiler chickens breeder will get revenue (production value) of IDR 130. By referring to this situation, broiler chickens in the research area are feasible and provide benefits even to be developed or run.

4. Conclusion

The conclusion of this study is broiler chicken farming in Beringin Village, STM Hilir Subdistrict, Deli Serdang District, which is economically feasible with the highest profit generated by respondent 5 of IDR 12.300.000, - and the lowest in respondent 5 is IDR 27.320.125 per period with an average profit of IDR 101.758.063 per period.

5. Recommendation

To increase farmer income, it is necessary to improve maintenance management and use production factors efficiently to increase business productivity.

6. References

[1] Badan Pusat Statistik. Provinsi Sumatera Utara, 2012-2015.

[2] Daniel, M. 2012. Pengantar Ilmu Ekonomi Peternakan. Penerbit Bumi Aksara, Jakarta.

[3] El-kabumaini, N. dan Tjetje, S, R. 2013. Yuk, beternak ayam pedaging dan petelur. Puri Pustaka. Bandung.

[4] Fanani, Z., 2013. *Evaluasi Usaha Peternakan Ayam Pedaging di Kabupaten Malang*, Fakultas Peternakan Universitas Brawijaya. Malang. .

[5] Franky Z. 2014. *Peraturan Pemerintah Republik Indonesia Nomor 44 tahun 1997 Tentang Kemitraan.*

[6] Galeriukm. 2012. Pola kemitraan untuk menjamin keberlangsungan bisnis ternak. http://galeriukm.web.id/unit-usaha/peternakan/pola-kemitraan-untuk-menjamin-keberlangsungan-bisnis-ternak.

[7] Gusasi. A dan Saade. M.A 2012. Analisis Pendapatan dan Efisiensi Ternak Ayam Potong pada Skala Usaha Kecil. Jurnal Agrisistem, Juni 2012 Vol 2 No.1Haryono,D., 2013, *Organisasi Produksi Usaha Ternak Ayam Pedaging Pola Kemitraan dan Non Kemitraan di Kecamatan Mojo warno Kabupaten Jombang*, Fakultas Peternakan UB. Malang.

[8] Kamil, M. 2013. *Strategi kemitraan dalam membangun pnf melalui pemberdayaan masyarakat. Badan Peneliti dan Pengembangan. Bandung.*

[9] Nasir, M., 2011. Metode Penelitian. Ghalia Indonesia. Jakarta.

IOP Conf. Series: Earth and Environmental Science **287** (2019) 012037 doi:10.1088/1755-1315/287/1/012037

[10] Nirwana, M. 2011. *Rahasia sukses memelihara ayam broiler di daerah tropis*. Penebar Swadaya. Jakarta.
[11] Rahardi, 2012. *Analisis pendapatan ayam broiler*. Kanisius. Jakarta.
[12] Rasyaf, M. 2013. XXIV. *Beternak ayam pedaging*. Penebar Swadaya. Depok.
[13] Soekartawi. 2011. *Analisis usahatani*. Universitas Indonesia. Jakarta.
[14] Sudjana, 2013. *Metode Statistik*. Tarsito. Bandung.
[15] Sugiarto, 2015. Teori dan Praktik Kemitraan Agribisnis. Penebar Swadaya, Jakarta.
[16] Sutama, 2012. Agrbisnis Berbasis Peternakan. Pustaka Wirausaha Muda, Bogor.
[17] Wahyudin E. 2014. *Sistem dan Pola Kemitraan di Indonesia*, Majalah Poultry Indonesia. No. 194. Jakarta.
[19] Yuwanta, T., 2014. *Dasar Ternak Unggas*. Edisi ke-5. Penerbit Kanisius. Yogyakarta.
[20] Zulian Y., 2011. *Manajemen Produksi dan Operasi*. FE UII. Yogyakarta.

The sustainability index of dairy cattle area in Padang Panjang City

James Hellyward[1], **Suyitman**[1], **and Arif Rachmat**[1]

[1)] Faculty of Animal Husbandry of Andalas University, Padang, West Sumatera, Indonesia.
Corresponding e-mail: jameshellyward@yahoo.com

Abstract. This study aims to analyze the sustainability status of dairy cattle based livestock areas in Padang Panjang City West Sumatera from the five dimensions of sustainability (ecology, economics, socio-cultural, technological-infrastructure, and legal-institutional) and identifies key influential factors in the future development of the system. This research uses Multidimensional Scaling (MDS) method of analysis which results are expressed in the form of sustainability index. The result of the index of sustainability using approach Rapid Appraisal of Dairy Cattle Breeding System yielded the value of 55.09. Based on these values, the sustainability index of dairy cattle breeding system in Padang Panjang City is categorized as "quite sustainable". An increase in the value of the sustainability index can be done by improving the 15 sensitive attributes of 45 attributes of the system's sustainability assessment. Analysis of dependency between factors using prospective analysis resulted in 7 (seven) key factors influencing the development of dairy farming system. The key factors are: (1) milk price, (2) feeding capacity, (3) feed industry, (4) utilization of livestock waste, (5) job socialization, (6) development of cooperative, and (7) microfinance (bank/credit). In order to improve the sustainability status for the future (long term), the scenario that needs to be done is a progressive-optimistic scenario by making a thorough improvement on all the sensitive attributes in enhancing the status of the region.
Keywords – status of sustainability ecological, economic, socio-cultural, technological infrastructure, and institutional-law dimensions.

1. Introduction

National milk self-sufficiency is currently still difficult to achieve, it is seen that more than 75 percent of milk supply across the archipelago is dominated by foreign farmers, and the remaining 25 percent for national milk supply comes from the contribution of farmers in the country (Directorate General of Livestock and Animal Health, 2015). One way that can be done to achieve national milk self-sufficiency is by increasing the provision of assistance to farmers and at the same time increasing dairy cattle [1]. The policy of developing dairy farming in West Sumatera is more focused on increasing local food-based population and integration with other businesses, increasing consumption and developing local marketing. Government support in the development of an integrated dairy cow industry must also be based on the development of farmers groups and need to be pursued in particular in order to increase market-oriented dairy farm income. Padang Panjang City is one of the development centers for dairy cattle in West Sumatera Province. The dairy cattle population in Padang Panjang ranks first, with 343 dairy cows, followed by Tanah Datar Regency with 183 dairy cows, and Padang City with 134 tails [2].

According to Livestock Services Office of West Sumatera data in 2010 the milk production of Padang Panjang city was 522,495 liters. The number of dairy cattle population in the city of Padang

Panjang experienced an increase from 2006 to 2010, from 244 to 343. Based on the increase in the number of dairy cattle population, it indicates that there is potential in the development of dairy cattle business in Padang Panjang City. One of the interesting information from them is the new population of dairy cows in the city of Padang Panjang which is around 343, producing around 1.200 liters / day, apparently the largest in Indonesia outside Java. However, when compared with dairy cattle farms in Sumedang, Ciawi, Cibungbulang, Bogor, West Java, it is indeed small. Especially when compared to New Zealand, farmers generally have more than 400 dairy cows per person.

However, Padang Panjang City with its dairy cattle business development, plus a number of supporting facilities lately, was once mentioned by the Deputy of the Republic of Indonesia Ministry of Agriculture, which has the potential to become a dairy cattle business development center in West Sumatera, especially because Padang Panjang City is on the road junction the heart of West Sumatera. The dairy business support facility in the city of Padang Panjang was according to the Head of the Padang Panjang Agriculture Service: First, the availability of dairy houses where milk production was collected from the 9 dairy cattle business groups in the city. At the milk house, there are labor mini testers for milk samples, milk pasteurization equipment, culling units, packing equipment and marketing containers. Second, has a calf maintenance / enlargement center (Rearing Unit) which is considered the best in the country, so it is visited a lot. Third, there is a national standard Puskeswan (animal health center) for cattle inspection. In 2016, a unit of broodstock by the Level I Provincial Government of West Sumatera Province is being built and the most exciting recently conducted by the Government of Indonesia - New Zealand in the field of dairy cattle business development in Indonesia, five consultants from New Zealand. by staff from the Ministry of Agriculture (Ministry of Agriculture) RI came to review the dairy farming business in Padang Panjang City in 2015.

To support the dairy farming business development program, it is necessary to develop a dairy farming business nationally. Sustainable development criteria need to be implemented to ensure the sustainability and optimal benefits of dairy farming that harmonizes economic, social and environmental sustainability [3]. That the criteria for sustainable development in principle involve the ecological, economic, socio-cultural and legal-institutional dimensions [4]. A technological aspect, so that the reference criteria for sustainable development include five dimensions, namely: (1) ecology, (2) economy, (3) socio-culture, (4) legal-institutional, and (5) technology [5]. In its development, the technological dimension is aligned with the dimensions of infrastructure [6]. Therefore, development sustainability criteria cover five dimensions, namely: (1) ecology, (2) economy, (3) socio-culture, (4) law-institutional, and (5) technologies.

The development of a farming system by applying the five dimensions of continuity requires modeling activities with a system approach. This is needed to describe the interrelationship between factors in the system. In addition, the resulting model is able to simulate the impact of interventions on factors, so as to be able to provide recommendations for better policy direction for the development of livestock systems in an effort to improve system performance. Modeling the dairy farming system in Padang Panjang by considering sustainability criteria is expected to improve system performance. Better system performance is expected to: (1) increase cow milk production, (2) increase farmer acceptance, (3) income distribution among farmers, (4) improve the quality of the local environment, (5) better management of livestock waste, (6) utilization of agricultural waste, and (7) the application of effective and efficient technology. Therefore, the sustainability analysis of the dairy farming system combined with an analysis of the dairy farming model is needed to provide convenience for stakeholders in determining the direction of the policy, so that optimal benefits can be obtained from the dairy farming business in Padang Panjang City. In addition, research related to the sustainability of the development of sustainable dairy farming in Padang Panjang has never been done before, so the results of this study are expected to be a reference for further related research and the West Sumatera Provincial Government in determining the direction of policy.

This study aims to assess the sustainability status of the dairy farming system in Padang Panjang in terms of each dimension of development sustainability, namely: ecological, economic, socio-cultural, technological-infrastructure, and legal-institutional dimensions, as well as identifying influential key factors that determine the sustainability of the development of the dairy farming system in the city of Padang Panjang (West Sumatera).

2. Materials and methods

2.1. Place and time research

The entire research area is in Padang Panjang City. Determination of the research location was chosen intentionally (purposive) with the consideration that the area was the center of dairy cow milk production in West Sumatera Province. This research was carried out from January 2018 to July 2018.

2.2. Research data

The data needed in this study are primary data and secondary data. Methods of data collection in the sustainability analysis of the development of dairy farms in the city of Padang Panjang were conducted through interviews, discussions, filling out questionnaires, and field surveys with respondents in the study area consisting of various experts and stakeholders related to the topic of this research.

2.3. Data analysis method

The research method used in general is descriptive method based on case studies analyzed using a system approach. The system approach is used to formulate policy directions and scenarios for the development of a sustainable dairy farming system in the city of Padang Panjang based on the five-dimensional sustainability criteria. Determination of sustainability index and status using the Rap-SIBUSAPE approach based on MDS (Multidimensional Scaling Analysis) techniques. Sensitive attributes that influence the value of the system sustainability index that is reviewed further are assessed based on the level of influence and dependence between factors through prospective analysis. The results of the analysis are key sustainability factors based on existing conditions. In the next stage, a needs analysis is needed from all stakeholders, so that the attributes of stakeholders' needs are obtained. These attributes are then assessed based on the level of influence and dependence between factors through prospective analysis. The results of these analyzed are key factors of stakeholders' needs.

2.4. Multidimensional Scaling Analysis (MDS)

The sustainability status of the farming system is expressed in the form of a sustainability index based on the Rap-SIBUSAPE approach by applying MDS techniques. The Rap-SIBUSAPE approach was modified from the Rapfish program (Rapid Assessment Techniques for Fisheries) developed by the Fisheries Center, University of British Columbia [7]. The MDS method is a statistical analysis technique that transforms every dimension and multidimensionality in the dimensions of the sustainability of the dairy farming system.

2.5. Prospective Analysis

Determination of key sustainability factors is carried out using a prospective analysis (Participatory Prospective Analysis). The analysis is used to determine important factors that affect the dairy farming system. Prospective analysis involves expert respondents to participate in order to find out, investigate, and anticipate changes to the system that are able to provide fast results [8].

3. Results and discussion

The results of Rap-SIBUSAPE analysis show that the Dairy Farm in Padang Panjang has an average index value of the sustainability of the dairy farming system of 55.09, so the sustainability status is categorized as quite sustainable. The results of Rap-SIBUSAPE analysis and leverage on each dimension of sustainability are as follows: ecological dimensions: 45.17 (less sustainable); economic dimension: 49.24 (less sustainable); socio-cultural dimensions: 48.89 (less sustainable); dimensions of infrastructure technology: 57.18 (quite sustainable); legal-institutional dimension: 74.98 (quite sustainable). The sustainability index value is generated based on an assessment of 45 attributes included in 5 dimensions of sustainability, including: ecological dimensions (10 attributes), economic dimensions (9 attributes), socio-cultural dimensions (9 attributes), technology-infrastructure dimensions (7 attributes), and legal-institutional dimensions (10 attributes).

3.1. Ecological dimension

The sustainability status of the ecological dimension is based on the assessment of 10 sustainability attributes. These sustainability attributes include: (1) sources of clean water, (2) conditions of groundwater utilization, (3) carrying capacity of feed, (4) types of animal feed, (5) sanitation of pens and milking equipment, (6) the quantity of waste (faeces and urine) remaining livestock per day, (7) availability of biogas installations, (8) utilization of cattle waste for organic fertilizer, (9) utilization of agricultural waste for animal feed, (10) distance of livestock business locations with residential areas. The results of the ordination of the ecological dimensions of RAP-SIBUSAPE, indicate that the sustainability index of Dairy Farming in Padang Panjang City is 45.17. Based on the results of the analysis, the sustainability status of Dairy Farming in Padang Panjang is categorized as less sustainable.

The unsustainable status of the Dairy Farming system in Padang Panjang City is due to the poor scores on some of the assessment attributes for the preparation of the sustainability index. Sustainability attributes that have poor scores include: (1) clean water sources, (2) feed carrying capacity, (3) sanitation of pens and milking equipment, (4) quantity of waste (feces and urine) remaining in the cage per day, (5) availability of waste / biogas storage installations, and (6) utilization of cattle waste for organic fertilizer. Most of the attributes that have bad scores have a strong relationship with aspects of livestock maintenance management, this shows that the management of livestock raising in Dairy Cattle Farming in Padang Panjang City is not good from an ecological perspective. The implementation of such maintenance management, if it continues to be maintained, will reduce the quality of the surrounding environment, especially the environmental carrying capacity.

3.2. Economic dimension

The sustainability status of the economic dimension is generated based on the assessment of 9 attributes of sustainability. These sustainability attributes include: (1) profit (profit) of dairy farming, (2) market for dairy products, (3) price of milk for the last 5 years, (4) place for farmers to sell milk, (5) ownership (beneficiaries of ownership), (6) level of subsidies to inputs, (7) transfer of profits, (8) types of superior commodities, (9) level of dependence of consumers.

The index value of the sustainability of the dairy farming system in the economic dimension of Dairy Cattle Farming in Padang Panjang City has a value of 49.24. Based on the sustainability index value, the sustainability status of Dairy Cattle Farming in Padang Panjang is categorized as less sustainable. The results of the leverage analysis of the sustainability of the economic dimension show that of the 9 attributes analyzed there are 3 sensitive attributes affecting the sustainability of the economic dimension. These three attributes are sensitive to sustainability, including: (1) transfer of profits, (2) level of subsidies to inputs, and (3) price of milk for the last 5 years. These three sensitive attributes are related to one another.

3.3. Socio-cultural dimensions

The sustainability status of the social dimension is generated based on the 9 attributes of sustainability. These sustainability attributes include: (1) socialization of work (individual or group), (2) the number of households of dairy farming workers, (3) the growth of dairy farming workers (last 5 years), (4) the frequency of conflicts relating to dairy cattle business, (5) alternative businesses other than dairy farming agribusiness, (6) time allocation for dairy cattle business, (7) population growth, (8) community role in management efforts, (9) frequency counseling and training.

Based on the RAP-SIBUSAPE analysis, it is known that the sustainability index value of the socio-cultural dimension of the dairy farming system in Dairy Cattle Farming in Padang Panjang City is 48.89. Based on the sustainability index value, the sustainability of the dairy farming system in the Dairy Cattle Farm in Padang Panjang is less sustainable. The results of the leverage analysis on the sustainability of the socio-cultural dimension show that of the 9 attributes analyzed there are 3 sensitive attributes that affect the sustainability of the socio-cultural dimension. The three attributes are sensitive to sustainability, among others: (1) alternative businesses, (2) frequency of counseling and training, and (3) job socialization.

The frequency of low intensity extension activities is only one time a year resulting in low levels of knowledge and skills of farmers in conducting dairy farming in a sustainable system. The low level of knowledge is reflected by the dependence on the involvement of the farmer's family in the implementation of dairy cattle cultivation, so that the efficiency in the input use of livestock business is low. The skills of farmers in the implementation of sustainable dairy farming systems are low.

The low level of farmer proficiency is reflected in the small number of business alternatives developed in addition to dairy farming, and does not even have an alternative business other than dairy farming. If the level of expertise of farmers is adequate, there are several business alternatives that can be done such as: building feed factories, animal medicine stores, yogurt making business, ice cream, butter, cheese, kefir, curd, frozen yogurt, making organic fertilizer from dirt or biourine cow urine, biogas making, cultivation of superior grass plants utilizing arbuscular mycorrhizal fungi (FMA) or other economically valuable plants. Alternative businesses, especially in this region are very limited, so alternative businesses other than dairy cattle breeding are difficult to develop, except business alternatives that are related to dairy cattle business from upstream to downstream.

3.4. Dimensions of technology-infrastructure

The sustainability status of the technology-infrastructure dimension is generated based on the assessment of 7 attributes of sustainability. These sustainability attributes include: (1) post-harvest handling, (2) feed technology, (3) deployment of poskeswan / IB places, (4) use of vitamins and probiotics for livestock, (5) availability of agribusiness facilities and infrastructure, (6) standardization of the quality of livestock products, (7) utilization of waste handling technology (biogas or organic fertilizer).

The results showed that the index value of the sustainability of the technology-infrastructure dimension in Dairy Cattle Farming in Padang Panjang City was 57.18. Based on the sustainability index value, the sustainability status of the technology-infrastructure dimension in Dairy Cattle Farming in Padang Panjang City is categorized as quite sustainable. The results of the analysis of the leverage of technology-infrastructure dimensions show that of the 7 attributes analyzed there is 1 sensitive attribute that influences the sustainability of the technology-infrastructure dimension. The sensitive attribute to sustainability is the use of waste handling technology (biogas or organic fertilizer).

The dairy farming business center in the Dairy Cattle Farming in Padang Panjang City has not fully utilized waste handling technology to be used as biogas or organic fertilizer. This is due to the quite high costs from the point of view of farmers in the application of the technology. In addition, the lack of maintenance (maintenance) activities in technology applications due to the limited knowledge of breeders is an obstacle that requires special attention.

3.5. Dimensions of institutional law

The sustainability status of the legal-institutional dimension is generated based on the assessment of 10 sustainability attributes. These sustainability attributes include: (1) the availability of formal management regulations, (2) land allotment zoning in accordance with the RTRW, (3) the intensity of illegal farming activities, (4) the development of cooperatives, (5) extension institutions agriculture, (6) transparency in policies or information delivery, (7) microfinance institutions (banks / credit), (8) livestock farmer groups, (9) institutional input of livestock businesses, and (10) institutional output of livestock businesses.

The sustainability index value of the legal-institutional dimension in Dairy Cattle Farming in Padang Panjang City is 74.98. Based on the sustainability index value, the Dairy Cattle Farm in Padang Panjang City is quite sustainable. The results of leverage analysis of legal-institutional dimensions show that from 10 attributes of sustainability assessment there are 3 sensitive attributes that affect sustainability. These three attributes are sensitive to sustainability, including: (1) microfinance institutions (banks / credit), (2) transparency of policies, and (3) development of cooperatives.

The existence of microfinance institutions that have the role of providing capital for dairy farming business such as cooperatives and banking institutions, is very sensitive to influencing the sustainability index of the legal-institutional dimension. The existence of microfinance institutions needs to get attention because of their function as credit channeling institutions. In addition, capital credit is the main factor to obtain good input for livestock business production. The better the input of livestock production, the greater the chance of better livestock business results.

The development of cooperatives has links with microfinance institutions. Cooperatives that are increasingly developing will be able to channel capital loans in larger amounts. The dairy cattle business cooperative in the city of Padang Panjang is considered not sufficiently developed based on the amount of milk received, the distribution of credit received by farmers, and the provision of available livestock input facilities. The problems that have been the concern of the cooperative in recent years include: (1) the number of members who are not active and do not understand the meaning of cooperating so that such members demand their rights more and ignore their obligations, (2) the amount of bad loans is caused by the many members who have difficulties pay credit installments, especially for members who are not active and their livestock are gone, (3) decrease in milk production due to the increasingly high price of beef so that farmers choose to cut their dairy cattle, (4) the number of farmers who sell milk outside the cooperative, (5) decreased volume of milk reception.

3.6. Key Factors for Developing Dairy Farming Systems in Padang Panjang City

The concept of development that harmonizes environmental, economic and social interests is a concept of development that has been accepted by all countries in the world. The concept of development is then called the concept of sustainable development with the aim of creating a sustainable condition as a result of the development process. The concept, has similarities with the concept of the development trilogy that was implemented during the reign of the New Order. The development trilogy requires that development is not only limited to creating economic growth, but also must pay attention to aspects
equity and the creation of national stability.

The concept of sustainable development aims to create a balance between economic growth (economic dimension), environmental conservation (ecological dimension), equity (socio-cultural dimension). Some opinions add the dimensions of technology-infrastructure (development and application of technology to better infrastructure), legal-institutional (compliance with the law and institutional functioning) for the implementation of sustainable development. The application of the concept of sustainable development in the real system requires a strong commitment of the main actors of the system (stakeholders) to ensure the success of development.

Prospective analysis is able to explore the possibilities of the future based on predetermined goals. The determination phase of sustainability index and status aims to provide an overview of the existing system of actual dairy farming systems. Prospective analysis aims to prepare strategic actions in the future by determining key factors that play an important role in the various possibilities that will occur in the future. These possible future variations, formulated in the form of a scenario for the development of a sustainable dairy farming system. Stages of analysis that need to be done in prospective analysis, namely: (1) identification of key factors in the future, (2) determining the

IOP Conf. Series: Earth and Environmental Science **287** (2019) 012038 doi:10.1088/1755-1315/287/1/012038

objectives and interests of the main actors, and (3) definitions and descriptions of possible future evolution as well as determining priority actions in accordance with the resources owned by the main actors and their implications for the system being studied.

The stages of determining the key factors for the development of the dairy farming system in the future, among others: (1) determination of key factors derived from the sensitive attributes of the sustainability index of existing actual dairy farming systems from leverage analysis of Rap-SIBUSAPE, (2) determination Key factors based on needs analysis (need analysis) sourced from the needs of stakeholders on the system that are reviewed through expert discussions with the help of questionnaires, (3) determine the key factors that influence the system based on a combination of existing condition and need analysis key factors.

Based on the results of leverage analysis on each dimension of sustainability, 15 sensitive attributes were obtained from 5 dimensions of sustainability, including:

1. Ecological dimension: (a) the quantity of waste (feces and urine) of livestock left in cages per day, (b) availability of waste / biogas storage plants, (c) feed carrying capacity, (d) utilization of cattle waste for organic fertilizer, and (e) sanitation of milking equipment and cages.
2. Economic dimensions: (a) profit transfers, (b) the level of subsidies to inputs, and (c) the price of milk in the last five years.
3. Socio-cultural dimensions: (a) business alternatives, (b) job socialization, and (c) frequency of counseling and training.
4. The technology-infrastructure dimension is only one attribute, namely the use of waste handling technology (biogas or organic fertilizer).
5. Legal-institutional dimensions: (a) microfinance institutions (cooperatives / credit), (b) transparency of policies, and (c) development of cooperatives.

These sensitive attributes are then assessed, so that the results obtained in the form of grouping of attributes in the four groups of quadrants are based on the level of influence and dependence of attributes on the system. Based on the assessment of the 15 sensitive attributes, 7 (seven) key factors were identified that had a strong influence on the system, namely: (1) milk prices, (2) feed carrying capacity, (3) feed industry, (4) utilization of livestock waste, (5) the nature of work (6) the development of cooperatives, (7) micro finance institutions (banks / credit).

4. Conclusion

1. This study resulted in the index value of the sustainability of the dairy farming system in the city of Padang Panjang of 55.09 (quite sustainable). The results of the analysis of each dimension of development shows that the legal-institutional dimension has the highest index value (74.98 / quite sustainable), then followed by the technology / infrastructure dimension (57.18 / quite sustainable), the economic dimension (49.24 / quite sustainable)), socio-cultural dimensions (48.89 / less sustainable), and ecological dimensions (45.17 / less sustainable). The results of the leverage analysis show that there are 15 sensitive attributes that influence the sustainability of this dairy farming system.
2. Based on the prospective analysis (level of influence of interest) the results of combining existing conditions (important factors derived from the sensitive attributes of leverage analysis) and need analysis (important factors derived from stakeholder needs) are known to seven important factors affecting the system, including: (1) five important factors have a high influence on the performance of the system with a low level of inter-factor dependency (milk prices, feed carrying capacity, feed industry, utilization of livestock waste, and the nature of work); and (2) two important factors have a high influence on the performance of the system with a high level of inter-factor dependency (the development of cooperatives and microfinance institutions). Seven important factors affecting the dairy farming system need to be managed properly and various states that may occur in the future to achieve the goal of creating a sustainable dairy farming system.

5. References

[1] Edwardi. 2016. Prospects of dairy farming in West Sumatera. Livestock Service Office of West Sumatera. Padang.

[2] West Sumatera Livestock Service. 2010. Population of dairy cattle in West Sumatera. Report of the Livestock Service Office. West Sumatera. Padang.

[3] Saragih, B. and T. Sipayung. 2002. Biological utilization in developmentalism and environmentalism. Paper Presented at the International Seminar on Natural Resources Accounting Environmental Economic Held in Yogyakarta, Indonesia, 29 April 2002. Yogyakarta.

[4] Munasinghe, M. 1993. Environmental Economic and Sustainable Development. The International Bank for Reconstruction and Development / The World Bank. Washington D.C.

[5] Mersyah, R. 2005. Design of sustainable beef cattle cultivation systems to support the implementation of regional autonomy in South Bengkulu Regency [Dissertation]. Bogor Agricultural University. Bogor.

[6] Suyitman. 2010. Model of the development of sustainable agropolitan area based on integrated beef cattle breeding in the Situbondo sub-district [Dissertation]. Bogor Agricultural University. Bogor.

[7] Fauzi, A. and S. Anna. 2005. Modeling of Fisheries and Ocean Resources for Policy Analysis. Gramedia Pustaka Utama. Jakarta.

[8] Bourgeois, R. and F. Jesus. 2004. Participatory Prospective Analysis, Exploring and Anticipating Challenges with Stakeholders. Center for Alleviation of Poverty through Secondary Crops Development in Asia and The Pacific and French Agricultural Research Center for International Development. Monograph (46): 1 - 29.

Acknowledgments

Thank you to the Directorate of Research and Community Service, Directorate General of Research and Development Strengthening - Ministry of Research, Technology and Higher Education, through the National Innovation Wave I Year 2018 Research Incentive System Program which has funded this research, according to the Research Contract No. 25 / INS-1/ PPK/E4/2018 and Andalas University.

Social and economic analysis of food crops and beef cattle sub systems in the dry land area of Kupang Regency, East Nusa Tenggara Province, Indonesia

Maximilian M J Kapa

Department of Agribusiness, Faculty of Agriculture, University of Nusa Cendana, Kupang, East Nusa Tenggara, Indonesia
*Corresponding author e-mail: maximilian.kapa@staf.undana.ac.id

Abstract. This study aims to analyze the system properties of food crop and beef cattle sub-systems in dry land area. The properties analyses are divided into two factors, namely the social and economic factors of both sub-systems. The study was conducted in Pakubaun Village in East Amarasi Sub District and Pukdale Village in East Kupang Sub District, Kupang Regency. The method applied was a survey method and the data was analyzed by using 5 (five) agro-ecosystem properties. Results of the research were the economic properties namely productivity of both subsystems were low to moderate, while, the stability of rice farming was low and the stability of the beef cattle enterprise was high. The sustainability of the two subsystems was high. Based on the social properties analysis, it was found that the equitability of food crops sub-system was low; on the contrary, the equitability of the beef cattle sub-system was high. From autonomy point of view, the results of this study indicated that the autonomy in both sub-systems was high. Based on agro-ecosystems properties, in general, the performance of both sub-systems was low, out of five properties only two properties described the high conditions.
Keywords – social economic analysis, food crops, beef cattle, dry land area, Kupang, East Nusa Tenggara.

1. Introduction

From the time of the new order to the reform era, the Indonesian government continued to pay important attention to the agricultural sector with the intention that villagers could achieve food and meat self-sufficiency. Self-sufficiency in food and meat programs in East Nusa Tenggara is approached with various development policies such as the development of food crops through the establishment of NTT as a corn province, and the development of livestock, especially beef cattle through the province's livestock program as a regional superior commodity [1]. However, these efforts have not been able to achieve the self-sufficiency program in this area. If explored further, it turns out that various factors are suspected to be obstacles to the development of the agricultural sector. Physical factors such as (1) geographic characteristics such as topographic patterns and climate behavior, (2) natural resource potential, and (3) procurement, utilization and development of natural resources. While non-physical factors such as (1) human resources, (2) institutional functions, and (3) farming technology.

Physical factors can be a driving factor in increasing farm productivity, but they can also be a factor inhibiting agricultural development in an area. On the other hand, non-physical factors, especially economic and social factors, knowledge and technological skills determine the choice of organizing a farming system. A combination of physical and non-physical factors determines the type of land use. While the selection of the farming system that will be carried out is determined by non-physical factors.

The above factors are also limiting factors which cause various problems in the development of dryland farming systems, especially the development of food crops, and livestock sub-systems in the dry land area of West Timor. Kupang Regency as a part of West Timor experiences problems that are identical to the problems of the farming system in West Timor based on dry land with short and uneven rainfall patterns. In accordance with the existing physical conditions, in Kupang Regency, there are three types of dominant farming sub-systems, namely the rice field sub-system, the upland system, and the livestock farm sub-systems [2] [3].

The problem is each subsystem has different social and economic characteristics. In terms of the concept of agroecosystem, these differences are thought to have an effect on the characteristics of the farming system, in this case, the productivity, stability, sustainability, and equitability of each farming sub-system. So far, existing studies are still partial and there were no integrated and comprehensive studies to assess the nature of the system in the dryland area.

2. Research objectives

This study aimed at determining and analyzing (1) the economic properties such as productivity, stability, and sustainability of food crops (rice enterprise) and livestock farm (local beef cattle enterprise) and (2) the social characteristics of the two dominant farming sub-systems (rice field and local beef cattle subsystems) in Kupang Regency of West Timor.

3. Research methods

This research was carried out in Pakubaun Village, East Amarasi Sub District, and Pukdale Village, East Kupang Sub District, Kupang Regency. Determination of sample farmers was done purposively, in this case, farmers who at least have operated two types of sub-systems from the three predominant predetermined sub-systems. The number of sample farmers in this study was 40 households in Pakubaun village and 45 families in Oenesu village.

Data collection methods for cultivation techniques and economics are carried out by interview techniques based on the questionnaire. Whereas, social aspects of data and information are obtained by interviewing key informants who are considered to understand and know it in each village. The data was then analyzed using the agro-ecosystems approach through the analysis of system properties (emergent properties) namely Productivity; Efficiency, Sustainability, Stability (those are categorized as economic systems), Equitability, and Autonomy (both are social systems properties) [4] [5] [6]. This analysis was applied to both sub-systems namely rice field subsystem and livestock (beef cattle) subsystem in the study area.

4. Results and discussion

4.1. The general background of the research area

Pakubaun village is one of 9 dry land villages located at an altitude of 419 m above sea level with a gentle slope. Administratively, this village is incorporated in the Amarasi sub-district. The total area of 11.91 km2 is accompanied by a sloping topography to wavy with a slope of 15-25 degrees and even part of a hilly area with a slope of more than 30 degrees [7]. While Pukdale village is a dryland village based on rice fields with a height of 15 m above sea level with a slope of <150. Pukdale Village is included in East Kupang Sub district with an area of 9.38 km2. The farming system in the two study areas is rice farming, dry field, and livestock enterprise, especially beef cattle [7]. In this study, only two subsystems namely rice farming and livestock especially beef cattle enterprise subsystem were taken into account.

4.2. Analysis of Rice Farming Systems

Generally, farmers in dryland areas operate several branches or sub-systems of farming in accordance with the capabilities and potential resources they control. Because working on more than one sub-system is highly needed high-level managerial skills from farmers to regulate the allocation of resources that have a certain combination pattern in order to provide better and more efficient results. The following is an analysis of each sub-system in the two study areas.

4.2.1. Sub-system rice field analysis

Rice fields cultivated by respondents in both study areas are rainfed so that they are only planted once a year. The average area of rice field per respondent in Pakubaun village is relatively small, namely 0.37 ha of the head of the family with an average planting area of 0.35 ha with an average production of 996 kg or 2.846 kg per ha. Meanwhile, in Pukdale Village, the average area of rice field ownership is 1.05 ha with an average planting area of 0.96 ha. The average production of rice per-farm was 1,490 kg or 2,021 kg per ha.

4.2.2. Livestock subsystem

Livestock raising activities, especially local beef cattle, cannot be separated from farming activities for people in rural areas in Kupang Regency. The activity of raising cattle is a hereditary activity or an inherited culture. The average ownership of cattle in Pakubaun village was 8.02 Livestock Unit (LU) while the average ownership of cattle in Pukdale Village was 12.02 LU. All respondents in both research areas are keeping cattle for breeding and fattening purposes.

4.3. Analysis of system properties

4.3.1. Economic characteristics

4.3.2. Productivity

Livestock raising activities, especially local beef cattle, cannot be separated from farming activities for people in rural areas in Kupang Regency. The activity of raising cattle is a hereditary activity or an inherited culture. The average ownership of cattle in Pakubaun village was 8.02 Livestock Unit (LU) while the average ownership of cattle in Pukdale Village was 12.02 LU. All respondents in both research areas are keeping cattle for breeding and fattening purposes.

Rice Farming Sub System. The results of the study showed the average rice production per ha in Pakubaun Village and Pukdale Village was 2,846 kg and 2,021 kg respectively. If the average selling price at the farm level in Pukdale was Rp 4,450, then the average revenue (Gross output) was Rp. 12,664,700 in Pukdale village. Meanwhile, the average price of rice at Pakubaun Village was Rp 3,500 per kg then the average gross income of farmer was Rp. 7,073,000. Looking at the results of rice production in both research villages, the level of productivity of lowland rice is low although most farmers, especially in Pukdale, have used superior variety of seeds. The low productivity of rice subsystems is considered to be influenced by several factors, among others, (1) types of rice fields which are rain-fed land, (2) unfavorable tillage systems, (3) the improper use of fertilizers and pesticides which are not in accordance with the recommended dosage, and (4) pest attracts and stem borer pests, especially in Pakubaun village

Livestock Sub System. The beef cattle business in the research area is still categorized as a non-commercial business, Therefore, there is no plan to sale so the sale in the form of breeders or fattening animals is not planned on a planned basis every year. Sales occur when there is an urgent need from the owner. From the results of this study, the sale of cattle in the year 2017 in Pukdale village totaled 2.75 LU with net income obtained Rp. 11,174,000. The results show that the productivity of beef cattle per LU is IDR 4,063,000. While, the sale of cattle in Pukdale Village per year is 2.06 LU with a total net income of Rp 8,912,000 or productivity per LU of Rp 4,300,000. This productivity value explains that the level of productivity of local beef cattle business in the study area is classified as moderate. Because, the level of productivity is actually still able to be increased if the farmers make changes to the maintenance system towards semi-intensive. In addition to the maintenance system, that factors to influence the low productivity of beef cattle in the study area include, (1) the environment of livestock keeping, (b) the carrying capacity of feed, (c) the lack of synchronization of government policies in the livestock sector [2] [8].

4.3.3. Stability

Stability is the consistency of products produced by a sub-system in this case how changes or fluctuations in production result from changes in environmental factors. Analysis of the stability of the sub-system of paddy farming is based on this classification. However, to know this, production data is needed in several time periods.

Rice Field Sub System. Data on rice production trends in the required timeframe is very difficult to obtain due to the low memory of respondents and the absence of good farming records in the two study areas. Therefore, the measurement of the stability of lowland rice farming in the two study areas was only based on the interview with respondents. During the interview, respondent farmers explained their experience about farm production.

Interviews with respondents indicated that the nature of rice farming in both Pakubaun and Pukdale villages was low. Rice production in both villages varies greatly every year and it depends on rainfall. When rainfall is good and timely, the production is quite high and vice versa if the rain is erratic causing low production.

Livestock subsystem. From the interviews with respondents, it was understood that the stability of livestock enterprise in the study area was quite high. As, the production obtained from the subsystems was not influenced by climate conditions and was more determined by the skills of farmers in managing the livestock enterprise.

4.3.4. Sustainability

Sustainability is the ability of a subsystem to maintain a certain level of production in the long run. In other words, sustainability is the prolongation of the productivity of a sub-system in the long run from various disturbances in the form of pressure or shock.

Rice Field Sub System. In accordance with the concept of sustainability above, the nature of the sustainability of the rice farming sub-system in Pukdale and Pakubaun villages is high. Factors that support the nature of the sustainability of the rice farming sub-system are, (1) the average area of wetland land ownership, (2) the suitability of agrarian, and (3) the suitability of land with land use, and (4) the motivation of farmers to cultivate rice is quite good [9]. In addition to these three factors, namely (1) support from traditional factors, (2) the availability of supporting institutions for rice farming activities such as farmer groups, and (3) government policies on increasing rice production [10].

Livestock Business Sub-system. As in the rice farm sub-system, the nature of the sustainability of beef cattle business in both Pakubaun Village in East Amarasi Sub-district and Pukdale Village in East Kupang District was high. This is reasonable because of various supporting factors. The sustainability of cattle business was strongly supported by various factors such as, (1) Status and extent of land severity, (2) high adaptability of cattle to the environment, (3) carrying capacity of animal feed, (4) availability of agricultural waste as a source of feed, (5) customs and culture of the community, and (6) government policies in the livestock sector [2] [11].

4.3.5. Social characteristics

4.3.6. Equitability

Equitability refers to the distribution of production of certain sub-systems evenly or equitably. In other words, equitability shows the extent to which the results of an agroecosystem are evenly distributed to the community.

Rice Field Sub System. In general, the equitability of the rice farming sub-system is low because most of the populations in the two villages do not have paddy fields. However, the results of observations and interviews with respondent farmers in Pakubaun and Pukdale villages showed that the equitability was quite high. Where 100 percent of respondent farmers in Pakubaun village own and cultivate their rice fields and 75 percent of sample farmers in Pukdale Village own and cultivate their paddy fields.

Local Beef Cattle Sub Systems. Although the beef cattle enterprise subsystem is integrated with rice farming as a source of labor, a source of fertilizer and also a source of income [3] [8] [11]. The results of this study proved the equitability of the sub-system of cattle enterprise was relatively low because cattle business requires a large initial capital which is rarely achieved by small farmers who generally lack capital.

4.3.7. Autonomy

Although the beef cattle enterprise sub-system is integrated with rice farming as a source of labor, a source of fertilizer and also a source of income [3] [8] [11]. The results of this study prove that the sub-system of cattle enterprise was relatively low because farmers are generally lack of capital.

Rice Farming Sub System. Referring to the previous Autonomy terminology, wetland rice farmers in the study area have various roles, namely as managers, workers (cultivators) and sometimes as farm laborers. As a manager, especially in decision-making because farmers themselves decide and choose farming branches (plants and livestock) what will be sought in accordance with the availability of resources. Likewise, if there are technologies or innovations such as superior varieties, fertilizers, strains or breeds of livestock or other production technologies. Based on this concept of thinking, rice farming in Pakubaun Village is high.

Local Beef Cattle Sub Systems. In contrast to rice farming, cattle enterprise has a high degree of autonomy, such as in the selection of livestock, maintenance methods and the motive for raising livestock. When asked what prompted them to raise cattle, all respondents in Pakubaun and Pukdale villages stated that cattle raising was continuing the habits of their parents, inherited from parents, and following neighbors or friends from the village as well as from neighboring villages.

5. Conclusion

Analysis of the economic and social properties of the two subsystems of farming, namely rice farming and livestock enterprise using the agroecosystem analysis approach are as follows:

(1) Economic properties:

a. The Productivity of the rice farming subsystem in the Pakubaun village is low because the yields obtained are low at 2.846 kg and 2.021 kg per ha, respectively. While the productivity of cattle subsystems was moderate to high in all research areas in both villages
b. The Stability of rice farming subsystems in the two villages was low. While, the stability of beef cattle subsystem in both study areas was high.
c. The sustainability property of rice farming subsystem in both study areas was categorized to be high. The sustainability category of livestock subsystems both in Pakubaun dan Pukdale was high as well.

(2) Social properties:

a. The equitability of the rice farming subsystems within the two villages was high. On the contrary, the cattle business subsystem in the two study areas was low. The low of stability occurred because the cattle husbandry system is extensive traditionally where the interventions the owner to their cattle husbandry is very limited.
b. The Autonomy category for rice farming subsystem in Pukdale was low, as the interventions of exogenous parties such as from agricultural extension worker and government policy through introducing inorganic fertilizers, exotic seeds, and other chemical matters.

6. References

[1] Anonymous, 2014. Peraturan Daerah Provinsi Nusa Tenggara Timur Nomor 1 Tahun 2014 Tentang Rencana Pembangunan Jangka Menengah Daerah (RPJMD) Provinsi Nusa Tenggara Timur Tahun 2013-2018. Pemerintah Daerah NTT.

[2] Kapa, M. M. J., 2017. Status Keberlanjutan Sistem Peternakan Sapi Potong Lokal di Daerah Lahan Kering di Tinjau dari Dimensi Ekologi dan Kelembagaan. Orasi Ilmiah pada Upacara Wisuda Magister, Profesi, dan Sarjana Universitas Nusa Cendana Tanggal 02 Desember 2017.

[3] Kapa, M. M. J. and Henuk, Y. L. 2018. Economic Performance of Local Beef Cattle Development under Rice Field in the Dryland Area. In: Proceeding of AAAP Congress 2018, 1-5 Agustus 2018, Kucing, Malaysia, p.85.

[4] Conway, G. R., 1987. The Properties of Agroecosystems. Agricultural Systems Journal. Vol. 24 (Issue 2): 95-117.

[5] Rambo, A. T. 1985. Applied Human Ecological Research on Asian Agricultural Systems. Paper Presented on at the EAPT/Murcep Workshop on Human Ecology Research on Asian Agricultural System. Paper presented at The EAPI/MURCEP Workshop on Human Ecology Research on Human Ekology, and Research on Agroecosystem, Nanjing, China, 9 – 18.

[6] Marten, G. G. 1988. Productivity, efficiency, stability, and sustainability as properties for agroecosystem assessment. Paper presented to the 'SUAN Workshop on Agroecosystem Analysis'. Khon Kaen, Thailand, January 6, 1986.

[7] BPS Kabupaten Kupang, 2017. Kabupaten Kupang dalam Angka 2017. Biro Pusat Statistik Kabupaten Kupang.

[8] Kapa, M. M. J., Henuk, Y.L., Hasnudi, and Suyadi. 2018. Contribution of local beef cattle production on farmer's income in the dryland farming of Kupang Regency, Indonesia. IOP Conf. Series: Earth and Environmental Science 122 (2018) 012118 doi :10.1088/1755-1315/122/1/012118.

[9] Dzikrillah, GF, Anwar, S., dan Sutjahjo, S.H., (2017). Analisis keberlanjutan Usahatani Padi Sawah di Kecamatan Soreang Kabupaten Bandung. Jurnal Pengelolaan Sumberdaya Alam dan Lingkungan Vol.7 (2): 107-113.

[10] Kurnia, Rosyani, dan Aulia (2018). Faktor-Faktor yang mempengaruhi Keberlanjutan petani berusahatani Padi Sawah (Studi Kasus di Desa Pulau Aro Kecamatan Tabie Ulu Kabupaten Merangin. https://repository.unja.ac.id/3914/1/D1B013078_LISA%20TRI%20KURNIA_JURNAL.pdf. Accessed: 10 Agustus 2018.

[11] Kapa, M. M. J., Soemarno, Bagyoyanuwiyadi, and Suyadi, 2017. Sustainability of Biologi Dimension of Local Beef Cattle Development in the Dryland Region, Indonesia. J. Econ. Sustain Dev. Vol.8(6): 102-108.

Cost of goods sold analysis of banana blossom chicken nuggets in the women farmers group "KWT Kebersamaan" in Padang Pariaman Regency

W Sartika and E Rahmi

Social Economic of Animal Science, Andalas University, Padang, West Sumatera, Indonesia.
E-mail address: winda.sartika@gmail.com

Abstract. Chicken nuggets are one of the products of meat processing technology that has good nutritional value and affordable prices compared to processed beef products. Despite having a fairly complete and good nutritional content, chicken nuggets contain high fat and low fiber. To increase the nutritional value of chicken nuggets, the Women Farmers Group "*KWT KEBERSAMAAN* " tried to make processed nugget with the ingredients of banana blossom and chicken meat. This study aims to analyze the cost of production and the cost of goods sold in banana blossom chicken nugget. Data used in this study were primary data from the company and secondary data obtained throught literature in line the tittle of the study. Method of calculating the cost of production and the cost of goods sold in this study is to use the full costing method and variable costing method. Results of this study showed that The cost of production obtained from the variable costing method is IDR 8,519/pack, while the full costing method costs IDR 10,719/pack. 200 gram of banana blossom-chicken nuggets per pack is sold at a price of IDR 10000, while the result of the research shows that the cost of goods sold is higher than the selling price set by the KWT Kebersamaan. The cost of goods sold using the variable costing method is IDR 13,074/pack and the full costing method obtains IDR 15,923/pack.
Keywords – cost of goods sold, banana blossom chicken nuggets.

1. Introduction

Chicken nuggets are one of the products of processed meat that has good nutritional value and is affordable, compared to processed beef products. Nutritional content of chicken nuggets consists of protein, fat, carbohydrates, and minerals. The protein is derived from chicken meat which consists of complete amino acids. Despite having a fairly complete and good nutritional content, chicken nuggets contain high fat and low fiber. [4] that the largest percentage of respondents who consumed chicken nuggets were between the ages of 20-25 years. This age range is a productive age range as they have their own income, and their needs of food vary. Thus, respondents can independently decide to consume chicken nuggets according to their needs and income. Additionally, productive age consumers tend to choose ready-to-eat and ready-to-be-cooked food products which promise no nutritional content reduction, including nugget products. As food technology continues to develop, changes in people's consumption patterns continue to occur. People, especially in urban areas, prefer to consume food products that are ready to eat and ready to be cooked, due to the high level of everyday community mobility [7].

Banana blossom is usually cooked and served as a kind of vegetable in a menu. In addition to its low price, banana blossom offers many health benefits, especially for those in a diet program, as it contains high fiber and low fat. Despite the high fiber content, banana blossom is relatively low on other nutritional content, especially protein. Therefore, there has to be some additional high-nutrition ingredients in the processing stage [1]. One way to supplement the nutritional content of banana blossoms is by mixing animal protein, such as chicken meat. Banana blossom nugget is a new product created by KWT Kebersamaan, with a background knowledge of children liking inexpensive fast food that has good nutritional value. Nugget products are still quite difficult to obtain in Padang Pariaman Regency, hence the large market opportunity to develop. A product survives in the market if they are liked or in demand by the community. In other words, the demand for the product should increase over time. The costs incurred to produce a product must be calculated meticulously, while currently, KWT Kebersamaan has not included all the cost components in the calculation of production costs, which has an impact on the selling price being not accordingly to the actual conditions. Determination of the selling price of this nugget product is only based on the common nugget prices in the market, making it difficult to calculate how much profit is obtained. Meanwhile, this business is expected to support and improve the financials of group members and the surrounding community.

2. Materials and methods

The research was conducted at the Female Farmer Group (KWT) Kebersamaan in Padang Pariaman Regency for 5 months in 2018. The selection of the research location was intentional, as KWT is the only group that has carried out the processing of banana blossom nuggets in Padang Pariaman Regency.

The research uses the case study method. [3] states that a case study focuses on research subjects related to specific or typical phases throughout a personality that aims to provide a detailed description of the background, traits, and characteristics that are typical in the case, or the status of the individual to form a general nature. Data are analyzed using quantitative descriptive analysis with full costing method and variable costing method.

The data used in this study is primary data from the Female Farmer Group (KWT) Kebersamaan, while the secondary, supporting data for the discussion is obtained from relevant Departments or Agencies. To determine the cost of production and the cost of goods sold for banana blossom nuggets, the cost components observed include:

a. Raw materials cost (IDR/kg/production),
b. Labor cost (IDR/kg/production),
c. Supporting materials cost (IDR/kg/production),
d. Overhead cost (IDR/production)
e. Marketing cost (IDR/production)

3. Results and discussion

Female Farmer Group (KWT) Kebersamaan is a group of female farmers in Padang Pariaman Regency who has processed banana blossom into various diversified products such as banana blossom jerky and banana blossom nuggets. KWT Kebersamaan's nuggets is a ready-to-eat food product that is highly favored by the public, therefore the high cost of raw materials production will have an impact on the selling price received by consumers. The high cost of feed ingredients and breeds of broiler chickens will have an impact on the sale of chicken meat and will also directly impact the high prices of products made from chicken meat. Nuggets are generally made from chicken meat with the addition of other complementary spices. But now, consumers can try nuggets with banana blossoms as the raw ingredient, in addition to chicken meat. [2] said that one of the shortcomings of fast food is the lack of vegetables. Therefore, fast food innovation needs the addition of vegetables to meet the required vegetable fiber content for health reasons. Some vegetables that have been added in the process of making nuggets include spinach, broccoli, carrots, and others. One of the vegetables that has the potential to be added to nuggets is Moringa oleifera L.

KWT Kebersamaan is able to produce 3 kg of banana blossom nuggets in one production. It will be marketed in 200-gram packages with a selling price of IDR 10000. This price is determined by calculating the price of raw materials used and divided by the number of nuggets produced. The calculation is still done with a very simple method and does not take into account labor costs or depreciation of the equipment used. The results of [6] study stated that the low selling price of a traditional business occurs because the business owner only uses intuition and makeshift calculations without classifying the costs incurred to determine production cost, while the cost of the products produced must be calculated thoroughly. Other factors include the lack of adequate financial records, causing the owner to experience difficulty in calculating the costs incurred, including in the classification.

Table 1. Raw material costs for Banana Blossom Nugget in one production.

No	Material	Unit	Price/ unit (IDR)	Total (IDR)
1	Chicken breast	Kg	50000	25000
2	Banana blossom	Kg	4000	8000
3	Tapioca flour	Kg	22000	1320
4	Egg	Egg	1500	7500

In one production of banana blossom-chicken nuggets, 1.5 kg of banana blossom and 0.5 kg of chicken meat were used. This mixture of raw materials will produce 3 kg of banana blossom nuggets per production. The banana blossom nuggets produced will be marketed in a plactic packaging, weighing 200 grams/pack.

Table 2. Overhead business costs of Banana Blossom Nugget.

Overhead cost	Variable Costing (IDR)	Full Costing (IDR)
1. Variable cost		
a. Gas Elpiji	13.000	13.000
b. Ingredients	3.960	3.960
c. Other materials	9.000	9.000
2. Fixed cost		
a. Tools depreciation cost	-	3.000
b. Electricity cost	-	20.000
c. Water cost	-	10.000
Total of overhead cost	25.960	58.960

The cost used in making banana blossom nuggets consist of fixed costs and variable costs. Variable costs will be included in the variable costing method while the fixed costs will be included in the full costing method calculation. Labor costs is a direct spending on labor, calculated as the result of business overhead costs. The processing of banana blossom-chicken nuggets involves 2 workers with a wage of IDR 30000/person/production.

Table 3. Cost of production and cost of goods sold per 1 kg of Banana Blossom-Chicken Nuggets.

Point	Variable Costing (Rp)	Full Costing (Rp)
1. Cost of Production	42.593	53.593
2. Cost of Goods Sold	65.371	79.671

Table 3 shows that the cost of production of banana blossom nuggets with full costing method is greater than that of the variable costing method. This is because the costs included in the calculation of variable costing are just the variable costs, while the fixed costs are not calculated. The cost of production and cost of goods sold in the table above are used to produce 3 kg of banana blossom nuggets, or equals to 15 packs with each package weighing 200 grams.

The cost of production with the variable costing method for each packaging is IDR 8,519/pack while the full costing method results in IDR 10,719/pack. 200 grams of banana blossom-chicken nuggets per pack is sold at IDR 10000 while based on the results of the study, it can be seen that the cost of goods sold using the variable costing method is IDR 13,074/pack, and the full costing method results in IDR 15,923/pack. It appears that the selling price offered by KWT Kebersamaan is much lower than the cost of goods sold calculated using the full costing and variable costing methods. According to [5], the selling price is the sum of the final value of goods generated from the production and other costs to produce an item, plus an amount of desired benefits.

4. Conclusion

The cost of production obtained from the variable costing method is IDR 8,519/pack, while the full costing method costs IDR 10,719/pack. 200 gram of banana blossom-chicken nuggets per pack is sold at a price of IDR 10000, while the result of the research shows that the cost of goods sold is higher than the selling price set by the KWT Kebersamaan. The cost of goods sold using the variable costing method is IDR 13,074/pack and the full costing method obtains IDR 15,923/pack.

5. References

[1] Aida, Y, Mamuaja, Ch. F dan Agustin AT. 2014. *Utilization of Inflorescence of Banana (Musa Paradisiaca) With The Addition Of Scad Fish (Decapterus sp) on Making Abon*. Jurnal Ilmu dan Teknologi Pangan, Vol. 2 No. 1 hal 20-26.

[2] Hastuti S, Suryawati S dan Maflahah I. 2015. Pengujian Sensoris Nugget Ayam Fortifikasi Daun Kelor. Jurnal AGROINTEK, Vol 9 No.1, Hal. 71-75.

[3] Nasir, M. 1998. Metode Penelitian. Penerbit Ghalia Indonesia. Jakarta.

[4] Prastiwi, WD, Santoso dan Marzuki. 2017. *Preference and Perception of Chicken Nugget As An Alternative of Chicken Meat Consumption in Secang, Magelang*. AGROMEDIA Vo. 35 No. 1 hal. 65-72.

[5] Waryanto, RBD dan Nasrulloh. 2014. Pengaruh penentuan Harga Pokok Produksi terhadap Harga Jual pada Industri Kerupuk (Studi Kasus di UD. INUN JAYA Sampang). Majalah Ekonomi. Vol. XVIII No. 2, Hal 109-119.

[6] Wauran, D. 2016. *Analysis determining the cost of product and the application of cost plus pricing method in the context of setting the selling price on restaurant soto rusuk ko' petrus branch megamas.* Jurnal EMBA. Vol. 4 No. 2 Hal. 652-661.

[7] Wulandari E, Suryaningsih L, Pratama A, Putra DS, dan Runtini N. 2016. *Effect of Tomatos Paste to Physicochemical and Sensory Characteristics Chicken Nuggets*. Jurnal Ilmu Ternak, Vol.16, No.2 hal. 95-99.

Acknowledgement

The authors acknowledged the Faculty of Animal Husbandry, Andalas University, who has funded this research using the 2018 Grant of Faculty of Animal Husbandry's PNBP.